Christine m
W9-API-356

5th edition

TEACHING ELEMENTARY SCHOOL MATHEMATICS

C. ALAN RIEDESEL

State University of New York, Buffalo

ALLYN and BACON

BOSTON LONDON TORONTO SYDNEY TOKYO SINGAPORE

Library of Congress Cataloging-in-Publication Data
Riedesel, C. Alan.
Teaching elementary school mathematics/C. Alan Riedesel.—5th ed.

p. cm.
Includes bibliographical references.
ISBN 0-13-892472-4
1. Mathematics—Study and teaching (Elementary) I. Title.
QA135.5.R5 1990 89-37123
372.7 044—dc20 CIP

Editorial/production supervision and
interior design: Mary Anne Shahidi
Cover design: Jeanette Jacobs
Manufacturing buyer: Peter Havens

to Ardeth
and our
5 + 2 + 2

Printed in the United States of America

10 9 8 7 6 5 4 3 95 94 93

ISBN 0-13-892472-4

CONTENTS

PREFACE

Teaching Elementary School Mathematics, fifth edition, is based on the same assumption as the first four editions: that there is a need for a book that focuses on illustrative situations that make use of the best ideas that concern teaching elementary school mathematics. The wide acceptance of the first four editions has borne out this assumption. Also, recent reports, such as *A Nation at Risk, What Is Fundamental and What Is Not, Everybody Counts,* and *Curriculum and Evaluation Standards for School Mathematics,* have stressed the need for a mathematically literate society. In particular, these reports emphasize the need for children to learn to discover, to think reflectively, and to solve problems. To meet these goals, teachers are needed who make use of developmental-discovery–oriented teaching methods.

Features of this edition include

1. Emphasis on the National Council of Teachers of Mathematics Standards.
2. Chapter sections designed to help the user assimilate the information and suggestions:
   (a) An overview for each chapter.
   (b) A new chapter on evaluation emphasizing the NCTM Standards
   (c) "Take Inventory," a list of pre-reading or post-reading questions that help students to assess their knowledge of the chapter material.
   (d) "Teacher Laboratory," exercises to give elementary mathematics teaching students opportunities to experiment with math ideas of the type contained in the chapters.
   (e) "Side Trips" within the chapter to encourage the reader to experiment with some interesting mathematics pleasantries.

(f) "Self-tests" at the end of the chapter to help review. (*Note:* the self-tests are more for thought than an exact true-false answer.)
(g) Suggestions for projects and further study.
3. Greater increase in calculator usage.
4. Greater emphasis on problem solving.
5. Illustrative children's textbook materials.

I have intentionally organized the chapters in a manner that fits the current organization of teaching in the elementary school. This organization is not always the most logical. However, I feel it will help the book be a good resource for the elementary school teacher. Reports of in-class teacher use of the book have borne out this approach.

Pre-service and in-service teachers in LAI 540 Improvement of Instruction in Elementary School Mathematics have made many valuable suggestions for revision. Also my experience with elementary school children in both mathematics and computer usage have allowed me to test out a number of ideas. I wish to thank Herbert F. Spitzer, formerly of the State University of Iowa, for arousing my interest in teaching approaches that require children to discover and think constructively. I would also like to thank Katharine D. Rasch, Maryville College; and Therese M. Kuhs, University of South Carolina, for their helpful suggestions.

Finally, I should like to suggest that in all of our thinking about teaching elementary school children, we take to heart these words of Dorothy Sayers: "... For the sole true end of education is simply this: to teach men how to learn for themselves; and whatever instruction fails to do this is effort spent in vain."

C. Alan Riedesel

chapter 1

TODAY'S CLASSROOM

OVERVIEW

Each generation of children and teachers has seen changes in the content of school courses and in the approach used to teach these courses. The present generation is not different; in fact, changes occur more rapidly now, owing to the speed of communication. During the 1950s and 1960s, there were significant changes in the content of the elementary school mathematics program; new content was added, and standard content was taught at an earlier grade level. During the 1970s, there was a slowdown in content change, and a good deal of attention was given to "system of delivery" of content.

In the late 1970s and early 1980s, the "back to basics" movement gave a heavy emphasis to drill and practice in mathematics. In the late 1980s, National Assessments Scores reflected this movement, indicating a need to increase emphasis upon teaching concepts and problem solving.

To improve mathematics programs, the National Council of Teachers of Mathematics issued their *Curriculum and Evaluation Standards for School Mathematics*.[1] (This document is explained in detail in later sections.)

The 1990s should be an exciting time for mathematics in the elementary school, what with the increased use of computers and calculators and the increased emphasis on estimation, problem solving, and in-depth thinking. For

[1]Commission on Standards for School Mathematics, *Curriculum and Evaluation Standards for School Mathematics* (Reston, Va.: National Council of Teachers of Mathematics, 1989).

one of the first times in history, improved education was a priority of both 1988 presidential candidates.

No matter what the emphasis, the single most important variable in the mathematics instructional program is the *teacher*. The way he or she guides the child in thinking and feeling is the crucial ingredient. Thus, it is of major importance for the teacher to be able to approach teaching mathematics with excitement and enthusiasm. Toward this end, the approach to teaching needs to focus on involving both children and teacher intellectually and emotionally. The procedures that follow in this book are designed to foster that type of involvement. Up to the present time, no single method has been found for teaching and organizing for teaching. There have been a number of specific theories and movements that have affected mathematics instruction in the elementary school, and each has made important contributions. But, it is the presupposition of this book that no one set of materials, no one sequencing, motivation, or model is appropriate for all the children in the United States. The teacher must be familiar with a variety of ideas so that he or she may use them when appropriate.

TAKE INVENTORY[2]

Can You:

1. Describe several strategies for teaching elementary school mathematics?
2. Give suggestions for procedures useful for any teaching strategy?
3. Describe the use of bulletin boards and games in the elementary mathematics program?
4. List and diagram three needs for curriculum construction in elementary school mathematics?
5. Sketch the history of mathematics teaching in the elementary school?
6. Describe the emphases of the National Council of Teachers of Mathematics (NCTM) Standards?

TEACHER LABORATORY

1. Think back to your elementary school years. How was mathematics taught? Outline a typical lesson. What was the role of the teacher? of the child?
2. How do current elementary school mathematics textbooks compare with those of a few years ago?
3. Obtain three elementary school mathematics textbooks: one published before 1960, one between 1965 and 1970, and the most recent you can find. How do they compare in approach and content?

[2]In each chapter, the "Take Inventory" section provides a "preview" of the chapter in question form. After studying the chapter, you should be able to answer the questions.

4. What do you believe to be the most current set of questions concerning elementary school mathematics teaching? Look through the index of this book. Where are these discussed?
5. Become familiar with this book. Look over the chapter features, such as "Side Trip," "Keeping Sharp," "Take Inventory," "Think About," and "Suggested References."
6. If possible, obtain a copy of *The Arithmetic Teacher*. What seems to be the major focus of the articles?
7. Recall your elementary school years. What types of mathematics lessons can you recall?

VISITING CONTEMPORARY CLASSROOMS

A picture of the role of the teacher of elementary school mathematics can be developed by looking into several classrooms of today. I have selected as an illustration the topic of division with fractions, not because it is crucial to the elementary school mathematics program but rather because it provides an opportunity to illustrate teaching strategies. (In fact, with the increased use of the metric system and hand-held calculators, division of fractions has become less and less important.) Three of these—Socratic questioning, group thinking, and pattern searching with laboratory work—are developmental in nature. The other, the explanatory method, is didactic in nature and involves the teacher's telling and showing.

Explanatory Method

IDEA 1

The teacher began by writing on the chalkboard

$$3 \div \frac{3}{8} = \square$$

The teacher said, "We're going to divide 3 by $\frac{3}{8}$. That is, we're going to find out how many $\frac{3}{8}$s are contained in 3. I'll show you several ways that we can look at the computation.

"First, look at the number line. I'll draw 3 and cut that into eighths. Now we can count forward or backward by $\frac{3}{8}$s:

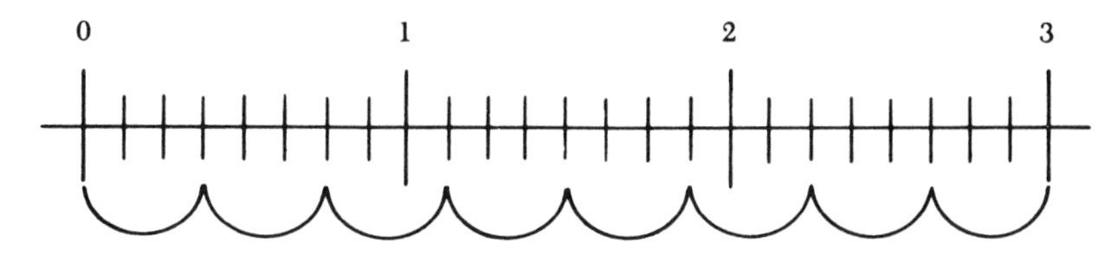

One, two, three, four, five, six, seven, eight. The answer is 8.

"We could also draw circles or squares and 'cut' them into eighths, then make $\frac{3}{8}$s groups":

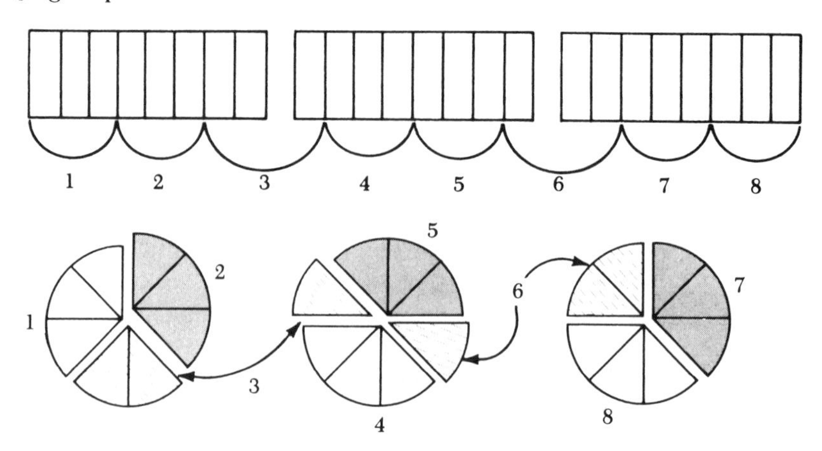

At this point in some classes, the teacher would provide the children with a number of exercises of the sort

$$4 \div \frac{1}{2} = \square \qquad 6 \div \frac{2}{3} = \square$$

The children would solve them by using the number line or drawings.

IDEA 2

The teacher then said, "Let's go back to our first sentence, $3 \div \frac{3}{8} = \square$. As I said, the problem is basically 'How many $\frac{3}{8}$s equal 3?' We could write it either of these two ways:

$$\frac{3}{8}\overline{)3} \qquad \text{or} \qquad 3 \text{ eighths}\overline{)3 \text{ whole units}}$$

"Remember when we divided

$$20\overline{)80} \qquad \text{and thought} \qquad 2 \text{ tens}\overline{)8 \text{ tens?}}$$

"Since 3 whole units can be expressed as 24 eighths, $\frac{3}{1} \div \frac{3}{8}$ can be expressed as $\frac{24}{8} \div \frac{3}{8}$, so we can use this idea:

$$\begin{array}{r} 8 \\ 3 \text{ eighths}\overline{)24 \text{ eighths}} \end{array}$$

or

$$\frac{24}{8} \div \frac{3}{8} = \frac{24 \div 3}{8 \div 8} = \frac{8}{1} = 8$$

This is called the common denominator method."

The teacher then answered questions concerning the procedure, and the children used this method until they were able to work well with it.

IDEA 3

When the children could handle the common denominator method, the teacher said, "Let's look at the mathematical sentence,

$$7 \div \frac{2}{3} = \square \qquad \text{or} \qquad \frac{7}{\frac{2}{3}} = \square$$

We could find our answer if we could make the denominator 1. We could do this by multiplying $\frac{2}{3}$ by $\frac{3}{2}$. Then we have to multiply the numerator by the same number $\frac{3}{2}/\frac{3}{2} = 1$. This is just multiplying by $\frac{3}{2}/\frac{3}{2}$ a name for 1."

$$\frac{\frac{3}{2}}{\frac{3}{2}} = 1$$

$$\frac{7 \times \frac{3}{2}}{\frac{2}{3} \times \frac{3}{2}} \rightarrow \frac{\frac{21}{2}}{1} \rightarrow 10\tfrac{1}{2}$$

The teacher explained the approach on two more examples and then helped the children as they used this method in working fractional divisions.

IDEA 4

After the children were able to handle the approach presented in Idea 3, the teacher wrote,

$$\frac{\frac{3}{8}}{\frac{1}{2}} \rightarrow \frac{\frac{3}{8} \times \frac{2}{1}}{\frac{1}{2} \times \frac{2}{1}} = \frac{\frac{6}{8}}{1} = \frac{6}{8}$$

"Notice that we could get the same answer by changing the $\frac{1}{2}$ to $\frac{2}{1}$ and multiplying":

$$\frac{3}{8} \div \frac{1}{2} = \frac{3}{8} \times \frac{2}{1} = \square$$

The children then worked exercises by using this approach.

IDEAS 5, 6, ETC.

The teacher continued to develop depth in fractional division by explaining the material in greater detail.

Developmental Patterns

FOUNDATION WORK (PATTERN SEARCHING)[3]

In the primary grades, the pupils should be given an opportunity to answer questions such as, "Alice has 5 apples. How many pieces will she have if she divides each one into halves?" The number line, actual objects, models, drawings, and diagrams prove useful in solving such problems.

[3]This lesson appears out of place in a division-of-fractions section, but it is necessary predevelopment for pupils.

In addition to the intuitive development of ideas concerning division, a knowledge of the multiplicative inverse of a rational (the reciprocal) is useful if a guided-discovery approach is to be taken. During the study of the multiplication of rationals, a teacher used the following procedure:

"Before you came in this morning, I drew a picture of three rectangular regions on the board and indicated their lengths and widths [see below]. Take a few minutes and find the area of each rectangular region.

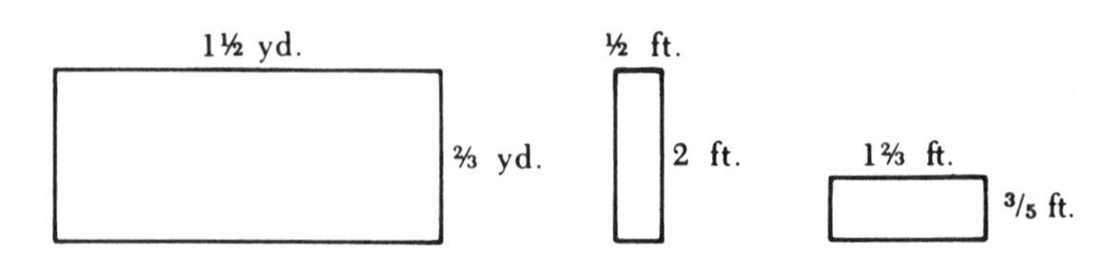

"Now, solve these":

1. $\frac{2}{5} \times \frac{5}{2} = \square$ 2. $\frac{8}{9} \times \frac{9}{8} = \square$

3. $\frac{7}{5} \times \frac{5}{7} = \square$ 4. $\frac{8}{11} \times \square = 1$

5. $\frac{9}{13} \times \square = 1$ 6. $\frac{7}{8} \times \square = 1$

7. What is the pattern? Why?

After a few minutes, four pupils were asked to show their solutions on the board. After the checking of solutions, several pupils wondered why each area was equal to one square unit. The teacher asked whether anyone could find a reason for this unusual result. Although some pupils immediately felt that they knew the answer, the teacher waited a few minutes to provide ample time for most of the class members to notice any similarity in the problems. Then one of the pupils explained, "In each problem, such as $\frac{3}{2} \times \frac{2}{3}$, the numerator of each fraction is the denominator of the other fraction. In such cases, the product is equal to 1."

By referring to a book suggested by the teacher, the class found that two rational numbers whose product equals 1 are the *reciprocals* of each other. They also found that these two numbers are considered the *multiplicative inverse* of each other because their product is always a name for the identity element for multiplication (1). Several exercises were given in which the pupils were to find the reciprocals of fractions and mixed forms.

GROUP THINKING WITH LABORATORY

The teacher who used the developmental pattern began the study of fractional division by giving each child a duplicated sheet containing the following problems:

1. Jill's club decided to sell $\frac{3}{8}$-pounders (bigger than quarter-pounders) at the carnival hamburger stand. They wanted to test them out ahead of time. If Jill brought in three pounds of hamburger, how many $\frac{3}{8}$-pounders could she make?

2. Peg wanted to make ribbon streamers $\frac{2}{3}$ of a yard long to decorate the hamburger stand. She had 6 yards of ribbon. How many streamers could she make from the 6 yards of ribbon?

The teacher read the first problem aloud and then said, "Solve the problem and then show that your answer is correct by solving it in another way. If you finish early, try to develop a diagram or explanation that you think would help someone having trouble with the problem." After the class members had started to work, the teacher moved about the room noting the various techniques they used and helped those having difficulty getting started by asking questions such as, "If you could make up a problem like this one using whole numbers, how would you solve it?" "Can you think of a math sentence you could use to solve the problem?" "What type of diagram could you use?"

When the majority of the class had finished the first problem and started on the second, the teacher asked several pupils to put their solutions on the chalkboard. When they had done this, the teacher said, "Let's stop for a minute and discuss the first problem. What was the mathematical sentence from the problem?"

The answer, "How many $\frac{3}{8}$s equal 3, or $3 \div \frac{3}{8} = N$," was given by a class member. Then the pupils who had written their solutions on the chalkboard explained their procedures. Some of the explanations follow:[4]

MATT: When you use the number line, you count $\frac{3}{8}$s each time; when you get to 3, you count the number of $\frac{3}{8}$s. She could make eight hamburgers.

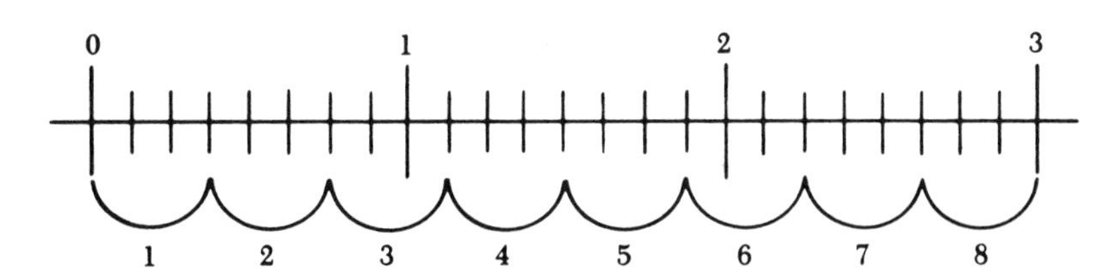

LARRY: Look at the diagram. We have three pounds of hamburgers. I've marked it off into eighths. I then marked off $\frac{3}{8}$s and counted them up. The mathematical question I'm answering is, "How many $\frac{3}{8}$s in 3?"

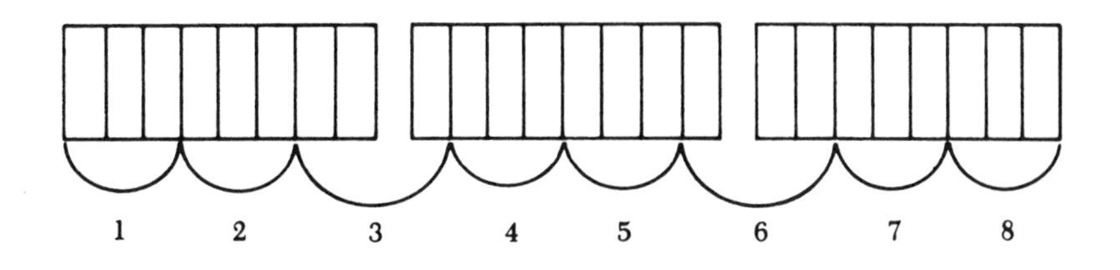

[4]It should be noted that all the solutions indicated are typical of a class that has had developmental teaching. Answers requiring the type of thought exhibited would not occur until the children were comfortable "thinking things out in math."

NAN: I used a slightly different drawing; eight hamburgers.

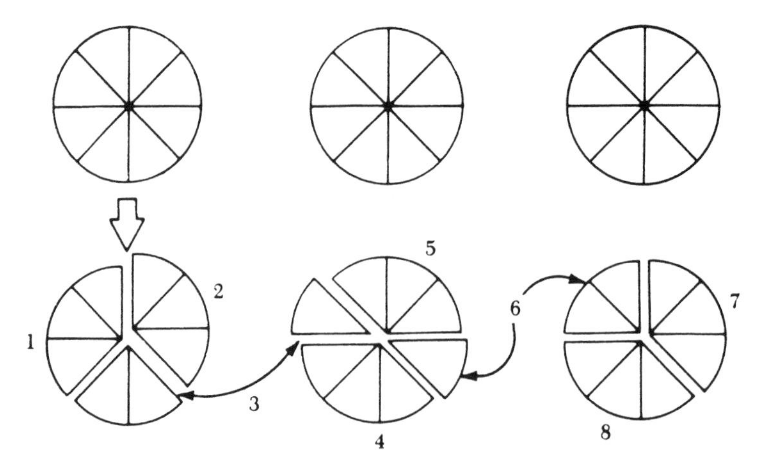

JANET: I added to solve the division:

$$\frac{3}{8} + \frac{3}{8} = \frac{6}{8} \qquad \frac{6}{8} + \frac{3}{8} = \frac{9}{8} \qquad \frac{9}{8} + \frac{3}{8} = \frac{12}{8}$$

1 ② ③ ④

$$\frac{12}{8} + \frac{3}{8} = \frac{15}{8} \qquad \frac{15}{8} + \frac{3}{8} = \frac{18}{8} \qquad \frac{18}{8} + \frac{3}{8} = \frac{21}{8}$$

⑤ ⑥ ⑦

$$\frac{21}{8} + \frac{3}{8} = \frac{24}{8} = 3$$

⑧

KATHY: There are eight $\frac{3}{8}$s. It could be solved by subtracting, too: $3 - \frac{3}{8}$, etc. The problem is, "How many $\frac{3}{8}$s equal 3?" I set the problem up this way: I didn't know how to divide by using a fraction and a whole number, so I changed 3 to $\frac{24}{8}$ and divided $\frac{3}{8}$ into $\frac{24}{8}$. The $\frac{24}{8}$ is another name for 3. Look:

$$\frac{3}{8}\overline{)3}$$

$$\begin{array}{r} 8 \\ \text{3 eighths}\overline{)\text{24 eighths}} \end{array}$$

KEN: Kathy's could be written:

$$3 \div \frac{3}{8}$$

$$\frac{24}{8} \div \frac{3}{8} = \frac{8}{1} = 8$$

The various methods used to solve the problem were discussed, and Kathy's method (the common-denominator method) was used to solve a number of problems. Some time was spent discussing why Kathy's method worked. Emphasis was given to the children's suggestions and evaluation of each other's contributions. A feeling of helpful cooperation was evident in the discussion. The children then worked division problems, using the method that they found most comfortable.

SOCRATIC THINKING

When the children were comfortable with fractional division, the teacher said, "Remember the other day we wrote the mathematical sentence $3 \div \frac{3}{8} = N$? Are there any other ways of writing that division?"

Several methods were suggested, and the teacher focused upon $\frac{3}{\frac{3}{8}}$.

TEACHER: We're now faced with a simplification-of-fractions problem. What denominator would you choose if you had to reduce 100 fractions before going home, but could choose any denominator?

TIM: 3.

JANE: 3 would be good for this problem, but the denominator that would always make the problem easy would be 1. Look:

$$\frac{8}{1}; \quad \frac{6}{1}; \quad \frac{5}{1}; \quad \frac{\frac{2}{3}}{1}$$

TEACHER: How could we get $\frac{3}{8}$ to be 1?

JACK: Remember when we worked those area problems and found the inverse? We would multiply by $\frac{8}{3}$.

TEACHER: Great! Look:

$$\frac{3}{\frac{3}{8} \times \frac{8}{3}}$$

Are we done?

MARY: No, we have to also multiply the 3 by $\frac{8}{3}$, like this:

$$\frac{3 \times \frac{8}{3}}{\frac{3}{8} \times \frac{8}{3}}$$

TEACHER: Why?

JOE: It's like multiplying by 1—

$$\frac{\frac{8}{3}}{\frac{8}{3}}$$

is another name for 1.

The remainder of the period was spent in working division by using the approach described above. After several days, a number of the children suggested that time would be saved by following the procedure,

$$\frac{5}{8} \div \frac{2}{3} \quad \text{rather than} \quad \frac{\frac{5}{8} \times \frac{3}{2}}{\frac{2}{3} \times \frac{3}{2}}$$

"Just multiply $\frac{5}{8} \times \frac{3}{2} = N$. We know the other will always be 1 when this approach is used."

The next day, the class compared its findings with the explanations contained in the textbook.

Laboratory Approach (used with Group Thinking)

The laboratory approach was not used in the development lessons presented on fractional division. The laboratory approach can be considered a developmental strategy and also an organizational procedure. Several laboratory lessons are presented below.

The activities might be made available to pupils on "laboratory sheets" and given to individuals and small groups from time to time as appropriate for individual needs and scheduled class work.

GROUP THINKING LABORATORY

What follows are a number of activities for use in a group thinking laboratory. The activities can be made available to children on "laboratory sheets" and given to individuals and small groups from time to time as appropriate for individual needs and scheduled class work.

I. Experiment with multiplication to develop insight in finding the product of numbers larger than 10.
Materials: Graph paper and scissors.
Procedure:
a. Cut out a rectangle 23 squares wide and 18 squares deep.
b. Find the number of squares in the rectangle. Use any method you wish.
c. What method did you use?
d. Try to find another method.
e. Cut out the following rectangles: 35 × 42, 21 × 33, 24 × 16. Use a different method to determine the number of squares in each. Possible methods are to count squares, perhaps count out a 10 × 10 square to represent 100, see how many times it "fits" on the rectangle, and count the remaining squares; to fold the rectangle carefully, count squares within one section, count sections, then add or multiply, and count any remaining squares.

II. Division experiments: to develop ideas about estimating quotients (can be adapted for division facts or multidigit division).
Materials: Pan balance, assorted rods in unit lengths 1–10.
Procedure: Place five 10-rods and six 1-rods in one of the balance pans. How many 7-rods will you need to put in the other pan to balance the scales? Guess, then check.

Materials: Graph paper
Procedure:

a. Mark off 144 squares to form a rectangle that will have 12 rows of squares. How many columns does the rectangle have?
b. Mark off 117 squares to form a rectangle that will have 9 rows of squares. How many columns does the rectangle have?

III. Experiment with triangles: for a group of two to four children.
Materials: 30 feet of string; masking tape if done indoors, small stakes if done outdoors.
Procedure:

a. Cut three pieces of string, and fasten the ends to the floor with tape to form a triangle. (Use stakes to secure the ends of the string if you are working outdoors.) Making a figure in this way with three pieces of string, would you always get a triangle no matter what the length of the pieces?
b. Find the middle points of the sides of the triangle by any method you wish. What method did you use?
c. Connect the midpoints with pieces of string and tape. What shapes do you see?
d. Compare the areas of the various shapes you have formed. How do you compare the areas with the original area?

DIFFERENCES IN APPROACH

It is obvious that no teacher makes exclusive use of explanatory procedures while another uses only developmental procedures. By the very nature of teaching, some aspects of each approach are used. However, many teachers tend to emphasize one of the two approaches. Also, not all superior teaching procedures are related to the issue of "teacher telling" as opposed to "pupil developmental thought." The comments that follow attempt to accomplish two things: (1) to point out differences between the "explanatory" and "developmental" procedures used in the preceding lessons and/or other lessons to follow and (2) to make several suggestions for teaching mathematics that are considered effective but not tied to either method.

1. The developmental patterns emphasize active learning. Rather than waiting for the teacher to tell them "what" or "how," the students attempt to develop a solution by themselves. If a pupil is not able to solve a problem, the teacher helps with a question or a comment designed to provide insight. Actively involved in the learning process, the student is more likely to be attentive. In the explanatory approach, the pupil's role requires listening rather than acting. Often, the pupil is not sure whether or not to listen to what the teacher is saying and thus may give the teacher's explanation halfhearted attention.
2. The developmental approach stresses building new knowledge on the foundations of experience, so it is of value to students in social situations as well as in academic endeavors. It also closely resembles the approach taken by mathematicians and scientists. Pupils taught by the explanatory method have a tendency to see if they can find a teacher or a book that will answer any questions they have. Thus, they are much more intellectually dependent in problem-solving situations.
3. Because developmental methods stress pupil thinking, the classroom is pupil-centered. The teacher must ask the right question or provide the right activity at the right time; to do this, it is necessary to understand the pupils. The explanatory approach fosters a classroom climate in which the teacher is the font

of all wisdom. Pupils wait to find out what the teacher thinks before they think or form opinions.

4. The developmental approach stresses a search for relationships and patterns and leads to an understanding of mathematical structure. Such insight is valuable at all levels of mathematics. Patterns pointed out by the teacher tend to be sensed rather than understood.

Now that we have listed the major differences between explanatory teaching and developmental procedures, let us focus on differences in the various developmental teaching strategies or approaches, as illustrated in Figure 1–1, Table 1–1, and Figure 1–2. It should be noted that these strategies are used most often in introductory lessons and that the line between them is a thin one. The various strategies should not be considered as pure or scientific but as methods of varying discovery-type teaching to meet the needs of specific mathematical topics, children, and teachers.

Figure 1–1 shows a general progression from pure telling to pure discovery teaching. The chart should not be considered at all absolute. Thus, it is

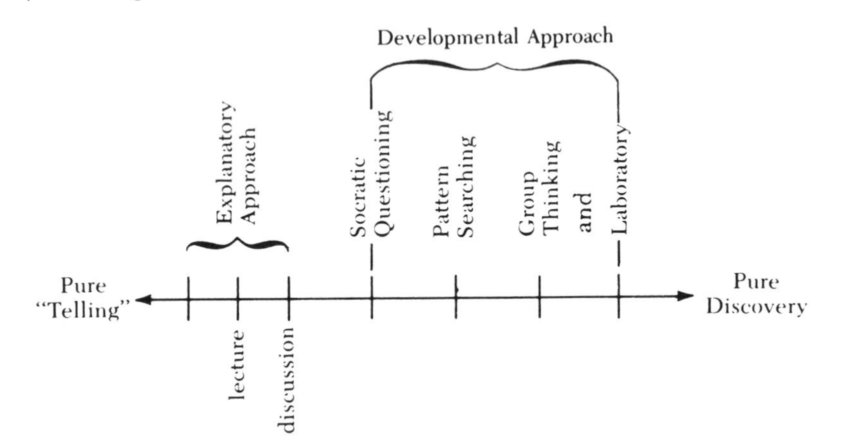

Figure 1–1 The Relationship of Developmental Teaching Strategies to Discovery

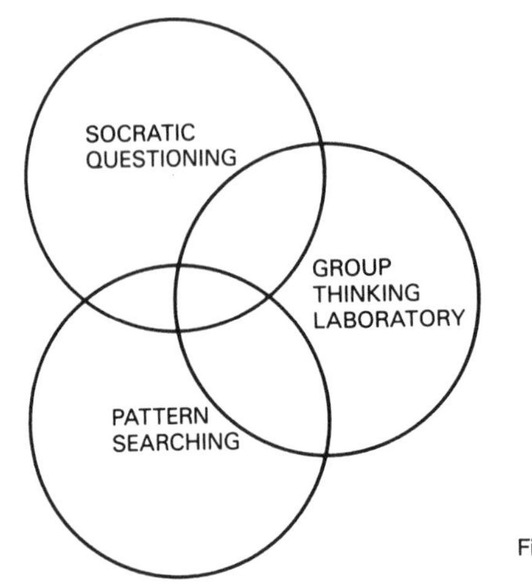

Figure 1–2 The Relationship of Teaching Strategies

Table 1–1. An Analysis of Specific Teaching Strategies

SOCRATIC QUESTIONING	*PATTERN SEARCHING*
1. Teacher directs children's discovery through a series of leading questions. 2. This is a large- or small-group activity under direct teacher guidance.	1. Several worked examples of the pattern to be discovered are presented. 2. Children work as individuals or in small groups. 3. Children discover a "rule" for the pattern and then test it.

GROUP THINKING AND LABORATORY
1. The teacher presents a problem that causes the children to use experience to discover the new mathematical idea. It may require exploration and activity. 2. Children exchange their ideas to synthesize and develop new ideas. 3. A variety of procedures are developed. 4. Manipulative materials are often used. 5. Often individual work is done first, then group work. 6. Children often work in small groups. 7. Often "science experiment" -type procedures are used.

quite possible for a "pattern-searching" strategy to be more developmentally oriented than a "group-thinking" or "laboratory" strategy. The degree of developmental orientation will depend upon the type of subject matter studied and the approach of the teacher.

Table 1–1 provides some general characteristics of the types of developmental teaching strategies described earlier in the chapter. Figure 1–2 reveals the overlap of the teaching strategies.

As you can see, there is a great deal of overlap between the group-thinking and laboratory approaches.

The single examples of the various teaching strategies and the charts are designed to introduce the reader to thinking about strategies of teaching elementary school mathematics. In the chapters that follow, many other classroom illustrations are given by using the various strategies. Thus, this brief introduction should be considered as explanatory; many examples will often be needed to clarify the strategies.

Several other important points need to be considered:

1. The best strategy to use often varies with the mathematical topic, the children to be taught, and the teaching style of the teacher.
2. The topic of fractional division does not lend itself with equal effectiveness to all the guided-discovery strategies suggested. However, a comparison of strategies is best accomplished by considering the same subject matter.
3. The teaching strategies considered are not "pure." The components of each strategy can be combined in several different ways.

SOME GENERAL PROCEDURES

There are a number of procedures that lend themselves to good teaching with any strategy. Some of the best follow:

1. *Verbal problems.* Such problem situations are effective for introducing new topics, for they provide the student with a physical-world model from which he or she is to abstract mathematical ideas. They also identify a use of the material; when students see a need or use for the phase of mathematics being taught, their motivation to learn is heightened.
2. *Multiple methods of solution.* The use of several methods of solving a problem or an exercise is helpful in several ways. (1) It allows for individual differences in approach and level of abstract thinking; one child can count to find the answer to an addition problem, while a more advanced pupil can think the answer. (2) It develops self-confidence; pupils are better able to attack new materials when they have developed several methods of approaching mathematical situations. (3) It leads to several solutions, encouraging the kind of pupil discussion and debate valuable for developing mathematical thinking.
3. *Materials.* A wide variety of manipulative, experimental, and/or environmental materials can be effectively used to make mathematics interesting, understandable, and relevant to children.
4. *"Learn to be their own teacher."* Certainly, discovery-oriented teaching is a major help; however, there are several techniques that can be used with a variety of teaching procedures (see, "What Is Metacognition?" in Chapter 2).
5. *Orally presented problems, computations, and challenges:* Teachers and researchers have found that spending 10 to 15 minutes daily on non–pencil-and-paper tasks (sometimes called mental arithmetic) greatly improves the problems-solving, computational, and mathematical thinking achievement of the children.
6. *Readiness.* Providing children with readiness for mathematical abstractions and readiness for new material should be built into the program 1 to 2 years before the topic is considered in a paper-and-pencil form. For example, simple orally presented work problems such as, "There are four of you at the table. If you each get two soda straws, how many do we need for the table?" can be used as kindergarten readiness for multiplication.
7. *Calculators and computers.* All children should have access to calculators, and there should be one computer per classroom.
8. *Enrichment.* Challenging problems, puzzles, historical materials, and computational algorithms should be provided for all children.

ROOTS OF TODAY'S ELEMENTARY SCHOOL MATHEMATICS CURRICULUM

Colonial Times

In early colonial days, schools existed only in towns and in the more densely settled districts. Where schools did exist, the study of mathematics was often not pursued and, even where it was taught, it usually consisted simply of counting and performing the fundamental operations with whole numbers. "Ciphering" was the principal mathematics taught in schools during colonial times, and the freshman mathematics course at Harvard College went no further than an eighth- or ninth-grade course of today. Very seldom was arithmetic taught to women.

The early textbooks, which were imported from England, were single copies for the "master"; the students copied problems and exercises from the master's book into copybooks. The first arithmetic textbook written by an American author and printed in the United States is believed to have been *Arithmetic, Vulgar and Decimal*, by Isaac Greenwood of Harvard College, in 1729. Greenwood's book received little publicity, and consequently, the first influential arithmetic book published in this country was that of Nicholas Pike, published in 1788, and was very popular during the late 1700s and early 1800s. Its major competition was a book written in England by Thomas Dilworth, first published in 1744. During the early 1800s, two American arithmetic books, *The School Master's Assistant*, by Nathan Daboll, and *The Scholar's Arithmetic*, by Daniel Adams, were introduced. These, along with Pike's book, dominated the field.

The approach taken by these early authors was (1) to state a rule and (2) to give several examples with explanation (steps in computation, not meaning). Their works dealt extensively with changing English money to U.S. money, since the transition from the old to the new currency caused merchants many problems. Study of verbal problems in such textbooks reveals much about the activities of the day. The following problems are typical:

> Supposing a man's income to be 2555 dollars a year; how much is that per day, there being 365 days in a year? Ans. 7 dollars.
>
> In 671 eagles, at 10 dollars each, how many shillings, three-pences, pence and farthings? Ans. 40260 shill, 161040 three-pences, and 1932480 qrs.[5]
>
> At the late Census, taken A.D. 1800, the number of Inhabitants in the New England States was as follows, viz. Newhampshire 183858; Massachusetts, 422845; Maine, 151719; Rhode Island, 69122; Connecticut, 251002; Vermont, 154465; what was the number of inhabitants at that time in New England? Ans. 1233011 inhabitants.[6]

1815–1900

In the period between 1815 and 1820, American educators began to revise their methods of teaching arithmetic. This reform was based on the philosophy of Joseph Pestalozzi, a Swiss educational reformer whose ideas were gaining great popularity in Europe and whose works in translation were beginning to appear in America. One of the first to adopt Pestalozzi's ideas was Warren Colburn, in his *First Lessons in Intellectual Arithmetic*, published in 1821. Rather than follow the then-typical question-and-answer approach, in which a question such as "What is arithmetic?" was asked and an answer such as "Arithmetic is the art or science of computing by numbers, either whole or in fraction" was given, Colburn asked questions such as "How many thumbs have you on your right hand?" "How many on your left?" "How many on both together?" The explanation was left up to the student. The preface to Colburn's first book contains the following suggestions for teaching arithmetic:

> Every combination commences with practical examples. Care has been taken to select such as will aptly illustrate the combination and assist the imagination of

[5]Note that the comma was not used in expressions such as 1932480.

[6]Daniel Adams, *The Scholar's Arithmetic of Federal Accounting* (Montpelier, Vt.: Wright & Sibley for John Prentiss, 1812).

> the pupil in performing it. In most instances, immediately after the practical, abstract examples are placed.... The examples are to be performed in the mind, or by means of sensible objects, such as beans, nuts, etc. The pupil should first perform the examples in his own way and then be made to observe and tell how he did them, and why he did them so.[7]

Colburn's inductive approach influenced the later writings of Adams, who included both reasoning and rules.

A. W. Grube of Germany, another follower of Pestalozzi, developed a program that became popular in the eastern United States in the middle 1800s. He made use of objects in his approach, but did not teach addition, subtraction, multiplication, and division in order. Instead, Grube worked with all four processes on very small numbers (usually 1 to 10) before proceeding to larger numbers. Under the Grube system, 3×3 was developed before $9 + 2$. An inductive approach to teaching was advocated by followers of Grube. In the light of the curriculum changes today, it is interesting to note that Grube recommended that geometry be a part of the mathematics program at all grade levels.

Until the middle 1800s, one book often served as the text for the entire arithmetic program. During the late 1800s, the number of books required to cover the curriculum increased from one or two to four or five. Often, the entire grade-school program was incorporated in two books on mental arithmetic (non-pencil and paper) and three books on written arithmetic. The majority of books at this time did not systematically introduce topics at the appropriate grade level.

Because many of the teachers of this period were unable to capture the spirit of either Colburn's or Grube's approach, these two reformers exerted far less influence than their ideas warranted. The majority of writers and teachers of the middle and late 1800s believed mathematics helped to train the mind (mental discipline), just as they believed Latin trained the mind. In his arithmetic books, which were very popular during this era, Joseph Ray took this view. Ray stated, "Two objects are to be accomplished in the study of arithmetic, viz., the acquisition of a science necessary to the business of life, and a thorough course of mental discipline."[8] While there were a few mathematicians who argued against "mental discipline," it was not rejected by the majority of mathematics educators until the early 1900s.

The first book specifically on the teaching of arithmetic was written by Edward Brooks. The book is well written and worth thorough study by the student of elementary school mathematics. The reader may observe from the preface, reprinted below, that had the suggestions of Brooks been adopted in 1880, the teaching of mathematics might have improved at a more rapid rate. Brooks stated,

[7]Warren Colburn, *First Lessons* (Boston: Houghton Mifflin, 1884), pp. 209–10. Quote taken from the reprint of the preface to the 1821 edition.

[8]Joseph Ray, *Practical Arithmetic, Ray's Arithmetic*, 3rd part, rev. ed. (Cincinnati: Winthrop B. Smith & Co., 1853), p. iii.

Progress in education is one of the most striking characteristics of this remarkable age. Never before was there so general an interest in the education of the people. The development of the intellectual resources of the nation has become an object of transcendent interest. Schools of all kinds and grades are multiplying in every section of the country; improved methods of training have been adopted; dull routine has given way to a healthy intellectual activity; instruction has become a science and teaching a profession.

This advance is reflected in and, to a certain extent, has been pioneered by, the improvements in the teaching of arithmetic. Fifty years ago, arithmetic was taught as a mere collection of rules to be committed to memory and applied mechanically to the solution of problems. No reason for operation was given, none were required; and it was the privilege of only the favored few even to realize that there is any thought in the processes. Amidst this darkness a star arose in the East; that star was the mental arithmetic of Warren Colburn. It caught the eyes of a few of the wise men of the schools, and led them to the adoption of methods of teaching that have lifted the mind from the slavery of dull routine to the freedom of independent thought. Through the influence of this little book, arithmetic was transformed from a dry collection of mechanical processes into a subject full of life and interest. The spirit of analysis, suggested and developed in it, runs today like a golden thread through the whole science, given simplicity and beauty to all its various parts....[9]

As you can see, the spirit of Brook's approach is in keeping with many ideas proposed today.

1900–1920

The turn of the century ushered in the practices of using a different textbook for every two grades and then a different one for every grade. Probably the most influential writer on mathematics education in the early 1900s was David Eugene Smith, who was exceptionally prolific. He wrote elementary, secondary, and college textbooks for students; books on the teaching of elementary school and secondary school mathematics; and standard works on the history of mathematics. In 1913, Smith made the following suggestions concerning content and method in arithmetic:

...(1) that pupils begin with content (having first felt some sensible reason for approaching the subject); (2) that they then pass to a use of symbols, to be handled automatically when expediency demands it, employing a particular form of expression only because that form best expresses the thought held; (3) that they be encouraged in flexibility of expression as well as of thinking, the former, however, always being controlled by the latter; and (4) that they be given many opportunities to exercise choice and judgment in applying the knowledge gained in life situations.[10]

Following the demise of "mental discipline," the stimulus-response (S-R) explanation of learning, often called connectionism and usually credited to Edward Thorndike, was in vogue. Drill procedures were strongly emphasized.

[9]Edward Brooks, *The Philosophy of Arithmetic* (Lancaster, Pa.: Normal Publishing, 1880), preface.

[10]David Eugene Smith, *The Teaching of Arithmetic* (Boston: Ginn, 1913), p. 51. Quoting W. W. Hart.

An example of the thinking of writers of the connectionist school in the field of elementary school mathematics is F. B. Knight's comment: "Theoretically, the main psychological basis is a behavioristic one, viewing skills and habits as fabrics of connection."[11] E. P. Cubberley states,

> In the field of methods, the new psychology has thrown much light on our teaching procedures. We know now the importance of learning of the formation of the right kind of mental connections, or bonds, and that right habit-formation is as important in the teaching of arithmetic as in action and conduct. We also have worked out definite procedures for the best types of practice and drill, and improvement can now proceed according to established rules.[12]

Extreme and often faulty application of connectionism caused many arithmetic programs to be little more than endless drill exercises. Pupils knew that $7 \times 5 = 35$ because their teacher and their textbook had told them so. They had little idea of why this was true.

Along with the S-R approach, writers such as Knight and L. J. Brueckner placed much emphasis upon the study of the relative difficulty of various computational materials and the identification of unit skills to be mastered. These studies led to practices such as introducing 7×3 a year before the introduction of 3×7. Such procedures vary greatly from the current emphasis upon the study of number relationships.

A modern advocate of an approach similar to the one outlined in the paragraphs above is B. F. Skinner.

1920–1955

From the 1920s through the middle 1950s, many educators advocated that the topics in arithmetic be selected only from material used in daily life. Prominent in the social-utility movement were such writers as Wilson, Brueckner, and F. E. Grossnickle. Wilson stated, "Limiting grade work in arithmetic to the mastery of the socially useful not only will remove a great burden from the backs of children but will contribute to better teaching and a better mental-hygiene program in the schoolroom"[13] The students of Wilson conducted extensive studies upon the use of arithmetic in the lives of children and adults. From these studies, Wilson made recommendations such as these: The only fractions to be taught should be the halves, fourths, thirds, eighths, and sixteenths; and only operations on whole numbers should be mastered. The emphasis was upon use, with little thought given to the understanding of mathematics.

In the middle 1930s, many mathematics educators placed less stress on the S-R theory and adopted the Gestalt theory. The Gestaltists emphasized

[11]National Society for the Study of Education, Part I, *Some Aspects of Modern Thought in Arithmetic;* Part II, *Research in Arithmetic* (Chicago: Distributed by the University of Chicago, 1930), p. 5.

[12]Ralph S. Newcomb, *Modern Methods of Teaching Arithmetic* (Boston: Houghton Mifflin, 1926), p. vii.

[13]From *Teaching the New Arithmetic*, 2nd ed., by Guy Wilson and others, p. 12. Copyright © 1951. Used by permission of McGraw-Hill Book Company.

insight, relationships, interpretations, and principles more than did the connectionists. One of the chief spokesmen for the "meaning theory," a field orientation, was William Brownell. He conducted significant studies that helped to establish the principle that pupil achievement in mathematics is better when the children understand the mathematical principles than when they learn only meaningless computational procedures. Over the years, evidence from further research has borne this out.

Brownell stated,

> "Meaningful" arithmetic, in contrast to "meaningless" arithmetic, refers to instruction which is deliberately planned to teach arithmetical meanings and to make arithmetic sensible to children through its mathematical relationships.... The meanings of arithmetic can be roughly grouped under a number of categories. I am suggesting four.
>
> 1. One group consists of a large list of basic concepts. Here, for example, are the meanings of whole numbers of common fractions, of decimal fractions, or percent, and most persons would say, of ratios and proportion....
> 2. A second group of arithmetical meanings includes understanding of the fundamental operations. Children must know when to add, when to subtract, when to multiply, and when to divide. They must possess this knowledge and they must also know what happens to the numbers used when a given operation is employed....
> 3. A third group of meanings is composed of the more important principles, relationships, and generalizations of arithmetic, of which the following are typical: When 0 is added to a number, the value of that number is unchanged. The product of two abstract factors remains the same regardless of which factor is used as multiplier. The numerator and denominator of a fraction may be divided by the same number without changing the value of the fraction.
> 4. A fourth group of meanings relates to the understanding of our decimal number system and its use in rationalizing our computational procedures and our algorithms.[14]

In an article that spoke to both social utility and meaning, Buckingham stated,

> The teacher who emphasizes the social aspects of arithmetic may say that she is giving meaning to numbers. I prefer to say that she is giving them significance. In my view, the only way to give numbers meaning is to treat them mathematically. I am suggesting, therefore, that we distinguish these two terms by allowing, broadly speaking, significance to be social and meaning to be mathematical. I hasten to say, however, that each idea supports the other.[15]

Since the time of Socrates, some educators have advocated the use of provocative questions accompanied by pupil thinking and "discovery." The thread of inductive procedures has appeared, disappeared, and reappeared in mathematics teaching. It can be noted from the previous material that Warren Colburn advocated an approach that somewhat resembled the current discovery approach, but the thread was lost again until the middle 1930s, when stud-

[14] William A. Brownell, "The Place of Meaning in the Teaching of Arithmetic," *Elementary School Journal*, 47 (January 1947); 256–65.

[15] B. R. Buckingham, "Significance, Meaning, Insight—These Three," *The Mathematics Teacher*, 31 (January 1938), 24–30.

ies by McConnell and Thiele once again brought attention to inductive or discovery procedures in elementary school mathematics.[16] Since 1938, Spitzer has worked on procedures that emphasize pupil discovery. His book on teaching arithmetic was probably the first to give practical procedures for the use of an inductive approach.[17]

Recently, a great deal of attention has been placed upon the studies of Jerome Bruner and Jean Piaget. Although the two differ in their interpretations of the developmental level of children, there is a great deal of similarity in their reporting of these stages of development of children's learning. Both stress pupil exploration and discovery as a key to learning.

1955–1970

The scientific advances of the twentieth century were heightened by the extensive governmental effort during World War II. During that time, the ordinary person became somewhat familiar with the large role that mathematics played in the development of the atomic bomb and of modern computer technology. Postwar interest in science and mathematics has been heightened by the space race. With the goal of continuing as the world's scientific leader, the U.S. government and various private foundations have become interested in aiding mathematicians and mathematics educators in the improvement of mathematics programs in the schools.

Many projects developed during the late 1950s. Some of the earliest and most significant were the School Mathematics Study Groups, the Madison Project, the University of Illinois Project, the Greater Cleveland Mathematics Program, the Minnesota School Mathematics Program (Minnemath), and the Stanford Projects (sets, geometry, and logic). In 1963, 29 mathematicians met in Cambridge, Massachusetts, to view the future of school mathematics. This conference developed *Goals for School Mathematics*, often referred to as the Cambridge report.[18]

The Cambridge conference has given rise to much discussion. To some, the report represents the ultimate limit to which the elementary school mathematics program might be "mathematized." To others, it represents a logical step in developing the mathematics curriculum. Some of the ideas from the Cambridge report have been tried in textbook materials, and the Central Mississippi Educational Regional Laboratory has experimented with activity packages for elementary children that use ideas from the report.

Each of the "modern math" programs for elementary school had features unique to it. Some of the programs placed a heavy reliance on "set" ter-

[16] T. R. McConnell, *Discovery Versus Authoritative Identification in the Learning of Children*, University of Iowa Studies in Education (Iowa City: State University of Iowa, 1934), Vol. IX, No. 5; C. L. Thiele, "The Contribution of Generalization to the Learning of Addition Facts," *Teachers College Contributions to Education*, No. 763 (New York: Bureau of Publications, Teachers College, Columbia University, 1938).

[17] Herbert F. Spitzer, *The Teaching of Arithmetic*, 1st ed. (Boston: Houghton Mifflin, 1948).

[18] *Goals for School Mathematics*, the Report of the Cambridge Conference on School Mathematics (Boston: Published for Educational Services, Inc., by Houghton Mifflin, 1963).

minology beginning early in the program; at least one did not mention "set." Some programs made use of a very precise vocabulary from kindergarten on; another avoided using mathematical terminology. Some introduced multiplication as a series of additions; others introduced it by the cross-product (Cartesian product) of sets. Most programs advocated a discovery approach, whereas a few emphasized the "show-and-tell" approach. With such differences in programs, it was difficult to answer the question, "What is modern mathematics?"

The careful student of elementary school mathematics is hardpressed to give a precise definition of modern mathematics. It would be difficult to find complete agreement between any two programs, but the following elements are common to the majority of new curriculums of the 1960s and early 1970s.

1. An increased emphasis upon the structure of mathematics, its laws and principles; a search for patterns to find sequences and order; insight into the logical and expanding development of mathematics.
2. Little emphasis upon the mathematics of "everyday life." The philosophy of the programs was, "If you know mathematics, you'll be able to solve problems."
3. An increased emphasis upon letting pupils figure things out for themselves. Such an approach has many names. It may be called discovery, guided discovery, developmental, or inductive, all of which emphasize the same basic goals.
4. An increased emphasis upon correct terminology. Correct names are used to identify mathematical ideas. If it appears unwise to use a particular word, rather than substitute an imprecise term, none would be used.
5. A readjustment of grade placement of topics. (The grade placement of most standard computational topics has gone back to what it was early in the century. During the 1930s, many topics were moved to higher grades.) There is a spirit of "Let's see what the child can learn."

Many "new mathematics" programs made a contribution to the learning of mathematics by children. Some, however, were too abstract for the majority of the children and were not effective at developing computational and problem-solving skill.

1970–1985

During the early 1970s, the focus turned from the development of new content to the development of laboratory-type approaches to teaching. Probably the best-known project of this type is the Nuffield Project, developed in England.[19] The Nuffield Project made use of the child-development theory of Jean Piaget.[20] The writings and curriculum work of Zoltan Dienes in Canada also focused on teaching approaches using laboratory materials.[21]

[19]Nuffield Mathematics Teaching Project, *I Do and I Understand* (New York: John Wiley, 1967).

[20]Association of Teachers of Mathematics, *Notes on Mathematics in Primary Schools* (London: Cambridge University, 1967).

[21]Zoltan P. Dienes, "Some Basic Processes Involved in Mathematics Learning." In R. Ashlock and W. L. Herman, Jr., *Current Research in Elementary School Mathematics* (New York: Macmillan, 1970).

As the decade moved along and the American economy slowed somewhat, there developed an emphasis on school accountability and proof of achievement. This movement went hand in hand with an emphasis upon behavioral objectives and a management system for teaching. Several "individualized programs" were developed, using small-step behavioral objectives and a system of management of these objectives. (Some of these ideas are mentioned in a later portion of the chapter, dealing with organizing for teaching.)

During the middle and late 1970s, a movement called "back to basics" was popular. The general tenor was on a return to memorization and drill and practice.

However, rather than help the program, this movement caused a reduction in scores on problem solving and concepts. To provide a comprehensive "basic skills" program, the National Council of Teachers of Mathematics (NCTM), in 1977, defined 10 basic skill areas: problem solving, applying mathematics in everyday situations, alertness to reasonableness of results, estimation and approximation, appropriate computational skills, geometry, measurement, reading, interpreting and constructing tables, charts, and graphs, using mathematics to predict; and computer literacy.[22] These suggestions lead to the later development of the NCTM Standards, which are described next.

Current Issues and Concerns

As we enter the 1990s, there is a clear-cut interest in a number of areas concerned with elementary school mathematics instruction. The Commission on Standards for School Mathematics of the National Council of Teachers of Mathematics has identified these five: (1) becoming a mathematical problem solver, (2) learning to communicate mathematically, (3) learning to reason mathematically, (4) valuing mathematics, and (5) becoming confident in one's ability to do mathematics.[23]

Becoming a Mathematical Problem Solver

Development of each child's mathematical problem solving ability is essential if he or she is to be a productive citizen. To develop these abilities requires the child's discovery and development of problem-solving strategies: working problems that take seconds, problems that take minutes, problems that take hours, problems that take days, and even long-term problems (Emphasis on this phase is given throughout the book with particular emphasis in Chapter 4.)

[22]"National Council of Supervisors of Mathematics Position Paper on Basic Skills," and "Ten Basic Skill Areas," *The Arithmetic Teacher*, Vol. 25, No. 1 (October 1977). Copyright © 1977 by the National Council of Teachers of Mathematics. Used by permission.

[23]*Curriculum and Evaluation Standards for School Mathematics*, (Reston, Va.: National Council of Teachers of Mathematics, 1989.)

Learning to Communicate Mathematically

Presently, mathematics is used to represent complex problems in business, economics, physics, etc. The almost universal use of the computer today illustrates the tremendous importance of the communication of ideas by all kinds of language: spoken, written, and mathematical. It is, in fact, almost impossible to deferentiate between "language" and "mathematics" in computer use. (See the question, "How can children learn to communicate mathematically? How should I handle mathematical vocabulary? in Chapter 2.)

Learning to Reason Mathematically

Children learn to make conjectures, gather evidence, build arguments supported by mathematical reasoning, and find correct answers. Development of these attributes requires doing a lot of mathematics. The discussions and suggestions in each of the chapters place emphasis upon mathematical reasoning. Remember: it is almost impossible to spend too much time working on this.

Valuing Mathematics

Mathematics should play an important role in children's lives, whether in school or out of school. Teachers in history or science should develop lessons about Pascal or Newton or the mathematical contributions of, for instance, the Mayans, Greeks, Babylonians, Egyptians, Romans.

Teaching mathematics by introducing each new topic with a meaningful problem to which the children can relate also contributes to their understanding of the value of mathematics. Some teachers have parents and others whose occupations use mathematics discuss their use of mathematics with the class.

Becoming Confident in One's Ability to do Mathematics

"Nothing succeeds like success." This is certainly true in mathematics. The attitude toward mathematics and the enthusiasm for teaching mathematics that the teacher brings to the class greatly affects children's confidences. Numerous studies have shown that if teachers and parents believe that a child "can do" mathematics, they can. In over 30 years of teaching mathematics, I have found that children always improve when they have two attitudes: First, they must *want* to do mathematics, and second, they must believe they *can* do mathematics. When these two attitudes occur, children learn mathematics and enjoy mathematics, problem solving, and mathematical discovery. A key person to establishing these attitudes is the teacher.

Carefully study the suggestions made by the commission concerning changes in curriculum and teaching. These changes are reflected in the approach taken in this book.

Summary of Changes in Content and Emphasis in K-4 Mathematics

TOPICS TO RECEIVE INCREASED EMPHASIS	*TOPICS TO RECEIVE REDUCED EMPHASIS*
Number	*Number*
–Number sense –Place value concepts –Meaning of fractions and decimals –Estimating quantities	–Early attention to reading, writing, and ordering symbolically
Operations and Computation	*Operations and Computation*
–Meaning of operations –Operation sense –Mental computation –Estimation and reasonableness of answers –Selection of an appropriate computational method –Use of calculators for complex computation –Thinking strategies for basic facts	–Complex paper-and-pencil computations –Isolated treatment of paper-and-pencil computations –Addition/subtraction without renaming –Isolated treatment of division facts –Long division –Long division without remainders –Paper-and-pencil fraction computation
Geometry and Measurement	*Geometry and Measurement*
–Properties of geometric figures –Geometric relationships –Spatial sense –Process of measuring –Concepts related to units of measurement –Actual measuring –Estimating measurements	–Naming geometric figures as a competency –Equivalencies between unit of measurement
Probability and Statistics	
–Collecting and organizing data –Exploration of chance	
Patterns and Relationships	
–Pattern recognition and description –Use of variables to express relationships	
Problem Solving	*Problem Solving*
–Word problems with a variety of structures –Use of everyday problems –Applications –Study of patterns and relationships –Problem-solving strategies	–Use of clue words to determine which operation to use
Instructional Practices	*Instructional Practices*
–Cooperative work –Discussing mathematics –Questioning –Justifying thinking –Writing about mathematics –Problem-solving approach to instruction –Use of calculators and computers	–Rote practice –Rote memorization of rules –One answer and one method –Use of worksheets –Written practice –Teaching by telling

Summary of Changes in Content and Emphasis in 5-8 Mathematics

TOPICS TO RECEIVE INCREASED ATTENTION	*TOPICS TO RECEIVE REDUCED ATTENTION*
Problem Solving	*Problem Solving*
–Open-ended problems and extended problem solving projects –Problem situations which allow student investigation and formulation of questions –Representing situations verbally, numerically, graphically, geometrically, or symbolically	–Routine, one-step problems –Problems categorized by types
Communication	*Communication*
–Discussing, writing, reading, and listening to other's mathematical ideas	–Seat work
Reasoning	*Reasoning*
–Spatial reasoning –Proportional reasoning –Reasoning from graphs –Inductive and deductive reasoning	–Relying on outside authority (teacher, answer key)
Connections	*Connections*
–Mathematics as an integrated whole –Applications of mathematics	–Isolated topics
Number/Operations/Computation	*Number/Operations/Computation*
–Number sense –Operation sense –Creating algorithms and procedures –Using estimation both in solving problems and in checking the reasonableness of results –Relationships among representations of and operations on whole numbers, fractions, decimals, integers, and rational numbers –Developing an understanding of ratio, proportion and percent	–Memorizing rules and algorithms –Paper-and-pencil computational facility –Exact forms of answers –Memorizing rules and procedures
Patterns and Functions	*Patterns and Functions*
–Developing and using tables, graphs, and rules to describe situations –Interpreting among various mathematical representations	
Algebra	*Algebra*
–Developing an understanding of variable, expression, and equation –Using a variety of methods to solve linear equations and to informally investigate inequalities and nonlinear equations	–Manipulation of symbols –Memorizing procedures and drill on equation solving

Summary of Changes in Content and Emphasis in 5-8 Mathematics (continued)

TOPICS TO RECEIVE INCREASED ATTENTION	TOPICS TO RECEIVE REDUCED ATTENTION
Statistics	*Statistics*
–Using statistical methods to describe, analyze, evaluate, and make decisions	–Memorizing formulas
Probability	*Probability*
–Create experimental and theoretical models of situations involving probabilities	–Memorizing formulas
Geometry	*Geometry*
–Developing an understanding of geometric objects and relationships –Using geometry in solving problems	–Memorizing geometric vocabulary
Measurement	*Measurement*
–Estimating and using measurement to solve problems	–Memorizing and manipulating formulas
Instructional Practices	*Instructional Practices*
–Actively involving students individually and in groups in exploring, conjecturing, analyzing, and applying mathematics in both a mathematical and a real-world context –Using appropriate technology for computation and exploration –Using concrete materials –Connecting mathematics to other subject areas and to the world outside the classroom –Connecting mathematical topics within mathematics –Teacher as facilitator	–Performing computations out of context –Paper-and-pencil work on computations –Topics in isolation –Teacher as dispenser of knowledge

Curriculum and Evaluation Standards for School Mathematics, Reston, Va.: National Council of Teachers of Mathematics, 1989, pp. 17, 74, 75.

ESTABLISHING A BALANCE

A number of factors interact to create curriculum change. Glennon and Callahan suggest that decisions on selection of content for inclusion in the elementary school mathematics program may be arrived at with help from three sources: (1) social theory (needs of society, sociological theory, or social-utility theory); (2) the needs of the subject theory (logical organization theory and meaning theory); and (3) psychological theory (needs of the individual, felt-needs theory, or expressed-needs theory).[24] They further suggest that any time these three theories are not well balanced, the curriculum suffers. In schools of

[24] V. J. Glennon and I. C. Callahan, *Elementary School Mathematics: A Guide to Current Research* (Washington, D.C.: Association for Supervision and Curriculum Development, NEA, 1975), pp. 1–6.

the Summerhill variety, the mental-health approach, which results in mathematical instruction only when the child expresses a need for it, may weaken the program.[25] We have noted how extreme application of the social-utility theory may have weakened the mathematics program for the 30 years prior to 1950. In the 1960s, the movement in the direction of the needs-of-the-subject approach may have been inappropriate for many children. It is hoped that the present trend is toward a balanced approach.

It may be helpful to consider the three extreme positions as the vertices of a triangle. A balance among the three theories could be considered to be a ring held in place by three springs (see the following figure). Thus, applying extra pressure to any one of the springs would move the curriculum out of balance. This diagram is helpful in considering individual differences in children or groups of children. For example, children with extreme psychological problems might need to have the program moved toward psychological theory; children from a disadvantaged area might benefit from a program moved toward the social theory; and children with strong academic goals might benefit most from a program moved toward the needs of the subject.

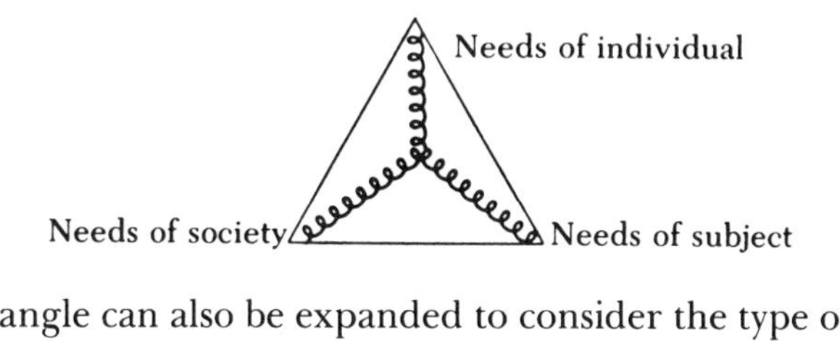

The triangle can also be expanded to consider the type of motivation to give to specific children or groups of children.

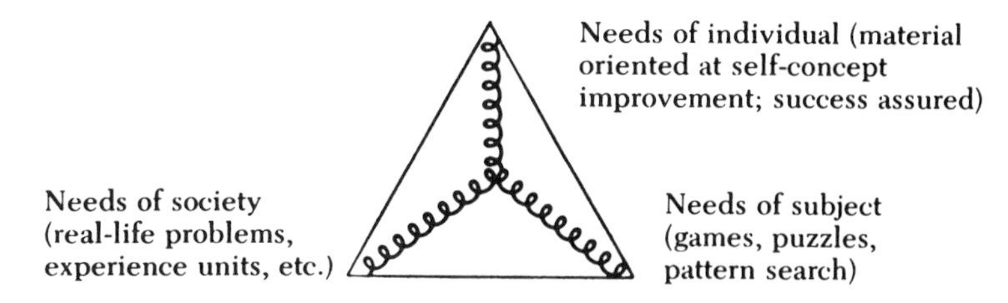

As can be noted, children with an academic orientation are often motivated by mathematics itself in the form of games, pattern searching, and puzzles. Children who see mathematics as of practical importance in the everyday world benefit from situations based on real-world problems. Children who are in need of an improved self-image are probably motivated best through material that ensures success and provides a feeling of achievement. It is true that most children benefit from a mixture of all three types of motivation, but specific groups of children should probably have a greater concentration of one type.

[25] A. S. Neill, *Summerhill* (New York: Hart, 1960).

SOCIAL AND EMOTIONAL CLIMATE

The following four suggestions for the selection of topics to be taught in elementary school mathematics are offered in the belief that their use will produce a healthy balance in the curriculum:

1. The topics taught should be mathematically sound.
2. The topics should be of a kind that, when presented, "make sense" to the student. The student should be able to see a reason for the study of a topic.
3. The topics should be teachable; that is, the difficulty should be such that, with effort, the student can understand the material.
4. The topics should lend themselves to mathematical exploration and pupil discovery.

KEEPING SHARP

Self-Test: True/False

_______ 1. In the explanatory method, the pupils can try out their own methods and explain them to the class.

_______ 2. Of the many variables affecting the mathematics program, the teacher is the most important.

_______ 3. The explanatory method is essentially didactic rather than developmental.

_______ 4. Under a developmental pattern of instruction, children are encouraged to experiment with different solutions.

_______ 5. Pupil discussion and analysis of solution strategies is an important aspect of the explanatory method.

_______ 6. The explanatory method usually encourages pupils to seek answers from an authority rather than create their own solutions.

_______ 7. Pattern searching and finding relationships are characteristic of developmental teaching.

_______ 8. In developing a mathematics curriculum, the only relevant concerns are the subject-matter needs and society's needs.

_______ 9. When the majority of class time is spent on developmental activities, students seem to perform better on achievement tests in problem-solving computation and concepts than when the time is spent on drill and practice.

_______ 10. Reading periodicals such as *The Arithmetic Teacher* has little value for elementary teachers.

_______ 11. Proper mathematics terminology is essential to understanding the concepts; thus, vocabulary should be developed prior to any conceptual work.

_______ 12. The theory of social utility dictates that the only math to be taught should be that which is most used in the daily life of adults.

Vocabulary

Socratic questioning	pattern searching	affective domain
explanatory method	social-utility theory	mastery

THINK ABOUT

1. Develop an introductory geometry lesson on the classification of shapes for your first grade, using the techniques of Socratic questioning.
2. Explain the three contributory needs in determining a math curriculum as described by Glennon and Callahan.
3. Discuss the advantages and disadvantages of the explanatory method and the developmental method of instruction.
4. Prepare a lesson on addition of whole numbers for a fourth grade. What method(s) did you use? Why?
5. List your goals in teaching mathematics. Get together with two or three friends to determine the goals each of you considers most important. Defend your views.
6. Look at the NCTM Standards material. How is this different from your childhood mathematics?
7. Think through your feelings about the use of calculators in school mathematics. How important are calculators in everyday life?
8. In your imagination, teach a lesson by using each of the lesson types described in this chapter.
9. Obtain elementary school mathematics textbooks; compare the grade placement of topics with those shown at the end of the chapter.
10. The American philosopher George Santayana said, "Those who do not study history are doomed to repeat it." What implications does this have for elementary mathematics teaching?

SUGGESTED REFERENCES

The Agenda in Action, 1983 Yearbook, Reston, Va.: National Council of Teachers of Mathematics, 1983.

COMMISSION ON STANDARDS FOR SCHOOL MATHEMATICS, "Let's Count," in *Curriculum and Evaluation Standards for Mathematics,* Reston, Va.: National Council of Teachers on Mathematics, 1989.

Estimation and Mental Computation, 1986 Yearbook, Reston, Va.: National Council of Teachers of Mathematics, 1986.

Everybody Counts, Washington, D.C.: National Academy Press, 1989.

POST, THOMAS R., *Teaching Mathematics in Grades k–8: Research Based Methods,* Boston, Mass.: Allyn and Bacon, 1988.

Research in Mathematics Education, Reston, Va.: National Council of Teachers of Mathematics, 1980.

Periodicals

The Arithmetic Teacher, National Council of Teachers of Mathematics, 1906 Association Drive, Reston, VA 22901.

Journal of Research in Mathematics Teaching, 1906 Association Drive, Reston, VA 22901.

Learning, Education Today Company, Inc., 530 University Avenue, Palo Alto, CA 94301.

Mathematics Student Journal, National Council of Teachers of Mathematics, 1906 Association Drive, Reston, VA 22091.

The Mathematics Teacher, National Council of Teachers of Mathematics, 1906 Association Drive, Reston, VA 22091.

Mathematics Teaching, Association of Teachers of Mathematics, Market Street Chambers, Neslon, Lancashire, BB9 7LN, England.

School Science and Mathematics, Central Association of Science and Mathematics Teachers, Inc., 535 Kendall Avenue, Kalamazoo, MI 49007.

chapter 2

QUESTIONS ON TEACHING ELEMENTARY SCHOOL MATHEMATICS

OVERVIEW

This chapter takes an approach somewhat different from the other chapters, inasmuch as in this chapter answers are provided to some questions students of the teaching of elementary school mathematics need to consider in the entire framework of mathematics teaching. Many of the topics are illustrated in later chapters; however, it is important at this point in the book to consider each question.

TEACHER LABORATORY

1. Read the chapter. What questions would you add to the list? Record your questions. Keep them on a sheet of paper in the back of the book. If they are not answered in the chapters that follow, perform the proper research to answer them.
2. Providing for individual differences is one of the greatest problems facing the elementary school mathematics teacher. Before reading this chapter, think about or write tentative answers to each of the following questions. Discuss your answers with others. Ask several children to comment on your suggestions.
   a. How can I interest children who are having difficulty with mathematics?

b. What are the advantages and disadvantages of a completely individualized program?
c. How many groups could I handle in a class of 30 children?
d. What are different worthwhile activities that could be occurring simultaneously in an elementary school mathematics classroom?
3. Work out a lesson designed to help a child to be his/her own teacher.

TAKE INVENTORY

Think about teaching elementary school mathematics, children's thinking, individual differences, and materials. What do you need to know?

How Can "Meaning" Be Maintained?

Often, teachers introduce new topics through exploratory procedures that lead to understanding of the mathematical idea and then proceed to teach efficient computational processing without constant stress on "meaning." It is a good idea that each time a topic is reintroduced later in a school year or in a different school year, the teacher review the ideas behind the processing and that periodically the teacher ask questions such as, "I noted that you all renamed in this subtraction situation without writing any changes in the numerals. Can you tell me what the basic principles of this process are?" "You've been inverting and multiplying when you divide fractions. Why does this work? What mathematical principles are involved?" "You multiplied the measure of the length by the measure of the width to find the area of this rectangle. Why?" Mathematical meanings, as well as computational procedures, need to be reviewed and reinforced.

What Procedures Help Children Remember the Mathematics Taught to Them?

Researchers have found a number of techniques that help children to remember mathematical material. Several rather easily used techniques are these:

1. Teach mathematical content in a meaningful manner, helping children to understand mathematical principles and computation.
2. Adjust the learning tasks to the appropriate achievement and intellectual level of the child.
3. Make use of intensive review procedures. (When several concepts have been taught, develop a day's work that involves probing review questions.)
4. Periodically retest. (Several weeks after you have given a quiz or test on material, repeat the quiz or test.)

5. Give children opportunities to diagnose their own errors. (Children can correct their own papers and then try to determine the reasons for mistakes they have made.)
6. Use guided-discovery teaching approaches.

How Much Time Should Be Devoted to Mathematics Teaching, and How Should This Time Be Used?

There has been little study concerning the optimum length of class time for elementary mathematics instruction. However, there are strong indications that an increased amount of time devoted to instruction results in significant increase in achievement. A teacher can feel relatively confident that a period from 55 to 60 minutes in length will produce achievement superior to that obtained in a period from 35 to 40 minutes in length.

There is some indication that many teachers spend the majority of mathematics class time on correction of assignments and on drill and practice activities. Research evidence strongly suggests that this is not the best use of mathematics class time. At least four studies have found that classes spending about 75 percent of the mathematics time on "developmental activities" score higher on achievement tests related to problem solving, computation, and concepts than those spending 75 percent of their time on drill-related activities. It is of particular importance that the children who spent most of their time working on developmental activities were better in computation than those spending most of their time practicing computation.

What are the developmental activities that should occupy well over half the instructional time in elementary school mathematics? Here are 10 basic skill areas: discussions of the whys and hows of the topic of study; pupil reports and explanations of the approaches that have been developed; pupil and teacher demonstrations of significant ideas being studied; small-group and individual handling, inspecting, analyzing, and arranging of visual and manipulable materials; individual and/or small-group exploration to find alternate solutions to mathematics exercises; solving and inventing puzzles and games related to the topic of study; and engaging in laboratory activities related to the topic of study.

Remember, there is usually no hurry to get to the final stage of development of an elementary school mathematics idea. In most cases, an increased amount of exploration time results in better understanding of the topic, better retention, and thus less need for drill. These suggestions do not mean that children should spend no mathematics time in practice, but that the amount should be reduced and an increased amount of time spent on developmental activities. See:

HOPKINS, C. D., "An Experiment on Use of Arithmetic Time in the Fifth Grade," *Dissertation Abstracts* 26 (1966): 5291.
JARVIS, OSCAR T., "Time Allotment Relationships to Pupil Achievement in Arithmetic," *Arithmetic Teacher* 10 (May 1963).
MILGRAM, JOEL, "Time Utilization in Arithmetic Teaching," *Arithmetic Teacher* 16 (March 1969).

SHIPP, DONALD E., and GEORGE H. DEER, "The Use of Class Time in Arithmetic," *Arithmetic Teacher* 7 (March 1960).
SHUSTER, ALBERT H., and FRED L. PIGGE, "Retention Efficiency of Meaningful Teaching," *Arithmetic Teacher* 12 (January 1965).
ZAHN, KARL G., "Use of Class Time in Eighth-Grade Arithmetic," *Arithmetic Teacher* 13 (February 1966).

How Can Children Learn to Communicate Mathematically? How Should I Handle Mathematical Vocabulary?

Children in elementary school should learn to talk about, listen to, and read and write about mathematics. The development of a child's power to use mathematics involves learning the signs, symbols, and terms of mathematics. In the past, this has often been difficult because of the approach used in teaching rather than the difficulty of the concepts.

Vocabulary. Let us use the development of mathematical vocabulary as an example. In many of the current elementary school mathematics materials, the vocabulary is introduced before the concept in considered. This causes difficulty. For example, one textbook series introduced "perimeter" by a heading at the top of the page that said, "Finding the Perimeter." This was followed by "When we work with closed figures of a line segment, it is often useful to know the total length of the segments." The sum of the lengths of all the line segments is called the 'perimeter.'" This would scare off any third grader that was average at either mathematics or reading.

By contrast, one teacher introduced "perimeter" with the following problem: "This bulletin board is six feet wide and two feet high. I want to put ribbon around our bulletin board. How much ribbon will I need? How could you find out?"

JOSE: Measure all around it (take tape) 6 feet and 2 feet and 6 feet and 2 feet. 16 feet in all.
CHRIS: 6 feet plus 2 feet plus 6 feet plus 2 feet. 16 feet in all.
TOM: $6 + 6 + 2 + 2 = 16$ feet.

After the children worked a number of problems involving perimeters, the teacher asked, "What have we been doing today?" The children indicated that all of the problems involved finding the distance around a shape or region. Then the teacher asked, "Do any of you know what the distance around a shape is called?"

In some classes, one or more of the children will know and respond with, "It's called the perimeter." In other classes no one will know, and the teacher can suggest checking the appropriate place in the textbook.

Note that this approach has several superior features: (a) The children are not frightened off by the early introduction of the vocabulary. For many children the teacher's saying, "Today we're going to find the perimeter," would make the lesson seem much harder than it actually is. (2) Learning the word after the idea has been developed and when there is a need to discuss "what we know" gives the children a feeling of accomplishment. Most children are interested in learning new words, and there is little reason ever to use an incorrect term. For maximum success in vocabulary development the following suggestions are offered:

1. Give pupils an opportunity to familiarize themselves with a mathematical concept before learning its name. Thus, pupils should have wide experience in using the commutative property of addition before naming it.
2. Avoid the "matching type" of vocabulary drill exercises. Concentrate on the correct use of vocabulary in situations in which it helps a discussion.
3. Remember, the mathematical idea is more important than the name.

Written communication. Superior teachers of elementary school mathematics have used short reports or themes as an aid in improving the mathematical understanding of the children.

Verbal communication. Small group discussions, as well as individual reports and total class discussion, should be used to improve the students' skill at communicating mathematics. Several techniques that have been effectively used follow:

1. Assigning an oral report on a topic such as "How I Add Fractions." Children react well to this. Use the suggestions given in standard books on elementary language arts instruction for suggestions.
2. Allow one student to help a fellow student who has been absent to learn the mathematics that was taught while he or she was gone. Note that care must be taken not to exploit some of the children as "teachers." However, one of the best ways to improve both mathematics and communication skills is to try to teach others.
3. Hold small group discussions on such topics as, "What is the best way to solve problems? Isn't there a best way?" At the end of discussion time, the small groups can share their findings with the entire class.
4. Have individuals from your class help children in lower grades with their mathematics.

WHAT TYPES OF LESSONS SHOULD BE USED?

Over a period of years, I have asked elementary school teachers to describe the typical mathematics lesson taught to elementary schoolchildren, both as they remember being taught and as they themselves teach. Over 80 percent of the replies fit this pattern:

1. Correct yesterday's work.
2. Open the textbook to a set of pages.
3. Explain the work contained there.
4. Assign the remainder of the pages for individual work.
5. Help class members with the work.

When I asked, "Are all the lessons the same? Are they different when introducing a topic, developing concepts, practicing, and so on?" the reply was, "No, the typical lesson is treated as a combination introduction and practice lesson." The brief paragraphs that follow are designed to alert the student of elementary school mathematics teaching to the need to develop different types of lessons. The lesson types fit into the flow chart in Figure 2–1.

Introductory lessons. Any one of the developmental strategies could be used to introduce a new topic. The introductory lesson should achieve most of the following goals: (1) provide a problem setting that would motivate the study of the topic and show its importance and relevance, (2) provide a setting in which the children can use their experience and learning to develop new concepts, (3) give the teacher an opportunity to observe the level of development of the children, and (4) lead into concept development or diagnosis.

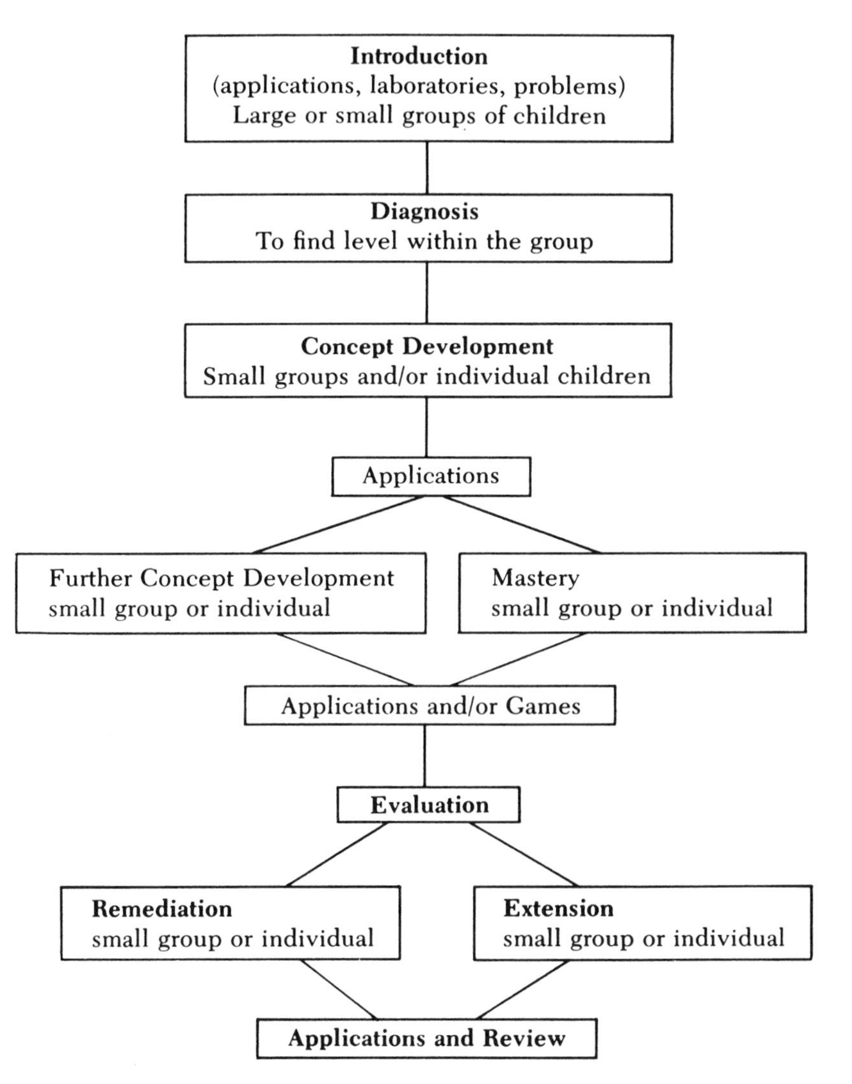

Figure 2–1 Types of Lessons

Diagnostic lessons. Diagnosis can be a portion of any other lesson or a particular lesson type. In the situation in which the teacher needs an in-depth look at the background of the children, a diagnostic lesson should be used. Such a lesson should contain features such as these: (1) It should gather information concerning the background, skills, and development of the children concerning a particular topic. (2) It should be developed in such a manner that the child does not feel the pressure of a "testing" situation. This can be accomplished through the use of games as a means of gathering information or through the use of diagnostic instruments that do not look like regular tests.

Concept-development lessons. Lessons of this type are designed to develop the mathematical meanings of the topic. Group-thinking strategies are often useful for this type of lesson. It is of great importance that adequate numbers of concept-development lessons occur before the move into mastery lessons.

Application lessons. An application lesson can occupy an entire class period or be part of another lesson type. The application lesson has the goal of furthering mathematical meaning and social relevance through "use" situations.

Mastery lessons. A mastery lesson is a form of drill or practice. The use of a variety of procedures is necessary to motivate such lessons. Games are useful as an important means of drill and practice. Master lessons should (1) strengthen mathematical ideas and meaning and/or (2) strengthen mathematical skills, performance, and speed.

Evaluation lessons. Periodically during the course of a unit a portion of a lesson should be given to evaluation of progress. Also, at the close of a unit, an overall evaluation should be made.

Review lessons. Periodically, a lesson should be developed to review previous teaching. Such a lesson could (1) review and reteach a topic studied earlier in the year or in previous years, (2) review topics concerned with the present unit, and (3) review the material and add depth to understanding and skill. *Note:* A review lesson can take the form of a mathematics theme in which a child "tells all about" the topic that is being studied. He or she can look at historical development and at various approaches.

Project lessons. Such lessons are often a part of a short interdisciplinary unit or they can take the form of lessons dealing with topics such as making measuring equipment, a mathematics field trip to a point in the community, student development of a mathematics fair, metric day, or student development of bulletin boards.

As in the case of teaching strategies, there is overlap from one lesson type to another. The important point is that the teacher should keep the described lesson types in mind in order to fit the lesson to needed content.

What Is Metacognition? What Is the Role of Metacognition in Elementary School Mathematics?

Metacognition deals with the knowledge and control a person has over one's thinking. In other words, do you have techniques that help you to learn, solve problems, and think? A great deal of interest in learning to learn and becoming your own teacher fit into the framework of metacognition.

Help in teaching elementary school mathematics. Teachers can give important help to the children they teach by aiding the children to be aware of the way in which he or she studies and learns mathematics. For example, knowing that, "I make computational errors when I hurry. Pulling out the numbers without knowing what a word problem is about works some of the time, but not enough to do well. If one way does not work, there is usually another method I can use to find and answer," can help a child to improve his or her learning.

Teachers can help students develop awareness of their best learning techniques by asking questions, conducting discussions, and rewarding children's thinking about thinking. Some questions you might on occasion ask are "What do you do when you see an unfamiliar problem? Why? Is it important to take one's time in working on a problem? When? What can you do when you're stuck with a computation and can't seem to get an answer? How often does it work? How can you keep track of what you're doing? What kind of errors do you usually make? Why? When is it useful to check your work? Why? What kind of problems are you best at? What kind of problems are you worst at? Do all problems take the same length of time? Why or why not? How many ways can problems be solved? When? Do you think you are good at mathematics? Why? How could you be better? What are things you could do to help yourself?"

In addition to helping children think through and improve their thinking strategies, the teacher can model good thinking behavior. Modeling requires that you let the children in on your thought processes. Too often we give children polished solutions rather than showing how we think and plan. A teacher who makes the classroom a place where both teacher and children are learning provides a setting for improving the metacognitive process of both.[1]

How Can We Improve Children's Attitudes Toward Mathematics?

In addition to the proper development of mathematical ideas, it is of prime importance that the feelings and attitudes of the child become a part of the planning for instruction. Figure 2–2 calls attention to the various forces that affect a child's "attitude" toward mathematics. The teacher should consider each of the areas for each child in the class.

[1]See: Garafalo, Joe, "Metacognition and School Mathematics." *The Arithmetic Teacher,* pp. 22, 23, May 1987; "Metacognitive Knowledge and Metacognition Processes: Important Influences on Mathematical Performances." *Research in Teaching in Developmental Education* 2 (April 1986): 34–39; and Schoenfield, Alan H., "What's All the Fuss about Metacognition?" In *Cognitive Science and Mathematics Education,* edited by Alan H. Schoenfield. Hillsdale, N.J.: Lawrence Erlbaum Associates, 1988.

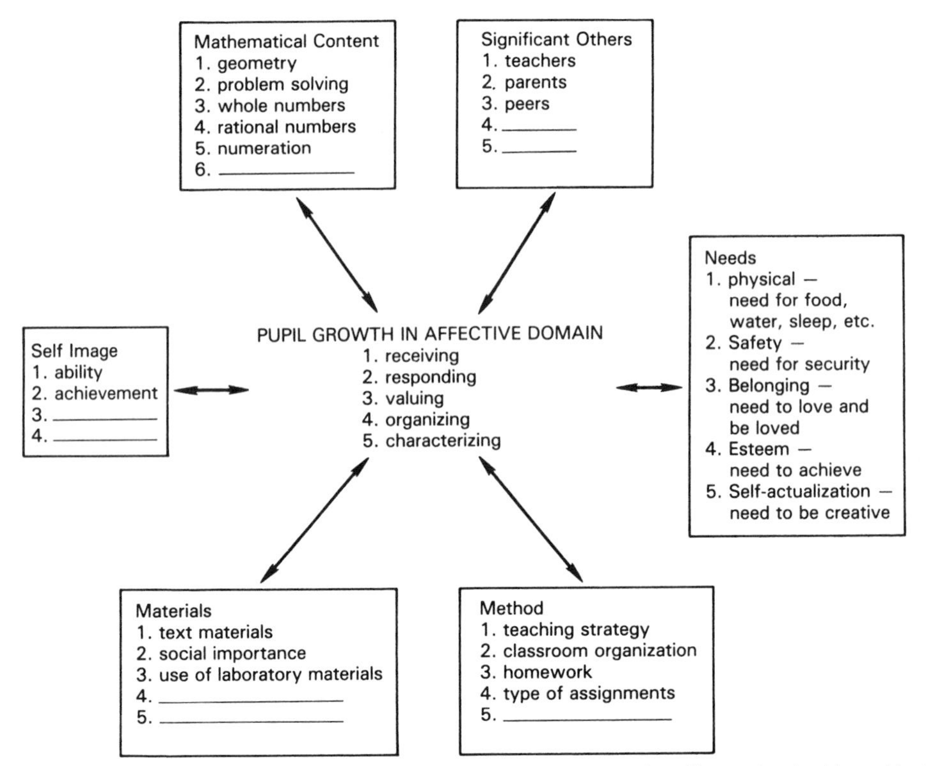

Note: The blank spaces are used to indicate that these lists are not complete. The reader should consider the addition of other categories.

Figure 2–2 Factors Affecting Pupils' Attitudes Toward Mathematics

It should be noted that none of the levels or the factors contributing to a child's achievement in the cognitive domain or to his "attitude" in the affective domain is absolute. They are provided to draw attention to some of the levels and ideas that should be considered in learning more about a child and his or her relationship to elementary school mathematics.

Another idea to consider in dealing with a child's feelings toward mathematics and in thinking about the teaching of mathematics to children is the "need orientation" of the child. Looking at the "needs" box we can see five types of needs that are listed in developmental order. That is, a child who is cold and hungry will not be interested in a mathematics activity designed to develop self-esteem. The wise teacher motivates children at their need level.

How Should Parents Be Involved in the School Mathematics Program?

Parents need to be kept informed of their child's mathematical progress and of the program in which he or she is involved. Several books on today's mathematics for parents have been written, and individual teachers and groups of teachers may well prepare periodic letters explaining the attack used on a particular topic. The interested parent may also take a vital part in the child's learning. It is suggested that groups of teachers at a given grade level work together in preparing letters or notes to parents, which can serve three major goals: (1) to inform parents of the content being studied, (2) to interpret the approach to that content being taken by the teacher (if the approach is different from that known by the parent, this provides an excellent opportunity for the teacher to explain the reason for the "new approach"), and (3) to suggest ways the parent can participate in aiding and encouraging the child.

HOW CAN BULLETIN BOARDS AND GAMES BE USED?

The Bulletin Board

Attractive and interesting science and social studies bulletin boards are found in most modern elementary school classrooms. However, a visitor to an elementary school classroom today would not be as likely to see one on mathematics. A mathematics bulletin board, properly handled, can do much to add interest to the program and challenge pupils' thinking. Several "idea books" are listed at the end of this chapter. Specific suggestions for bulletin boards are given below.

1. Use mathematics for bulletin boards to stimulate interest, to introduce a new topic, for enrichment, as a summary of a unit, to display mathematics themes, to elicit pupil discussion, and to illustrate number or geometric ideas.
2. Many mathematics bulletin boards should be handled by the teacher; others should be handled by pupils. I suggest that the teacher handle bulletin boards that are used to motivate a mathematics unit and that pupils take care of the boards that explain or summarize. When the teacher maintains high standards in bulletin-board construction, the pupils will do the same. It is rather easy to get an idea of the caliber of bulletin boards constructed by the teacher by looking at those of the pupils.
3. A wide variety of materials can be used for bulletin boards, and the boards can be used in various ways. For example, one can actually mount measuring instruments, use cloth or burlap for backing, or connect the bulletin board to the mathematics table with yarn or arrows.
4. Use captions that *ask questions.* A question is much more of an attention getter than a statement is. A question also forces the reader to do some thinking. For example, compare *Equivalent Fractions* with *What Are Equivalent Fractions?* or *Points* with *What Do You Know About Points?*
5. Bulletin-board material can be filed for future use. The teacher can make an easy-to-use file by sketching a picture of the board on a large manila envelope and then putting the materials in the envelope when they have been removed from the board.

Games

Games are an important part of the life of almost every child and every adult. Thus, they can be a very important vehicle in teaching elementary school mathematics. Here is a brief listing of suggestions concerning games and their uses:

1. Games can be used for practice, review, diagnostic testing, and concept development. It is of major importance that careful study be given to matching the game with the objective.
2. Use games for specific purposes rather than as time fillers.
3. Keep in mind the interest of the children in selecting games.
4. The majority of games should have self-correcting features.
5. The game should have enough of the element of chance so that it allows the student with less information to keep interested but also to see a need to improve his or her skill.
6. Normally it is best to use games that require little time to complete. It is probably better to play a game three times in a half hour than to require a half hour for its completion.
7. Check periodically to see whether the games are accomplishing the objectives for which they are being used.

WHAT ABOUT THE SLOW LEARNER?

Slow learners in mathematics often

1. Possess little motivation or drive. They are willing to just sit.
2. Enjoy "drill-type" work.
3. Possess short attention spans. They are easily distracted.
4. Possess a less-than-average memory. They may forget material from day to day.
5. Are low in verbal comprehension. They have difficulty in interpreting written material.
6. Are unable to be original in their thinking.
7. Are physically weaker, socially less secure, and emotionally less stable than the average.

These traits vary from pupil to pupil, but awareness of them should serve the teacher as a guideline in observation. In addition, the power of motivation should not be overlooked. It is possible for the slower pupils to do very fine work on a topic in which they are greatly interested or that they believe they have a chance of mastering.

The suggestions that follow are designed to improve the interest of the slow learner, although many of them apply equally well to the average and the above-average student.

1. Show interest yourself. Interest inventories reveal that arithmetic rates at or near the top as a favorite subject of children, but teachers do not rate it as high. Below-average pupils often reflect the enthusiasm or lack of enthusiasm of the teacher.
2. Often, the bright pupils are provided with attractive supplemental materials. Provide appropriate supplemental materials of equal attractiveness for the below average.
3. During a class discussion, use the "levels-of-depth" approach to discussion, in which the teacher allows the pupils who have used a very simple means of arriving at an answer to express their views first. Then he or she builds on their explanations by getting suggestions next from the average pupils, and last, the suggestions of the bright. By such an approach, all ability levels are able to make a contribution to the class consideration of a topic in mathematics.

4. Slow learners have few opportunities to present unique contributions to the class. During the course of the year, work out a "special" presentation with one of the slower pupils. "Knowing something first" is often a good incentive for continued work.
5. Provide situations that "make sense" to the slow learner. Bright mathematics pupils are often interested in the abstract puzzle aspects of mathematics. Although some slow learners are also interested in these aspects, in general it is important for them to see a use for the mathematics they study.
6. Know the interests of the pupils, and capitalize on them in the mathematics class. Many boys who are below average in arithmetic understand baseball percentages very well.
7. Be impartial. Show that you like each pupil and that you are interested in the work of each, even if the person will never be a good mathematician. Be positive rather than negative. Almost all slow learners react much better to "You're coming along better; let's see if we can do well on the next lesson" than to "If you don't improve, you're going to fail mathematics."
8. Point out the value of praise to the parents. Encourage them to take pride in some of the good work a slow learner accomplishes.
9. Be patient. When a pupil is trying very hard and working as fast as possible, it is very discouraging to hear, "Hurry up; quit loafing!"
10. Last and probably *most important,* give the slow learner a feeling of being important to you and of being liked as a person.

WHAT ABOUT THE GIFTED?

Giftedness is very difficult to define. The trait may be the result of various combinations of high intelligence, creativity, interest, motivation or drive, and curiosity. Further, the child may be gifted in many areas or traits or may have a gift or talent for particular subject matter.

Gifted students in mathematics often

1. Are more persistent in pursuing a task than others.
2. Are able to perceive mathematical patterns, structures, and relationships.
3. Possess a greater amount of intellectual curiosity and imagination.
4. Are physically stronger, socially more secure, and emotionally more stable than average.
5. Possess a superior vocabulary.
6. Are able to transfer mathematical learning to new or novel situations that have not previously been taught.
7. Are able to remember longer what they have learned.
8. Are flexible in their thinking.
9. Are able to discover new principles from the principles they already know.
10. Are able to think and work abstractly and enjoy working with abstractions.
11. Have a high verbal comprehension and are able to communicate mathematical ideas to others.
12. Are able to objectively analyze the strengths and weaknesses in their own mathematical thinking.
13. Are bored with "drill-type" work.[2]

[2]See: Riley, Jane, and Nancy Carlson, *Help for Parents of Gifted and Talented Children,* (Carthage, Ill.: Good Apple, Inc., 1984); and *Providing Opportunities for the Mathematically Gifted* (Reston, Va.: National Council of Teachers of Mathematics, 1987).

Should superior students be used to help the slower pupils? And if so, how? Probably the best answer to this question is "Sometimes." On occasion, both the fast pupils and the slow pupils can gain from working together. The fast pupil has to understand the mathematical content well enough to explain it clearly to the slower pupil, and often the thought processes the fast pupil goes through in verbalizing an explanation are helpful to the fast as well as to the slower pupil. If the "team-learning" approach is sometimes used in a classroom, there will be occasions on which the teacher can profitably team a slow student with a fast student.

This approach can be, and often is, overdone. The better student should not have the responsibility of being a tutor to a slower student. A rule of thumb might be this: Use this technique in situations in which you believe both parties will derive benefit from the exchange.

WHAT ARE SOME ORGANIZATIONAL PLANS FOR PROVIDING FOR INDIVIDUAL DIFFERENCES?

One Plan

A sixth-grade teacher found that many of the word problems presented in the textbook were not challenging the better students, and the poorer students were experiencing extreme difficulty with them. He devised a period-length problem-solving test made up of typical sixth-grade problems. The test was tape recorded for the benefit of pupils with reading difficulty, and each pupil also received a duplicated copy of the problems. The test was administered and corrected by the teacher.

In the next class period, the teacher began by saying, "Yesterday we took a test in problem solving designed to help you gauge your present problem-solving ability and to serve as a basis on which to check the improvement you make. I've developed two sets of problem exercises. The problems on the yellow sheet of paper are the more difficult ones. Those on the white sheet are not so difficult. I am giving you either a yellow or a white sheet on the basis of the score you made on the tests. Those who made below the median (halfway score) will get the white sheets; the others will get the yellow ones. After you've worked a while with the problems, we can more easily decide which sheet is best for you. Some of you may want to work with both sheets."

The first two problems were the same on both sheets. These were used as a basis for discussion and to acquaint the pupils with the materials. Pupils then began to work, with the teacher going about the room offering encouragement and help to those in need. After the pupils had been at work for a little while, the teacher called their attention to the fact that the last exercise, headed "How's Your P.Q.?" (problem quotient), was the most difficult on the page and was to be worked only if the pupil wished to do so. The teacher also suggested that when the students had finished all the exercises, they might compare their answers first with a companion and then with a check sheet on the teacher's desk.

During the weeks that followed, the teacher made use of the supplementary-worksheet plan when working on problem solving. At times, all the students were assigned a few problems from the text and their choice of problem-solving sheets.

Problem-Solving Lesson: Using Drawings and Diagrams (Above Average)

Read the problem carefully, and then make a drawing or diagram to use in solving the problem. Try to check your answer by using another method of solution.

1. Two people depart both from one place and both go the same road; the one travels 12 miles every day, the other 17 miles every day; how far are they distant the 5th day after departure? (Taken from *The Scholar's Arithmetic*, by Adams, 1812.)
2. Mr. Kramer and Mr. Black are going to put new fence around their farms. Mr. Kramer's farm is in the shape of a square that is 1 mile long and 1 mile wide, and Mr. Black's farm is a rectangle 1 mile long and 1/2 mile wide. How many times as much fence will be needed by Mr. Kramer then by Mr. Black?
3. Mary is cutting 4-inch-by-5-inch cards to use as tickets for the school play. How many cards can she cut from a sheet of cardboard that is 12 inches by 20 inches?
4. A circular fish pond has a border of red and white bricks. If it has 20 red bricks spaced evenly around the edge and follows a pattern of two red bricks, one white brick, two red bricks, etc., how many white bricks are there?

HOW'S YOUR P.Q.?

5. If a brick balances evenly with a 3/4 lb weight and 3/4 of a brick, what is the weight of the whole brick?

Problem-Solving Lesson: Using Drawings and Diagrams (Below Average)

Read the problem carefully, and then make a drawing or diagram to use in solving the problem. Try to check your work by using another method of solution.

1. Alice wants to divide 3 candy bars equally among 4 people. What fraction of a candy bar will each person's share be?
2. Clyde and his father set out for the mountains. In 3 hours they had gone 140 miles. They still had 28 miles to go. How far was it to the mountains?
3. One Friday George rode his bicycle $\frac{3}{10}$ mile to the store, $\frac{7}{10}$ mile to the YMCA, and $\frac{9}{10}$ mile to school. How far did he ride that day?
4. In the morning the snow behind Rachel's house was $6\frac{1}{2}$ inches deep. It snowed during the day, so that in the evening the snow was 12 inches deep. How much had it snowed during the day?

HOW'S YOUR P.Q.?

5. Jack said to his friend Bill, "I met a group of boys practicing marching. There were 2 boys in front of a boy and 2 boys behind a boy and there was a boy in the middle. How many boys did I see?" Try using a drawing to help you answer Jack's question.

Several features of the approach suggested can be noted:

1. Materials on the same topic but with varying levels of difficulty were provided. Note that the "How's Your P.Q.?" problem provided an opportunity for both the fast student and the slow student to work with challenging supplemental problems.
2. The introductory test demonstrated to the pupils who were having difficulty a need for the study of problem solving. One of the most important aspects in improving a pupil's achievement is to have the pupil see a need for an intensive study.
3. The pupils were allowed leeway in choosing material of varying levels of difficulty. Teachers using such materials have found that pupils who are experiencing some difficulty with problem solving will often ask to work both sets of materials. They are encouraged to get the majority of exercises on the easier sheets correct and to feel a sense of achievement if they are able to get any of those on the more difficult sheets correct.
4. The opportunity to check answers upon completion of the work provided for a correction of errors while the material was still fresh in the pupil's mind.
5. Class unity was still maintained, since the class discussed some of the problems worked in common. Class discussion also gave the pupils a chance to hear the explanations from their classmates. Often, another pupil is able to explain an idea in a manner that is understandable to the slow learner. These explanations also allow the able pupil to clarify in his or her own mind his or her method of solution.
6. The procedure provided the teacher with an opportunity to work with individual pupils. During this time, the teacher can also suggest materials for individual class members. In such cases, one or two pages from a supplementary book or worksheet may help to correct difficulty experienced by only one pupil.
7. A program such as the one described allows for a study in depth by the above-average pupil. This procedure has applications in all areas of the mathematics curriculum. The fact that pupils with varying degrees of mathematical ability can work on materials at several levels of abstractness allows the class members to move from topic to topic together. Some pupils will use concrete materials, some drawings, and some the standard algorithms, and some will work beyond the level of the standard form. Examples of pupils working at these levels are given in the chapters that follow.

Using Stations

Many teachers have found that the use of the "station" approach for a portion of each mathematics unit is helpful in making provisions for individual differences.

The station approach involves providing one to five sets of materials that may be at various levels of difficulty. Each station provides the challenge and the materials necessary to complete the work. Children often work at stations in pairs.

Classroom Organization

Assume that you have four groups of learners needing specific skill help. These groups have been formed on the basis of their lack of success in performing adequately on the pretests for some skill. You are planning to provide each individual group with 15 minutes of direct instruction. Your task is not only to plan the four skill lessons but also to plan activities that can be used by the learners not directly involved in the groups you are teaching.

While the teacher gives the 15-minute lesson on a particular skill to the first group of children needing it, the other children are occupied by working with individual materials (laboratory or otherwise). In addition, corrected papers and special assignments can be used as a part of the independent activities to be carried out by each child when he or she is not involved in a skill group. A bulletin board might list specific directions regarding independent activities the children can perform. Recreational reading, listening activities, or practice-material packets based on mathematics skills taught previously are other ideas for independent tasks.

The most effective classroom organization cannot take place overnight; it will involve several weeks of careful planning. The following issues will need careful consideration: (1) administering of the pretests for the skill objectives, (2) recording pretest performance on your recordkeeping device, and (3) planning for independent activity that will involve the rest of the class meaningfully while you are working with the skill groups.

Enrichment and Remedial Laboratories

Some schools provides special instruction once or twice a week for pupils above or below average in mathematics achievement. These special sessions are usually 30 to 50 minutes long. Although the pattern or organization varies, usually some form of team teaching is used. For example, in a school containing four fifth grades with a total of 120 fifth graders, one teacher may work on enrichment materials with the top students, a second teacher on remedial work with students in need of a particular kind of remedial aid, and the other two teachers with the average achievers on an appropriate topic. In other schools, a special mathematics consultant may work with enrichment materials with all classes. Still another pattern is for a teacher with a special interest in mathematics to work with gifted children from several rooms during the period in which music or art is taught in his or her homeroom. The success of such programs has varied with the enthusiasm and capabilities of the teachers involved.

Team Learning

On occasion, the idea of "team learning" may be employed. In the team-learning situation, two or three children of about the same ability level work on materials as partners. The children study the materials together or work with the teacher.

This procedure has a great deal of merit if wisely used. There are many laboratory lessons, game situations, and practice materials that can be effectively handled through the team-learning approach. It should be noted that the pupils are able to make their own discoveries.

Some teachers have found that it is effective to begin some lessons with children working individually and then move into a team-learning situation in which the children compare their findings.

Individualizing Instruction

Currently, many educators are calling for almost complete individualizing of instruction. If this means that the children are to be working on materials that are appropriate for their level of maturity and interest, this idea is very good. If, however, the goal is to have each child in a classroom working on a different set of materials, the idea is probably unrealistic, since only highly self-motivated pupils can maintain interest in mathematical study without the stimulation of discussion, arguments, and other exchange with peers and teachers.

Plans aimed at total individualization, often called individually prescribed instruction, have yet to produce consistently better achievement than "typical" large-group and small-group instruction. This may be due to a lack of superior teaching materials or a lack of sustained interest in working alone.

Another danger in complete individualization is the lack of direct attention that the child would receive from the teacher. In a class of 30 children, how much time would each child receive if the class lasted 40 minutes?

It should be noted, however, that materials designed for complete individualization can be very effectively used at points in the program where working individually is important and necessary.

Variations

It should be remembered that there is no *one* perfect procedure for providing each child with *the* mathematics he or she needs. The teacher must act as a conductor calling on almost all the various procedures for dealing with individual differences. Too often in the past, teachers or school systems have settled upon one procedure as a panacea, only to completely reject the program a few years later.

The flow chart and table in Figure 2–3 are designed to give the teacher an idea of the decisions that must be made for each unit of mathematics instruction. The teacher must consider the content to be developed, the materials that can be used for instruction, and the best approach to teaching that particular topic, all in the light of the individual children. With these variables in mind, the teacher then decides which mode of instruction (large groups, small groups, individual) will be best for the lesson, group of lessons, or unit.

How Can All Children Have Enrichment Mathematics?

Not only must the mathematics program be appropriate in difficulty for all children, it must provide in-depth or enrichment experiences for all children, including the mentally retarded and the handicapped as well as the gifted child. Enrichment can be thought of as a combination of *breadth and depth* expe-

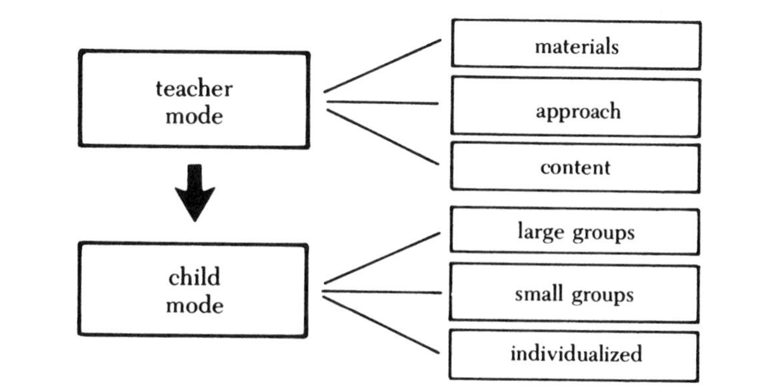

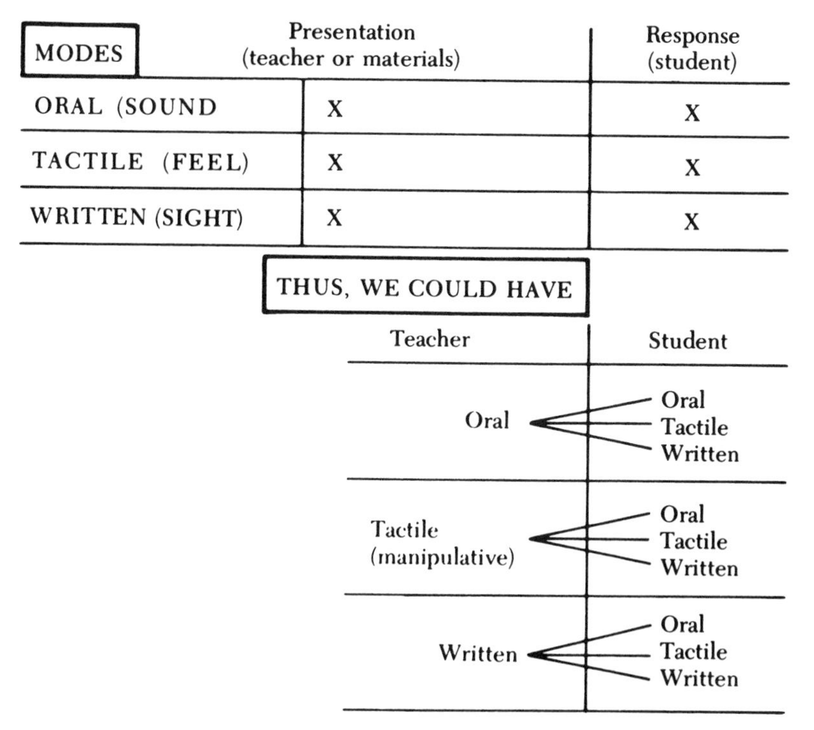

MODES	Presentation (teacher or materials)	Response (student)
ORAL (SOUND	X	X
TACTILE (FEEL)	X	X
WRITTEN (SIGHT)	X	X

Figure 2–3 Teaching Decisions

riences: breadth in that the materials used for enrichment will be broader in scope than the typical textbook program while being related to the program and depth in that enrichment materials may be designed to "dig deeper" into the mathematical ideas than is typical. Several "breadth and depth" suggestions follow:

1. Emphasize the multiple solution to number operations and verbal problems. The inventive pupil may develop several alternate solutions to an exercise while the slower student is developing one. Such a procedure not only adds to the learning situations but is also a valuable means of balancing the length of time required for the class to finish an exercise.

2. The use of small groups using a laboratory strategy often provides an opportunity to work with other groups on specific difficulties.
3. The use of a mathematics corner that contains appropriate manipulative materials, project cards, and work space is a great aid in organizing for group instruction.
4. Make use of historical materials. Such materials help to reveal the number system as a human invention and to develop cultural appreciation of it.
5. Make use of a mathematics bulletin board. This should act as a motivator for a unit. In addition, the bulletin board should always include several challenging problems, puzzles, or exercises developed at several different levels of difficulty so that every pupil can work at least one.
6. Provide supplemental worksheets containing challenging material. It should be noted that for many years, supplemental puzzle-type exercises have been used by the better teachers. It is suggested that a somewhat different stress be placed on these exercises; that is, not only should figuring out an answer be stressed, but the pupil should be directed to figure out what mathematical principle is involved in the solution.
7. Provide books and booklets that emphasize the use of mathematics in various careers (for example, National Council of Teachers of Mathematics, *Careers in Mathematics*). Pupils are often amazed at the need for mathematics in apparently non-mathematical occupations.
8. At the end of units, make use of "review study questions." In addition to providing a general review of understanding, some of these questions should encourage investigation and exploration. Here are examples of such questions:
   Grade Three (addition). How could you rename one of the 8s in $8 + 8$ so that a friend who knew only addition combinations with sums of 10 or less could probably understand? [Answer: $8 + (2 + 6) = 10 + 6 = 16$] Use of the associative principle.
   Grade Five (multiplication of fractions). What is the smallest number and the largest number that could be the product of a little less than 3 times a little more than 4? *Note:* A little less than 3 would be $>2\frac{1}{2}$, but <3; a little more than 4 would be >4, but $<4\frac{1}{2}$. [Answer: 10 plus to $13\frac{1}{2}$ minus] Smallest: $2\frac{1}{2} \times 4 = 10$. Largest: $3 \times 4\frac{1}{2} = 13\frac{1}{2}$.

What is Typically Taught at the Various Grade Levels in the Elementary School?

Each state—and to an extent, each school district—is responsible for establishing the grade placement of topics. Also, the teacher needs to consider the students' maturity and intellectual ability. However, there is a great deal of commonality in textbook series, state curricula, and school curricula. The content summary table following is typical of the current grade placement of mathematical topics.

How Do I Use a Textbook? How Can I Evaluate Textbooks?

THE ROLE OF THE TEXTBOOK

The textbook is an indispensable item in the operation of a successful elementary school mathematics program. However, too often it becomes the only instruction material used in the mathematics program. This is unfortunate

Content Summary Table

At all grade levels, Open Court's Real Math places appropriate emphasis on real applications, thinking skills and problem-solving strategies, mental arithmetic, estimation, approximation, measurement with metric units, and topics in geometry, probability, and statistics. In addition to the topics listed below, substantial review is built into each grade level.

LEVEL K	*LEVEL 1*	*LEVEL 2*	*LEVEL 3*	*LEVEL 4*	*LEVEL 5*	*LEVEL 6*	*LEVEL 7*	*LEVEL 8*
Numbers (cardinal and ordinal) through 10: counting; writing numerals; measurement using nonstandard units; one-to-one matching; preparation for adding and subtracting.	Number 0 through 100; addition and subtraction concepts; basic addition facts (through 10 + 10); measurement with nonstandard units; introductory work with multiplication, fractions, recording data, maps, and inequalities.	Numbers through 10,000; basic addition and subtraction facts; multidigit addition and subtraction algorithms; introduction to multiplication and division; measurement with standard units; fractions of area and fractions of numbers; reading maps.	Numbers through 1,000,000 and beyond; fractions and decimals; multiplication and division; multiplication facts through 10×10; multidigit multiplication algorithms; measurement; introduction to graphing and functions; adding and subtracting decimals.	General multidigit multiplication algorithm; division by a 1-digit divisor; digit divisor; addition and subtraction of common fractions; rounding and approximation; linear, composite, and inverse functions—and graphing; multiplying decimals and whole numbers; introduction to mixed numbers.	Multidigit division algorithm; use of handheld calculator; rounding; linear, composite, and inverse functions—and graphing; introduction of negative numbers; rates, ratios, and percents; relation of fractions and decimals; addition, subtraction, and multiplication of fractions, mixed numbers, decimals; division with decimal dividends and quotients.	All operations with whole numbers, fractions, decimals; some operations with negative numbers; compass-and-straightedge constructions; nonlinear functions—and graphing; exponents; use of handheld calculator.	All operations with rational numbers; computational shortcuts; computer literacy; use of handheld calculator; exponents; paper folding and solids; squares and square roots; use of percents, variables; linear and nonlinear functions—and graphing.	All operations with real numbers; solving equations algebraically and graphically; statistical tools, such as mean, mode, median, range, and clipped range; analyzing standardized tests; use of handheld calculator; computer literacy.

From *Open Court/Real Math,* undated pamphlet. Courtesy of Open Court Publishing Co., LaSalle, Illinois.

and may happen because the teacher lacks a clear-cut philosophy concerning the use of materials. The following suggestions for the use of materials are in keeping with the approach to the teaching of elementary school mathematics developed in this book:

1. Normally, an introduction to a new topic should involve a pretext development that allows for pupil thinking and discovery. Suggestions for this type of introduction is detailed throughout the book. For introductory lessons the text serves as a reference "to check to see if the book agrees with our statements." It also provides extension material for further development of the topic.
2. Textbook materials normally contain fewer errors and ambiguities than teacher-prepared duplicated exercises. Therefore, they provide the single best source for practice material, once the basic understanding of a mathematical idea has been developed. In addition to the textbook used by all of the pupils, the teacher should have special exercises for below-average and above-average pupils. Well-developed teacher-made exercises are of great importance.
3. If a teacher uses a variety of teaching approaches, the textbook will have to be supplemented with laboratory cards, games, and special enrichment and remedial materials.
4. There are now series of textbooks specially designed for the lowest 30 to 50 percent of the mathematics pupils. Thus, it is often desirable to use two or more textbook series with the children in a typical elementary school classroom.
5. The majority of superior textbooks provide a teacher's manual with suggestions for appropriate instructional equipment, reference books, games, and other activities. These suggestions should aid the teacher in presenting a varied mathematics program. The teacher should make a careful evaluation of the merits of various suggestions because many books provide more suggestions for materials than are appropriate for a single day's study of mathematics. Also, some of the suggestions are of dubious value.
6. If the sequence of topics in the adopted textbook is in keeping with the curriculum of a school, it is normally best to follow that sequence. This does not mean that every page needs to be studied by the class. Most books provide more practice material than is necessary for the average student. The teacher is responsible for using discretion in selecting how much of the material on a given topic will be of value to the class.
7. The early chapters of many texts are a review of the previous year's mathematics program. In situations where this is the case, the teacher should devise a variety of pretext reintroductions to improve motivation on the topic and avoid making children feel that "we're doing the same old thing."
8. In the intermediate grades, where pupils can readily read the material, the textbook is usually of greater value than in the primary grades. The teacher of primary-grade pupils will have, at best, a mediocre program if only the textbook is used.
9. Textbooks vary in quality and approach. Often an adopted series has many superior materials and some that are not of high caliber. The teacher can often receive valuable ideas for classroom instruction from the teacher's manuals of several other series.
10. While the textbook plays a very important part in the elementary school mathematics program, it is only one source of needed materials. A superior mathematics program will use many others.

SELECTION OF THE TEXTBOOK

The textbook selected for a school system should be in accord with the philosophy, curriculum, and goals of the elementary school mathematics program. Because teachers use the adopted textbook series extensively, great care should be taken in its selection. The vignette that follows presents a possible plan for textbook selection.

The superintendent called together the administrative personnel who worked with elementary school mathematics, principals, supervisors, and curriculum consultants, and charged the mathematics consultant (or an elementary principal interested in mathematics) with the responsibility of setting up a committee to select an elementary school mathematics textbook series. In addition, the superintendent provided each of the participants with a policy statement concerning ethics in textbook adoptions. The remainder of the meeting was turned over to the consultant, who was to act as the coordinator for the textbook selection. The coordinator asked that principals recommend teachers who were interested in mathematics curriculum improvement.

From the list of interested teachers, five teachers were appointed to the committee. They represented the early primary, primary, and intermediate grades. The high school mathematics chair was available for consultation concerning coordination (K–12). The selection committee chair wrote to each textbook publisher who published elementary school mathematics textbooks (the list of publishers was obtained from the American Textbook Publishers Institute, 1 Madison Avenue, New York) and also obtained copies of the current professional books concerned with the teaching of elementary school mathematics.

When the textbooks and professional books arrived, the chair called a meeting of the committee. At this meeting the committee members were informed that (1) a total of 10 half-days would be made available for study meetings and other work (substitute teachers would be hired to free the committee members), (2) clerical help needed for typing and duplication was being made available, (3) the entire staff of teachers was being informed concerning the membership of the committee and was being asked to make suggestions and to consult with the committee (the committee was free to use the other teachers as advisors), (4) the selection was to be made in 6 months, and (5) the new course of study was to be considered as a guide to help in textbook selection.

Because almost 20 textbook series were submitted for consideration, the committee agreed that the first efforts should be devoted to eliminating those series that did not fit the goals of the school district and to narrowing the field to six. To accomplish this task, each member was asked to review the textbooks at the grade level at which they taught and to select six they thought merited further consideration. The chair selected a grade level that was not represented by committee members for study.

The committee members were given the district policy statement concerning textbooks and the following tips on textbook selection.

WHAT A GOOD TEXTBOOK SHOULD DO (IN KEEPING WITH THE COURSE OF STUDY FOR THE SCHOOL DISTRICT).

1. Provide teacher direction and guidance, not dictation and limitation.
2. Provide basic material for study and review.
3. Be interestingly written and well organized.
4. Be free from mathematical inaccuracies.
5. Reflect the latest research in teaching methods and use an exploratory-discovery approach to teaching.
6. Provide a common frame of reference.
7. Show careful development of mathematical principles and generalizations.
8. Provide gradation of topics according to difficulty.
9. Provide appropriate diagrams, charts, and illustrations.
10. Provide a spiral arrangement of topics that promotes continuity from grade to grade.
11. Provide good problem situations and material emphasizing the usefulness of mathematics.
12. Provide variation in content to meet individual needs.
13. Provide interesting and challenging practice exercises to develop basic skills.

ERRORS TO BE AVOIDED

1. Don't let a single error prejudice you against the entire series.
2. In final textbook selection, keep in mind all grades, not just your own.
3. Consider only the materials that are available.
4. Do not let minor conflicts between a series and the course of study (particularly in grade placement) rule out a series. It may be necessary to use only portions of the adopted series, or it may be necessary to change the course of study.
5. Take care in using textbook selection guides provided by a publisher of mathematics textbooks. These are often slanted toward that particular series.

FORMAT AND CONTENT CONSIDERATIONS TO BE KEPT IN MIND.

Format. Cover design, style of type, size of type, layout of pages, consistent placement of page numbers, use of color, eye appeal, binding, glossary, table of contents, index, appendix, illustrations.

Content. Development of meaning of mathematical operations and of number systems (whole numbers, rational numbers, integers), problems, tests, study suggestions, exercises.

Two weeks later the committee met, and each member submitted the names of the six textbooks that were felt would best fit the needs of the school's mathematics program. From those lists, six text series were suggested for further consideration. Committee members then went about rating the series by using the specific topics suggested above. Less weight was given to format than to content. Specifically, the committee members were asked to rate each item on a five-point scale from 1 (included but poor) to 5 (superior). For example, the textbook series might rate 3 on verbal problems and 1 on tests.

When this rating was completed, the six textbook publishers were asked to send in representatives who were to give a presentation to the committee concerning specific topics. The representatives were to be ready to answer questions submitted by the committee. After meeting with all of the representatives, the committee compiled rating points and selected a textbook series.

After the selection of the series, a plan was formulated to educate the teachers in the use of the series. Also, plans were laid to evaluate the series after a year of use.

KEEPING SHARP

Self-Test: True/False

_______ 1. Children are easily categorized by their learning problems.

_______ 2. Individualization of instruction has proven to be the best means of providing for individual differences.

_______ 3. Children with specific learning disabilities are almost always placed in special classes.

_______ 4. Motivation is probably the single most important procedure for improving achievement in mathematics.

_______ 5. A station approach does not require many special materials.

_______ 6. Only very standard materials should be used with slow learners.

_______ 7. When dealing with a child having difficulty in mathematics, if one approach does not work, another should be tried.

_______ 8. The calculator is a good device for use with EMR students.

_______ 9. Acceleration is the best method of providing for the gifted.

_______ 10. The slow learner can engage in the "breadth" type of mathematical experiences.

_______ 11. You cannot have a good mathematics program if you have a poor textbook.

_______ 12. The role of parents should be an important part of your planning.

THINK ABOUT

1. Find a school, college, or university that has several elementary mathematics textbook series. Study them for the scope and sequence of the material taught and the provision for individual differences.
2. Take one textbook series and apply the criterion for textbook selection given in the chapter.
3. Design a letter to a parent explaining one phase of the elementary mathematics program.
4. Why do we group pupils for instruction according to their chronological age (typical first-grader is 6 years old) when we know that the ability to learn is more closely related to mental age than to chronological age? Think of reasons for and against.

5. To provide suitable content for study by the superior pupils in a fifth-grade class, sixth-grade textbooks must be used. What are the chief objections to such a procedure?
6. If content from the upper grades (see question above) is not to be used, what is to be the source of the arithmetic content for superior pupils?
7. Why has within-class grouping for instruction in arithmetic not been as popular as it has been in reading?
8. What features of the developmental method of instruction in arithmetic make it desirable as a method of providing for individual differences in ability within a class?
9. Should fifth-grade pupils who are doing remedial work be left with the class or put into a separate class? For what reason?
10. How much of a problem is the grade placement (assignment to grades) of exercises in an arithmetic enrichment program?

SUGGESTED REFERENCES

ASHLOCK, ROBERT B., *Error Patterns in Computation,* Columbus, Ohio: Charles Merrill, 1984.

BARATTA-LORTON, ROBERT, *Mathematics: A Way of Thinking,* Menlo Park, Calif.: Addison-Wesley, 1977.

BAROODY, AUTHUR J., *Children's Mathematical Thinking,* New York: Teachers College Press, 1987.

BITTER, GARY G., MARY M. HATFIELD, and NANCY T. EDWARDS, *Mathematics Methods for the Elementary and Middle School,* Boston, Mass.: Allyn and Bacon, 1989.

Developing Computational Skills, 1978 Yearbook, chap. 1–5 and 14, Reston, Va.: National Council of Teachers of Mathematics, 1978.

Estimation of Mental Computation, 1986 Yearbook, chap. 19, Reston, Va.: National Council of Teachers of Mathematics, 1986.

Mathematics for the Middle Grades (5–9), 1982 Yearbook, Reston, Va.: National Council of Teachers of Mathematics, 1982.

Providing Opportunities for the Mathematically Gifted, Reston, Va.: National Council of Teachers of Mathematics, 1987.

RILEY, JANE, and NANCY CARLSON, *Help for Parents of Gifted and Talented Children,* Carthage, Illinois: Good Apple, Inc., 1984.

THIESSEN, DIANE, MARGRET WILD, DONALD D. PAIGE, and DIANE L. BAUM, *Elementary Mathematical Methods,* 3d. ed., New York: Macmillan, 1989.

WIEBE, JAMES H., *Teaching Elementary Mathematics in a Technological Age,* chap. 10, Scottsdale, Ariz.: Gorsuch Scarisbrick, 1988.

chapter 3

EARLY CHILDHOOD COUNTING AND PLACE VALUE

OVERVIEW

"Bill has more candy than I do."
"How many more pennies do I need to buy a candy bar?"
"Aunt Nancy gave Tom a nickel and me a penny. Who can buy the most?"
"I can count to 10. Want to hear me?"
"Jack is bigger than I am. Mother is bigger than Jack. Dad is the biggest one in our family."

Early in the life of a child, parents or brothers and sisters begin to acquaint him or her with mathematics. Many times each day, the child hears references to time, temperature, the number of objects, days of the week, shapes of objects, and other mathematical situations.

Preschool children vary in the contacts they have with mathematics. Studies reveal that the average beginning kindergarten child recognizes or is able to understand, at least to some extent, (1) rational counting (counting with meaning) beyond 10 by ones; (2) sets of one, two, three, and four objects; (3) situations requiring the use of *largest, smallest, tallest, longest, inside, beside, most, closest,* and *farthest;* (4) the calendar, foot rule, and quarts; (5) the telling of time on the full hour; (6) the geometric figures of a circle and a square; (7) the solution of simple, orally presented problems involving addition and subtraction; and (8) an object cut into halves or thirds.

It should be remembered that although many beginning kindergarten children have lower skills than those listed, many have more. The teacher of young children should enrich the pupils' backgrounds, in addition to correcting any mathematical misconceptions they have.

TEACHER LABORATORY

Materials

Shapes (see accompanying figure) in four colors.

Experiment:

1. What different attributes (characteristics) do the shapes have?
2. With at least two other persons, play this game: With the set of colored shapes, you are to pick in turn a shape that is different from the first shape, played in two—and only two—ways. For example, the first person might pick up a large/red/triangle; the next person might then pick up a small/blue/triangle (different in two—and only two— ways). The game would continue until a player is unable to find a piece that meets the criteria.
3. Make up a similar game that early schoolchildren could use in learning to classify objects.
4. Make up a set of materials with different attributes (characteristics) of real-world objects familiar to kindergarten and first-grade children.
5. Go to a library, and look through several children's books. Select one that has some ideas involving numbers. Develop ideas for using the book in mathematics lessons.
6. It is often suggested that the mathematics program for young children should be directly related to their "real" world. Try to think of an everyday activity or interest of the children for each letter of the alphabet. Then compare your list with others'.
7. Develop a set of 10 problem situations involving mathematical ideas that you might present to a 5-year-old child. Find one or two children of that age and try them out. Revise, and try them out on one or two other children.
8. Have several friends write large numbers, such as 876,549, as you state them orally. Analyze the writing. Which numerals are least legible? Compare your findings with those reported in the chapter.

TAKE INVENTORY

Can You:

1. Describe the role of structure versus incidental learning in the pre-first-grade elementary mathematics program?
2. List several suggestions for the teaching of ideas associated with sets at the elementary school level?
3. Describe the role of pattern searching in pre-first-grade mathematics?
4. List three counting experiences that can be used to improve children's mathematics concepts?
5. Give four suggestions for teaching children to read numerals?
6. Suggest three materials that can be used to improve children's ideas of number and comparing number?
7. Give four suggestions for improving children's concepts concerned with place value?

STRUCTURE VERSUS INCIDENTAL LEARNING

Great variation in mathematical content and mathematical emphasis can be found at the preschool and kindergarten levels. An observer visiting kindergartens throughout the nation would find some situations where the teacher carefully avoids any reference to numbers, others where the program consists of rote counting and an occasional use of the classroom store, and still others where the instruction in mathematics is very systematic and rather abstract.

The content of mathematics for the kindergarten should meet the same criteria demanded for the grades, but should have a somewhat different emphasis. At the kindergarten level, the pupils should be given wide opportunities to learn mathematics, but without extensive study for mastery. The kindergarten program should provide the pupils with ideas and experiences upon which they can build. Thus, it should be (1) understandable, (2) accurate, (3) designed to make sense to the learner, (4) geared to give the pupils many opportunities to explore ideas without pressure for mastery, and (5) related to the physical world. As the studies of Piaget and others have shown, it does little good to have children practice saying, "Four plus five equals nine," or writing the symbols for such an addition, when they have not had a great deal of experience with number ideas in the form of objects and pictures of objects that make the ideas meaningful.

By the end of kindergarten, the children should have experience leading to

1. Classifying objects on the basis of size, shape, and color.
2. Counting rationally to 10 and beyond; experience counting to 100 by using tens and one.

3. Recognizing the numerals to 10; recording numbers as tens and ones.
4. Developing a useful vocabulary of mathematically oriented words based upon experience (see Chapter 2): such as few, many, big, little, up, down, more than, less than, short, tall, light, heavy.
5. Comparing shapes leading to identification of rectangles, squares, triangles, balls, cubes, boxes, etc.
6. Copying, extending, and inventing patterns with blocks, pictures, and shapes.
7. Identifying by counting sets of one more than, one less than, two more than, two less than, etc.
8. Matching numerals to sets from 1 to 10 and beyond.
9. Using nonstandard units of lengths and weights to compare objects (see Chapter 14).
10. Solving a variety of word problems involving readiness for addition, subtraction, multiplication, division, fractional, and decimal work.
11. Developing ideas concerned with place value (positional notation).

The emphasis for the pre-first-grade program should be to provide the children with a wide variety of experiences with little stress upon mastery. Those pre-kindergarten and kindergarten programs that stress mastery have begun a pattern of failure for some students while beginning a pattern of boredom for others. Mathematics for these children should stress discovery and thinking and, above all, be exciting.

A LOOK AT KINDERGARTEN TOPICS

One of the most complete kindergarten programs was developed at the University of Wisconsin.[1] What follows presents an overview of this type of program, with teaching suggestions concerning a number of specific topics.

Some Basic Processes

The following basic processes involve the use of actual experimental situations involving the child in active learning:

1. *Describing.* The process of characterizing an object, set, or event in terms of its attributes. (See Figure 3–1.)
2. *Classifying.* The process of sorting objects, sets, or representations on the basis of one or more attributes. (See Figure 3–2.)
3. *Comparing and ordering.* The process of determining whether two objects, sets, events, or their representations are the same or different on a specified attribute. (See Figure 3–3.)
4. *Equalizing.* Equalizing is the process of making two objects, sets, or representations the same for an attribute. This can be done by taking away from a "larger" or putting on a "smaller." (See Figure 3–4.)
5. *Joining and separating.* The process of putting together two objects, sets, or representations that have an attribute in common. (See Figure 3–5.)

[1]The Wisconsin Research and Development Center for Cognitive Learning, *Resource Manual, Topics 1–40, Developing Mathematical Processes* (Chicago, Ill.: Rand McNally, 1974), pp. 21–26.

What would you tell
a friend about Silly?

He's purple! . . . long
legs . . . spots on
his hat . . . fuzzy . . .
stripes on his shoes.
He's happy . . .

How are these different?
How are they alike?

One's blue, the other's
green. One's longer . . .
one's higher. One goes
faster . . . They both have
four wheels . . . two doors . . .
windows.

What do you know
about this solid?

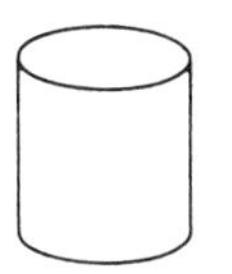

It rolls . . . it slides . . .
it stacks . . . it's three links
tall. (and later) It's
eight centimeters tall . . .
it's nine centimeters
around . . . the area of top
is about nine square
centimeters . . . it weighs
two ounces . . . its volume is . . .

Figure 3–1 Chart for Describing

Put all the long ones together and all the
short ones together.

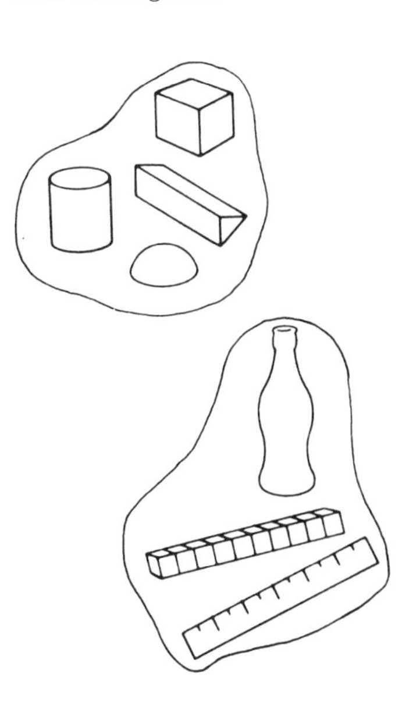

or

Put the ones that are alike together.

Figure 3–2 Chart for Classifying

Are they the same height?

no

Are there as many cubes as links?

no

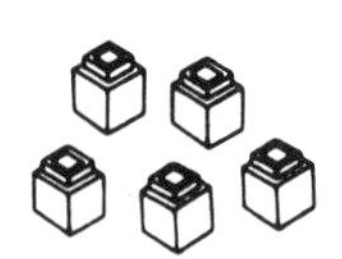

Are they the same weight?

Do these have the same shape?

yes

Figure 3–3 Chart for Comparing

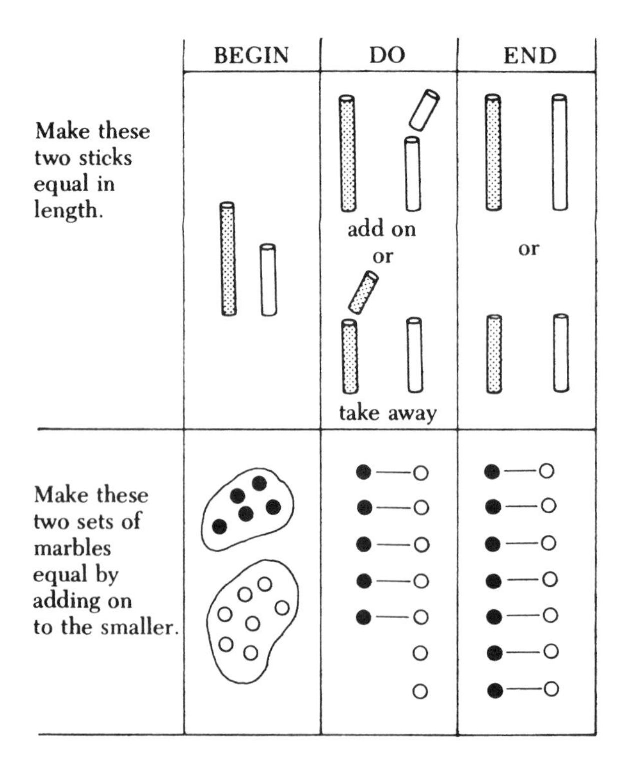

Figure 3–4 Chart for Equalizing

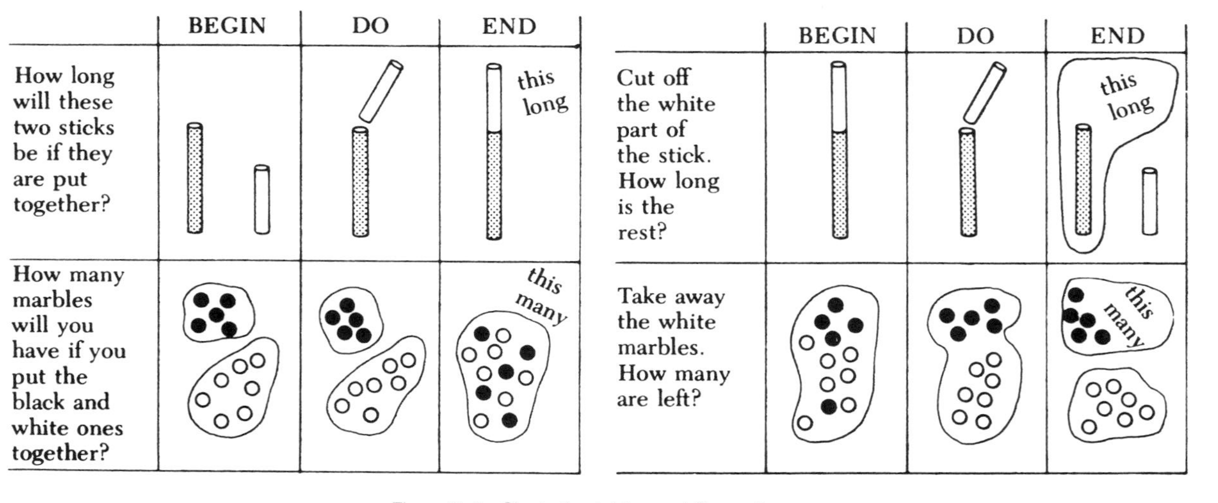

Figure 3–5 Charts for Joining and Separating

6. *Grouping and partitioning.* The process of arranging a set of objects into equal groups of a specified size, with the possibility of one additional group of any leftovers. (See Figure 3–6.)
7. *Representing and validating.* The process that enables a child to progress from solving problems directly (the real situation) to solving them more abstractly. (See Figure 3–7.)

EARLY ELEMENTARY TOPICS

CLASSIFYING

An alert teacher can use situations that arise in normal classroom activities to develop ideas concerned with identifying, describing, and classifying sets. *Group thinking* and *Socratic questioning* can be used to develop some of the basic concepts, and then a *laboratory* strategy can be used to extend and solidify the ideas.

After children had finished some free play with blocks, a teacher said, "Let's put the blocks away in boxes (or on shelves) so that it will be easy for us to find particular blocks we want next time we use them. How should we decide which blocks go on which shelf?" The children suggested that the blocks could be grouped into several sets depending on how they were to be used. One child said they should be grouped by size, another said to group them by color, and still another said it would be better to group them by both size and shape. The children decided that it would be most helpful if they had the blocks grouped by their shape. Taking this suggestion, the teacher used questioning to bring out the properties of set membership, questions such as, "Do these blocks belong in the same set?" (yes) and "Why?" (because they both have six sides). The teacher continued the same type of questioning to bring out comparison of the attributes of members of a set. In the days that followed, activities were provided that gave children opportunities to classify set members by color, size, shape, and other properties, such as cold and hot, shiny and dull, and boys and girls. The classifying activities, in addition to being inherently useful, provide readiness for future study of logic.

There are many objects that can be used for this purpose, such as leaves, sea shells, buttons, toy cars and trucks, dolls, nuts and bolts, paper clips, or rubber bands. It was previously mentioned that laboratory-type games provide valuable experience in classifying attributes. The materials that can be used for such games, as well as several of the games themselves, are described below:

I can top that. Materials: a typical classroom. Three or four children play. Directions: The first child names an object in the room. The next child names a larger object. This continues until no larger object can be named.

Shape and color. Materials: attribute pieces such as those shown in the "Teacher Laboratory" at the beginning of the chapter. Directions: Each child has a bag of attribute pieces. The leader takes a piece out of his or her bag and holds it up. The other players reach into their bags and, without looking, try to

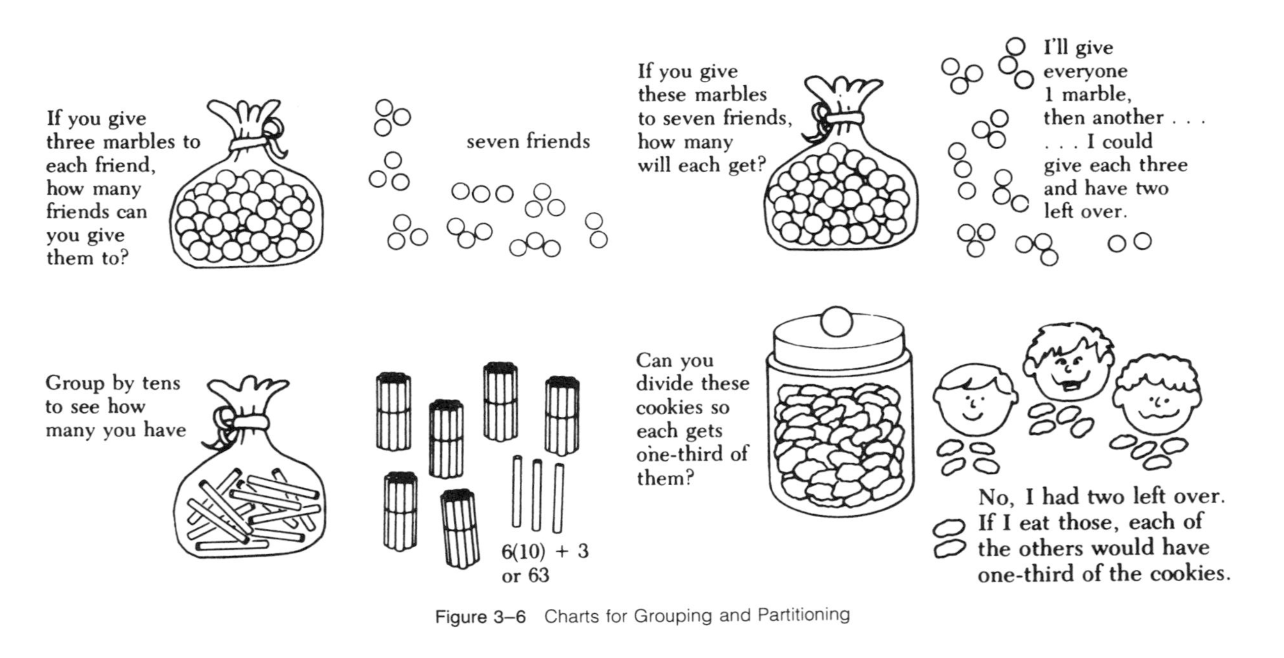

Figure 3–6 Charts for Grouping and Partitioning

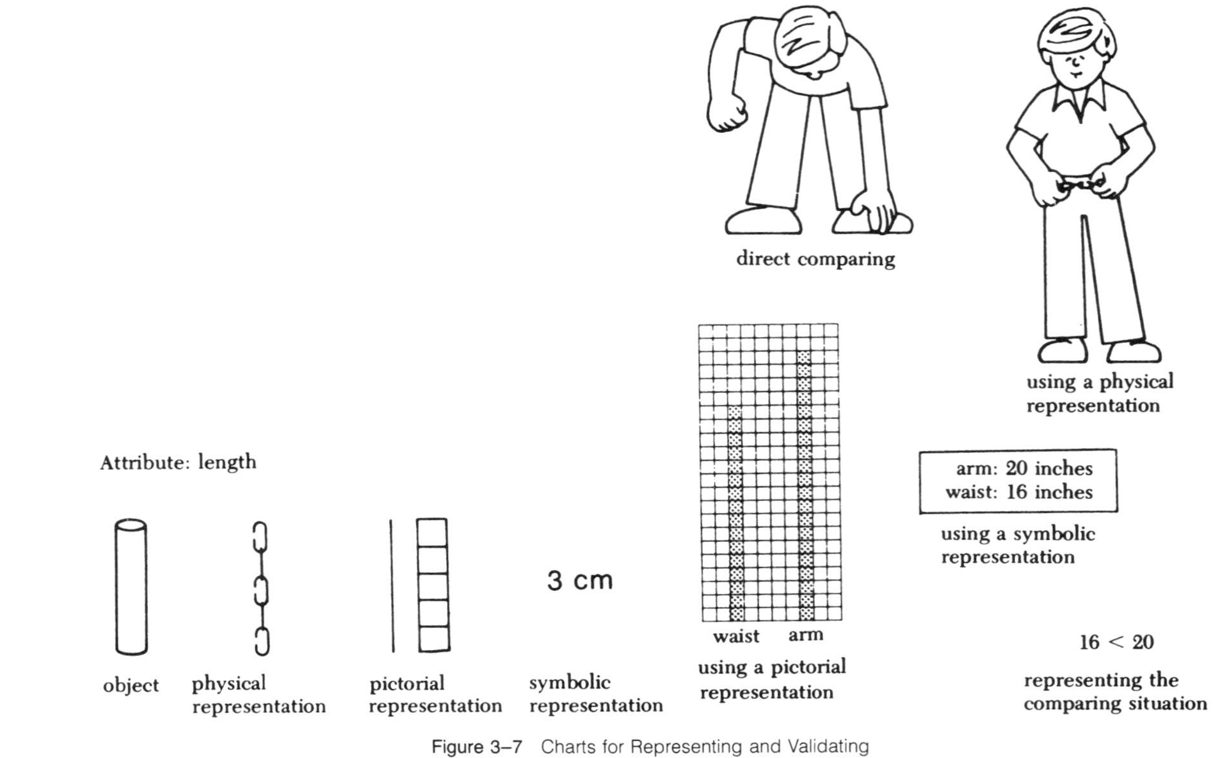

Figure 3–7 Charts for Representing and Validating

draw out a piece of the same shape. They then look at the color of the piece. If it matches the leader's piece, they call, "Color and shape." The first child to call the match takes the leader's piece. The first child to get four pieces becomes leader; the pieces are then redistributed, and the activity starts with the new leader.

Alike in one way. Materials: attribute pieces. Directions: Three or four children sit on the floor with their attribute pieces in front of them. The first player places any piece in the center of the group. The next child is to place another piece on top of the first that is like it in one (and only one) way. For example, one child sets out a small blue circle. The next child then sets out a large yellow circle. A child who makes a mistake loses a turn. The player with the fewest attribute pieces left when no one can play is the winner.

Different in one way. Materials: attribute pieces. Directions: The game is very similar to "Alike in one way." Children take turns setting out pieces that are different in one (and only one) way.

SIDE TRIP

Try the following activities as your friends arrive for class in the next few days:

1. Decide on two categories by which you will be able to sort all the students so there will be no students left out. Either list those in each group or sufficiently describe the group so that anyone else could determine its members.
2. Predetermine three categories or groups and repeat activity 1. Are there questions about group membership? Do some students fit in two groups? Can you change your groups so that each student truly belongs in only one group and all students have a group?
3. Try it again with four categories. Did your experiences with three categories help?
4. Discuss what you did with five or six classmates. Try to remember how you did your sorting, questions, and problems.
   a. What activity was easiest? Why?
   b. What was needed as more categories were added?
   c. Would the sorting be more or less difficult if the groups could overlap? (Some students could belong to more than one.)
   d. Is it better to predetermine categories or to wait until you see those to be sorted?

Try your own. Materials: colored crayons and 3×5 cards. Directions: Children make their own sets of attribute pieces by drawing cartoon characters. For the first try, this activity usually takes some guidance by the teacher. Later, children can develop their own sets of pictures and play the games in the same way as those described earlier.

There are many variations on these games. Often it is a good learning experience for the teacher and groups of the children to try to invent such games. In addition, there are a number of different environmental materials that can be used for attribute games; for example, pencils in four colors, of long and short lengths, with erasers and no erasers, and sharpened and not sharpened.

In addition to these types of activities, books and materials designed for reading readiness can serve double duty for use in classifying activities. Also, catalogs and magazines provide pictures that can be classified by using various properties.

COMPARING SETS

Problem situations and activities in which children have an opportunity to make comparison of objects provide the basis for developing this skill; for example, "Who has more maple leaves, Bill or Nan?"

THE EMPTY SET

Through the use of games such as beanbag toss and imaginative problems, the idea of the empty set can be developed. Children enjoy "wild" examples of empty sets, such as the set of "elephants in the room."

Patterns

The search for the development of patterns is basic to mathematical thinking at all levels. At the early school level, work with patterns aids in the basic understanding of numerousness, counting, and sequencing. The study of patterns also provides background for application of functions and number theory.

Primary children should have experience dealing with matching by color, size, shape, and kind. Such activities can be sequenced from simple matching of objects of the same color or the same shape to combinations of color, size, and shape.

Several types of pattern study are listed below. Following each idea is a possible activity to use with children. In most cases, a *laboratory* strategy can be used effectively. However, if the children have difficulty with pattern ideas, *Socratic questioning* should prove effective. After a topic has been explored in a laboratory setting, then *pattern searching* can be used for transfer-type lessons.

Creating a pattern. Children were given paste and strips of paper in three colors. The teacher discussed the idea of "pattern" with the children. Each child then was to create a "rope" pattern (making a paper chain by using some type of repeating pattern). When the children were finished, each told about his or her pattern. *Note:* The same type of activity could be developed with making necklaces, Indian headbands, flower-planting patterns, beads, and macaroni shapes.

Reproducing a pattern. Two or four children working in a group could try to make each other's pattern as an extension of the lesson suggested in the preceding paragraph, or, from a textbook page or a duplicated sheet, children could cut out pictures of objects and make a pattern like the one shown on the page. This could be part of lessons on following instructions in making birthday decorations and so on.

Finding the next object. The teacher formed a pattern of crayons, repeating the pattern twice. The children were then to take turns putting the next crayon in the pattern in the proper place. After some experience, the children played a "follow the pattern" game. For this game, one child formed a pattern, repeating it twice. The other two children playing continued the pattern in turn. After the materials were all used, another child formed a pattern, and the game continued.

Extending a pattern. Children can be given a pattern started with beads, paper chains, and the like and directed to continue the pattern to complete the desired length of material.

Transferring shape, size, and color, For this type of activity, a pattern is formed on the basis of one attribute, such as color (for example, red, red, blue, red, red, blue). The children are then to use another attribute, such as shape, to form the same pattern (for example, circle, circle, triangle, circle, circle, triangle).

Describing patterns. At a more mature level, children can be provided with experience in describing a physical-object pattern in terms of symbols. For example, a pattern of colored flowers that are red, yellow, and blue can be described by using R, Y, and B.

Counting

Many beginning schoolchildren possess a background of counting, both rote (calling of the number name in order, but without meaning) and rational (identifying the "how many" or "which one" aspect of numbers). The beginning use of numbers should be developed in a manner that makes use of the background of these students without causing those without counting background to flounder. The role of counting in the kindergarten and first-grade programs cannot be overemphasized.

COUNTING EXPERIENCES FOR CHILDREN

In developing counting experiences, one teacher began by asking, "Will those of you who have birthdays this week (or today or tomorrow, if there are a great number in a given week) raise your hand?" Three children raised their hands. The teacher asked, "How many birthday hats will we have to make?" A large number of children raised their hands. When called upon, one child said, "We'll have to make three hats." The teacher questioned the group as to how they knew the number of hats was three. Explanations included, "I counted one, two, three"; "My sister is 3 years old, and we hold up three fingers to tell how old she is"; and "You can match three fingers with the children that raised their hand."

During the next few days, the teacher used a number of questions that required the determination of the number of objects in a set. The class engaged in experiences such as counting the number of pupils who needed milk at each table (usually somewhere between five and eight), chairs needed to seat some members of the class, dogs in a picture on the bulletin board, pennies brought by each pupil for a charity drive, children absent, shells in a collection, days before holidays, and books in the library.

The chart depicted in Figure 3–8 was used as a reference set to give the number name to another set. The development of the chart was a total class project that was directed by the teacher's questions. At this time, no attempt was made to identify the numerals. Children who could read the numerals were free to refer to their names.

As the children continued to develop their understanding of the number of objects in a set, the teacher made use of songs, problems, games, and finger play to provide further experience with using the number names. Here are some of the situations used:

1. How long do you think it is until Valentine's Day? Let's look on the calendar and count the number of days.
2. I noticed that many of you have picked beautifully colored leaves on the way to school. Let's count to see how many we have.

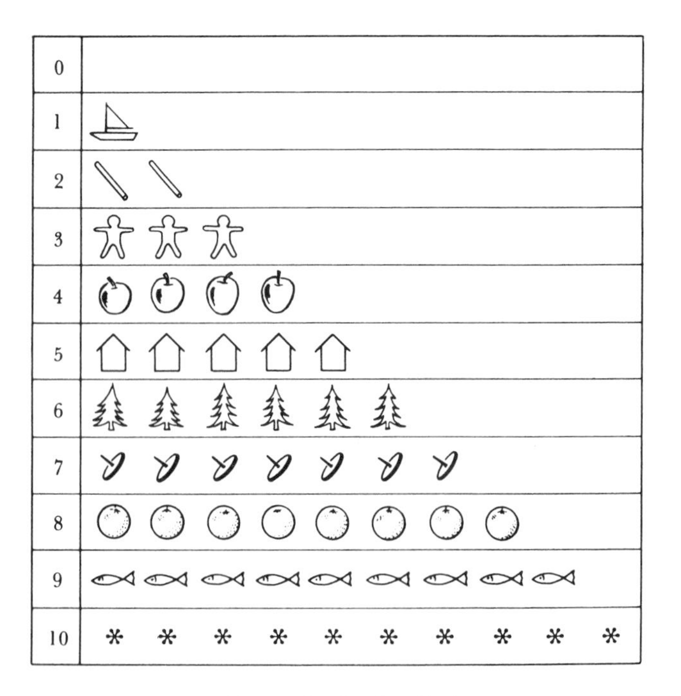

Figure 3–8 Counting Chart

3. Bob has built a very nice train with the blocks. Count to find the number of cars in the train.
4. Let's see how many minutes it will take all of us to get our boots on to go outside for recess. Count along with me. I'll write the numeral on the board, and then tomorrow we will see if we can beat today's record.
5. As we sing "This Old Man," hold up your fingers to show the number we're singing about. Let's start: "This old man, he played one, he played nic nac on my thumb," etc.
6. As I read *One Fish, Two Fish, Red Fish, Blue Fish,* hold up the number of fingers that show the number of fish talked about in the story.
7. Simon says clap your hands four times; simon says jump up and down six times.

COUNTING BEYOND 10

Because many children starting school can count beyond 10 either by rote or rationally (with understanding), the teacher's main concern may be one of emphasizing the idea of "Eleven is the word we use to mean ten and one more," and so on. Pupils often have greater difficulty with the "teens" than with larger numbers. This is probably because of the inconsistency of *eleven* and *twelve,* compared to the rest of the "teen" values, and the reverse sequences as compared with the number names beyond. For example, we call 36 *thirty-six,* but call 16 *sixteen,* not *teen six.* Also, if a pupil learns the names of the even tens (twenty, thirty, forty), he can be led to discover the pattern of reading the tens value and the ones value. For example, if the pupil knows that 22 is read *twenty-two,* and knows that 40 is *forty,* he or she can reason that 42 would be read as *forty-two.*

Because of the inconsistency of the names for numbers between 10 and 20, it is advantageous for pupils to first learn to count to 10 and then learn to count to 100 by tens. With such a procedure, the pupil would count, "Ten, two tens, three tens, four tens, . . . one hundred." This procedure has merit for the pupil who does not count beyond 10 when starting school. Emphasis on the base of 10 also develops the concept of grouping by 10s, which is of major importance in developing place-value concepts.

ORDINAL USAGE IN COUNTING

Children have many early uses for the position of an object in a linear sequence. For example, they discuss such things as, "Who is first in line?" "I'm the third person to bat." "We're going on our vacation on the twenty-first of the month." "I finished fourth in the race." The teacher can make use of ordinals in situations involving the location of pages in a book, the daily use of calendars, and the line pupils form for games. Pupils experience little difficulty with this "which one?" aspect of numbers when it is used naturally. The calendar, the sequence of pages in a book, and the number line can be used as aids in development of ordinal-number ideas.

COUNTING IN THE UPPER GRADES

Counting is one of the best background experiences for addition, subtraction, multiplication, and division. It is also one of the best means of checking the accuracy of an answer to a computational exercise. Because of counting's many uses, pupils who have very good facility in it usually have less difficulty with later number work. The teacher can provide valuable counting experience at every grade level. Here are some suggestions for counting in the upper grades:

GRADE 3

1. Tom and Bill were arguing. Tom said that if you count by fours starting with 5, one of the numbers you would count would be 21. Was he right?
2. Can you count backwards by twos? Let's try. We'll start at 16.

GRADE 4

1. Tom and Bill were working on the chart of numerals I've started on the board. Can you tell me what numeral will appear above the 15? What numeral will be three rows above the 12?

13	14	15	16	17	18
7	8	9	10	11	12
1	2	3	4	5	6

2. A friend told me that she didn't think it would be possible to count by using fractions. What do you think? Let's try counting by one-thirds. Ready—$\frac{1}{3}$, $\frac{2}{3}$, $\frac{3}{3}$ [pause]—what's another name for $\frac{3}{3}$?

GRADE 5

1. Some of you seem to be having trouble with addition. Counting practice will often improve your addition. Start at 23 and count by eights to the first number you reach that is beyond 60. What is that number? Count from 75 by nines to the first number you reach beyond 100.
2. How well do you know fractions? How well do you know how to count? Let's see. Start at 8 and count by $\frac{3}{8}$s to the first number you reach that is beyond 10.

GRADE 6

1. Counting by using fractions often helps you to improve your ability to add and subtract fractions. Try these:
   a. Start at 8 and count by $\frac{5}{8}$s until you count a number larger than 10.
   b. Start at 15 and count backward by $\frac{3}{8}$s until you reach a number smaller than 10.

Reading Numerals

When most of the pupils showed that they understood the names of the numbers for sets containing up to 10 objects, the teacher began work on the reading of numerals. This work was designed not to have every student "master" the reading of numerals but to develop background for further work. There was little pressure placed on the slow students, while the faster students were given an opportunity to work at a high level of proficiency.

FIRST DAY

The teacher used *group thinking* and began by asking, "How many fish do we have in our fishbowl?" The pupils responded that there were four fish in the bowl. The teacher asked one of the pupils to place four felt fish on the flannel board. Then the teacher commented, "Larry has placed a set of four fish on the flannel board. Do any of you know the symbol we use for the number four?"

Because of experiences with sets, counting, and the use of the chart in Figure 3–8, many of the pupils knew the symbol for the number four. The teacher gave each pupil a set of plastic numerals (they could be paper ones) and asked the children if they could hold up the one that indicated four.

Next, the teacher placed three flannel cutouts of children on the flannel board and asked, "Do you know how we could write three?" One of the pupils went to the board and wrote the numeral 3, explaining that his older brother had helped him learn to write. The teacher placed a felt numeral 3 on the flannel board, and then asked a number of children to show sets that contained three objects. During the discussion that followed, the teacher's questions brought out the idea that the various sets that contained three objects were not the same set, although they were alike in that each contained three objects. Discussion also emphasized that the "3" was a way of writing three, just as the teacher might write "Bill" on the chalkboard but did not actually put the boy Bill on the chalkboard.

SECOND DAY

The teacher gave each child a set of *laboratory* materials to be used for individual exploration and checking of the numerals. Each child was given this large card:

3	7	5	1	8
6	2	9	4	0

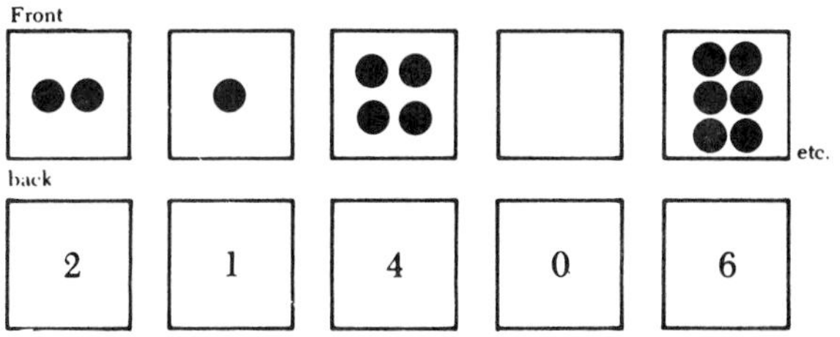

Front	●●	●	●● ●●		●● ●● ●●	etc.
back	2	1	4	0	6	

and the smaller cards shown above.

The children were instructed to place all their small cards with the dots up (it was noted that one card did not have any dots). Then the teacher said, "Experiment and see if you can match each card with the numeral that is used for the number of dots on it. Set the card on your large card. After you have filled your large card, you can see if you are right by turning each card over and checking to see if the numerals match."

This activity is particularly useful, since it allows for individual differences. Children who have not yet learned the numerals associate them with pictured objects for meaningful learning, while children with some experience with numerals can use a "guess-and-test" approach to find out if they are right. For all children, the activity is self-correcting. Thus, the child with little numeral identification experience is not faced with a paper filled with correction marks.

Comparing Numbers: Introducing Relations

After children understand the relationship of "more than," "less than," and "the same number as" between two sets and are able to associate numerals with sets, the teacher can begin to develop the children's ability to compare numbers. Comparing the number associated with sets of varying sizes is not only a valuable mathematical endeavor, but it also has many valuable social applications. In fact, a discussion between or among preschoolers very frequently involves "Do I have as much as?" or "Do I have more?"

The teacher can use a *Socratic-questioning* approach to begin work with comparing numbers by saying, "I've placed some figures on the flannel board that represent the pieces of candy Jeff brought to school [4] and the pieces of candy Alice brought to school [5]. Does Jeff's set have more pieces or fewer than Alice's? How could we tell without matching?" One child suggested that they count the number in each set and then decide which was the larger number.

The teacher then had one of the children place the appropriate felt numeral below each set and said, "In Florida, they have alligators with very large mouths. Let's pretend that this alligator [the teacher held up a cardboard alligator with a wide-open mouth] named Freddy likes candy. Which of the sets of candy will he eat?" The children responded that he would eat the set of five rather than the set of four. The teacher then placed the alligator head between the felt numerals 4 and 5 and asked, "Which way will the alligator face?" (Toward the 5, because there will be more.) "I have a symbol that I can place between the two numbers that will tell us that Alice's set has a greater number of pieces of candy than Jeff's." (She held up a red felt < and then turned it to >.) "Which way do you think the symbol should point? Will the alligator help us?" A short discussion followed in which the children noted that the symbol (the greater-than–less-than symbol) would be the same as the alligator's mouth; that is, the open end would always go toward the greater number. Then, two sets with the same number of candies were placed on the flannel board. Through directed questions, the teacher developed the notion that the alligator would decide to eat either of the sets and thus might keep his mouth shut while deciding, thus developing the = symbol.[2]

The next day, the teacher used a laboratory approach to further thinking concerning *greater than, less than,* and *equal to.* Each group of three children was given a duplicated alligator to cut out, a lab sheet, a simple beam balance, and tagged sets of washers. The children were directed to use the balance to determine which side had more (a greater number of) washers and write the appropriate sign on the lab sheet. It was suggested that they could use the alligator to help in writing the < or > sign.

A GAME

When the children have had some experience with the idea groups, two children can play this simple game for reinforcement. If needed, they can use the balance to check. Rules: Each two children are given a red and blue block with numerals marked on each side (masking tape works effectively) and an alligator or cardboard < sign. Each child rolls a block and, in turn, faces the alligator or sign in the correct direction. Each time the child is correct, one point is scored. Instead of blocks, two number wheels and a spinner can be used. One of the best and most nonbiased spinners can be made by using a paper clip and a sharpened pencil (see Figure 3–9).

EQUALITY AND EQUILVALENCE

When you deal with sets and numbers, the ideas of equality and equivalence become important. In traditional mathematics programs, the difference

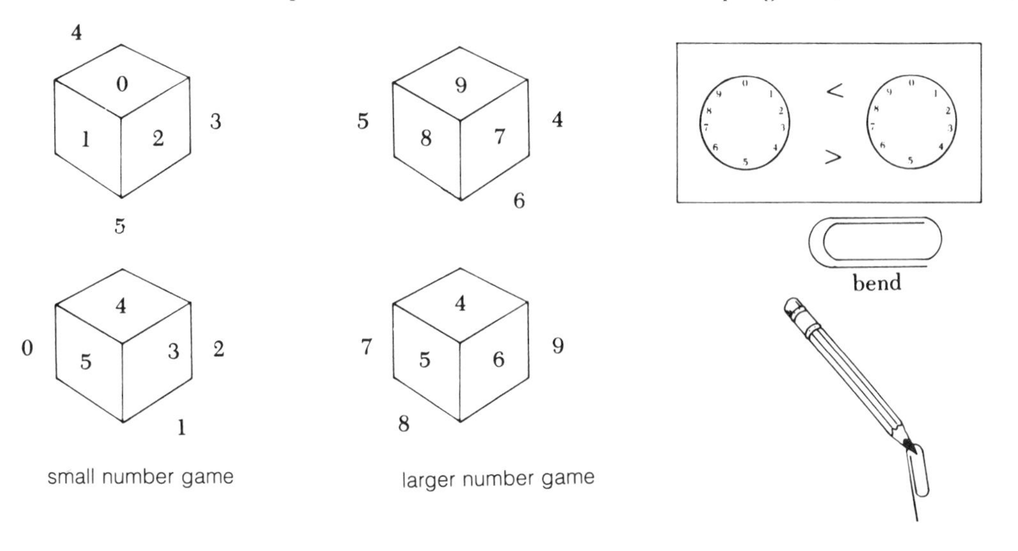

Figure 3–9 Game Equipment

[2] If in a lesson similar to this the children do not think of the variety of approaches suggested, use Socratic questioning to bring out such a variety. The knowledge that there are many ways of finding an answer usually gives the child greater confidence when undertaking a new task.

between equality and equivalence was often hazy. Occasionally, a teacher said, "Although 4 + 5 is equivalent to 9, not equal to 9, we write the sign of equality (=) rather than the sign of equivalence (~) to save time." In this case, the teacher was having trouble making a distinction between sets and numbers. Two sets may be only equivalent, but the number property of the two sets is the same; therefore, they represent equal numbers. Thus, {Tom, Mike, Alice} and {Bob, John, Harry} are equivalent sets, since they can be matched in one-to-one correspondence. However, N\{Tom, Mike, Alice\} = N\{ Bob, John, Harry\} = 3, because the number or idea of threeness is conveyed by both sets. It should be remembered that two sets are equal if they have the same elements, that two sets are equivalent if their elements may be matched in one-to-one correspondence, and that numbers have many names (6, 3 + 3, 8 − 2, and 5 + 1 are all different names for the same number).

Writing Numerals

While numeral writing often appears to be a very mundane part of the curriculum, there are several reasons why correct numeral writing should be stressed:

1. In the majority of schools, the first handwriting a child is taught involves numeral writing.
2. Current use of numerals in addresses (ZIP codes), telephone exchanges (237-2090 rather than AD 7-2090), credit cards, bank accounts, and social security numbers on income tax returns requires a high level of legibility. Incorrect formation of numerals can cause costly delays in delivering mail and in processing data.
3. Illegible numeral writing causes errors that nationally cost victims well over a million dollars. Many persons have had experience with the difficulties that arise when one of the persons using a joint checking account writes a numeral illegibly.

The need to write numerals arises early in the elementary school program. The first need to write numerals usually occurs when the student needs indicate a page number, to record date or the temperature, or to keep score for a game.

The pupils have had experience in reading the number names and thus refer to sample numerals posted on the bulletin board. Several use situations should precede a systematic numeral writing program.

Traditionally, the handwriting period has focused upon "what's wrong" with "your" numeral writing. Comments on "what's right" usually produce better results. The teacher may begin a numeral writing period in the following manner: "Yesterday I asked each of you to write on your papers the numeral that would tell me how many pictures you had collected of objects that begin with the same sound as 'dog.' I was very pleased with the way that many of you wrote the numerals. I've copied some of the numerals you wrote on these large pieces of oak tag, and I've photocopied a sheet containing the correct forms of the numerals for you to refer to. Compare the way in which you wrote the numerals with the board and with the sheet I've given you. (pause) Now, let's see if we can write the numerals even better. I'm going to count and then stop. You write the numeral that would represent the number that occurs next. If you aren't sure which number would be next, raise your hand, and I'll be happy to give you some help. All right, is everyone clear on the instructions? Let's begin 1, 2, 3, 4, 5, 6, (pause), 8, 9, etc.

"Now that we've finished, take a look at the numerals you've written and compare them with your chart or the chart on the board. I'd like to know how you began to write each numeral. Please draw an arrow like this (→) to indicate where you started each numeral." When the class members had indicated the direction in which they had begun each numeral, the teacher set up a chart such as the one below, and the pupils compared their writing with the chart.

1 2 3 4 5 6 7 8 9 0

Detailed accounts of the research done on numeral writing and of the many ideas that classroom teachers have found helpful in teaching numeral writing are beyond the scope of this book. However, the following comments are offered as an aid to teachers:

1. The strongest factor in good numeral writing is the student's motivation. Most children can write numerals legibly (if not beautifully) if they have a real desire to do so. Thus the teacher should be constantly on the lookout for situations that will positively motivate the student to increased effort in numeral writing.
2. Most handwriting programs begin with the student's writing very large numerals. In such cases, the student usually draws rather than writes the numerals. While such drawing practice is helpful, the student should also have an opportunity right from the beginning of instruction to write numerals of standard size.
3. While different pupils experience difficulty with various numerals, research reveals that the three numerals that cause the greatest difficulty are 5, 0, and 2. The numerals 3, 6, 7, 8, and 9 cause some difficulties.
4. A large pencil is often suggested for primary work in handwriting. Research does not support the contention that it is important for primary school children to use large pencils.[3]
5. There should be a reintroduction of numeral writing at each grade level. This reintroduction should present a challenge. The teacher might use such approaches as comparing the numerals of today with those used in the past.

A Comparison of Numerals[4]

	1	2	3	4	5	6	7	8	9	10
Sanskrit	१	२	३	४	५	६	७	८	९	०
Arabic	١	٢	٣	٤	٥	٦	٧	٨	٩	٠
Later Arabic	1	2	3	4	5	6	7	8	9	0

[3]Virgil E. Herrick, "Handwriting and Children's Writing," *Elementary English,* 37, (April, 1960). 248–58.

[4]From *History of Mathematics* by D. E. Smith, Vol. II, pp. 70–71. Copyright 1953 by Eva May Luce Smith. Published by Dover Publications, and reprinted through permission of the publisher.

6. Normally, short intensive periods of instruction are preferable to longer periods. If a teacher plans to spend 60 minutes in numeral writing, four 15-minute periods would produce better results than two 30-minute periods.
7. Normally, instruction begins in grade one with manuscript numerals. Cursive numerals usually are introduced in grade two or three. There is little evidence for or against this procedure.

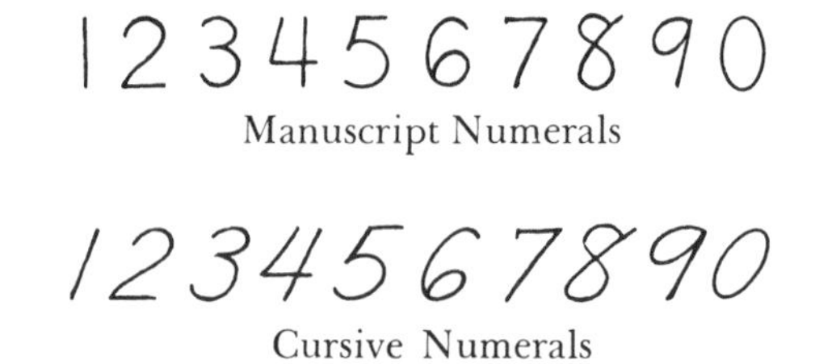

Manuscript Numerals

Cursive Numerals

8. Paper placement for manuscript writing is at right angles to the desk. When cursive writing begins, the paper should be placed as illustrated below. *Note:* When the paper is placed in the same manner for both left-handed and right-handed pupils, the left-handed pupil will usually develop a hook. The teacher should carefully check to see that the left-handed pupil has placed the paper correctly.

Manuscript

Cursive

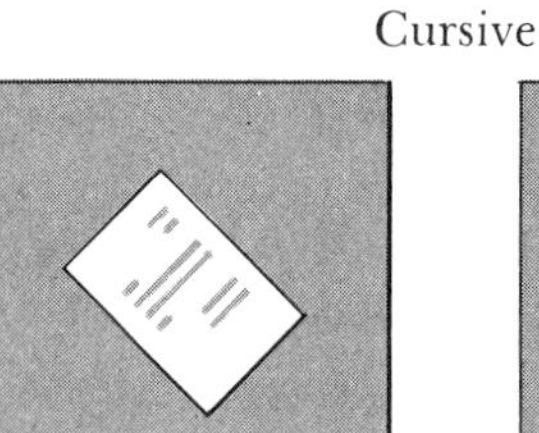

right-handed

left-handed

PROBLEM SOLVING AT THE PREREADING LEVEL

Many teachers feel that the solution of the verbal problem is one of the most difficult portions of the elementary school mathematics curriculum. They find that, in many cases, pupils have difficulty with problems and do not enjoy working them. This difficulty may well be caused by the delay in starting the problem-solving program. The majority of commercial mathematics programs have very little verbal problem solving until late in the second grade. This lack of problem-solving material usually stems from the assumptions (1) that problem solving cannot or should not begin until the pupil can read the problems and/or (2) that problem solving as such is unimportant to the mathematics program.

A careful study of the two assumptions will reveal little sound evidence to support them. First, it is not necessary for the pupil to have reading facility to work with verbal problems. These problems can be presented orally by the teacher and solved by the use of the concrete sets of objects. Oral presentation is in keeping with the typical problem situations that children and adults encounter, for most physical-world problem situations are presented in an oral rather than a written context. Second, work in verbal problem solving at the early primary level is consistent with the modern emphasis on sets and concrete materials. A problem situation concerning things (sets) provides pupils with an opportunity to work with sets in a situation where they see a need for their use.

In addition to orally presented problems, nonverbal, or low-verbal problems with limited vocabulary in cartoon sequence can be developed. The kindergarten or first-grade teacher can isolate many simple problem situations from occurrences in the room and contrived situations that he or she has developed.

How would the problem situation be used? The following classroom description is typical where orally presented problems form a major portion of early mathematics instruction.

This problem was read to the children: "Mark has been telling me about his fishing trip. His father caught four fish, and Mark caught three fish. How many did they catch in all?" Each pupil was directed to try to find the answer and then be able to show that it was correct. Each pupil had a set of plastic counters and paper and pencil. Some pupils began to use the counters to find an answer, while others thought for a moment and raised their hands. The following explanations are typical of the different levels of thinking that occurred:

JIM: I drew four fish on my paper because Mark's father caught four fish. Then I drew three fish on my paper because Mark caught three fish. Then I counted all the fish. They caught seven fish.

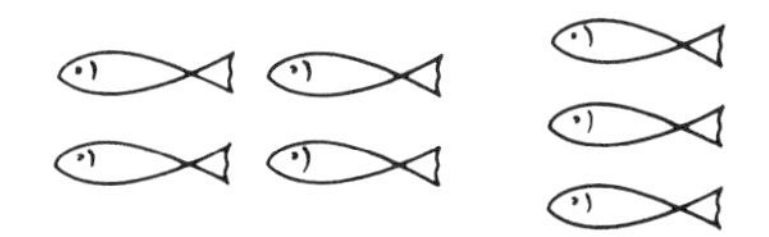

NANCY: I laid down four counters to show the fish that Mark's father caught, and then I laid down three counters to show the fish that Mark caught. I counted all the fish and got an answer of seven fish caught.

TIM: I know that Mark's father caught four fish and that Mark caught three fish. I said to myself four—five, six, seven. I knew I could start with four and then count three more.

JILL: I just know that four and three combine to equal seven.

More difficult problems may be worked by total-group or small-group consideration. Such total class experiences should involve problems that are significant to the members of the class. Problems involving the union or separation of sets are excellent background for the later study of addition and subtraction.

PLACE VALUE

Place value is one of the most important ideas that elementary school students need to understand and master. However, place value is not a topic that lends itself to isolated study. Knowledge of place value is important to an understanding of the algorithms of addition, subtraction, multiplication, and division and to the later development of decimals and percentages. A lack of understanding of such procedures as regrouping in addition and subtraction originates from a lack of understanding of place value. Further, positional notation is at the heart of the workings of computers and calculators.

"Ken is anxious to fill in his savings-stamp book." (The teacher pointed to the drawing on the board.) "How many stamps will he need to fill the first

Ken's Stamp Book

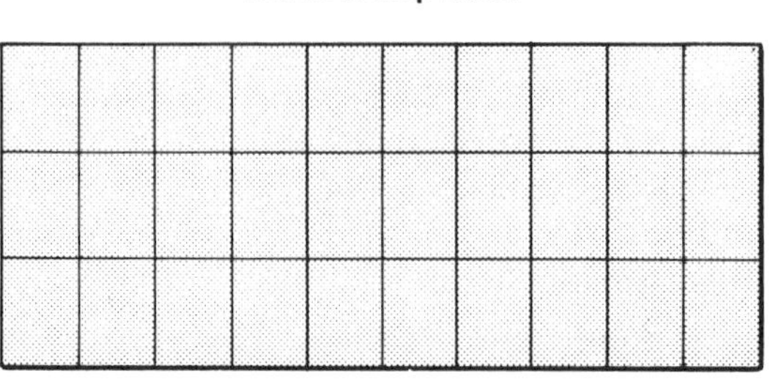

row?" (The pupils counted and answered, "Ten." "How many tens will he need to cover the page?" (Pupils responded, "Three tens.")

"How many groups of 10 stamps has he pasted in the book?" (Pupils responded, "Two tens.") "How many stamps in all?" (Pupils responded, "Two tens and four ones," or "Twenty-four stamps.")

"How many markers are there on the hundred board?" Several children immediately raised their hands and answered, "Thirty-four." The teacher asked, "How did you know so fast? You didn't have time to count them." The pupils responded that they knew there were 10 markers in each row on the hundred board. This meant that there were three tens and four ones, or 34 markers.

"I've given each of you a number of Popsicle® sticks and some rubber bands. Group the Popsicle® sticks so that I or one of your friends can easily tell how many sticks there are." The teacher had made no attempt to give the same number of sticks to every child. The majority of the class grouped the sticks into bundles of 10 each. Some of the pupils used bundles of other sizes. The pupils then discussed the grouping of the sticks. They found that if the sticks were in groups such as 8, 11, or 12, it was still necessary to count the entire number of sticks. After discussion, the children agreed that any size group could be used, but that grouping by tens and ones helped them tell how many sticks were on another desk.

Then, after giving each pupil a dittoed sheet containing several place-value frames, the teacher referred to a frame drawn on the board and asked, "How could I represent this number of sticks on the place-value frame? Show this on your place-value frame."

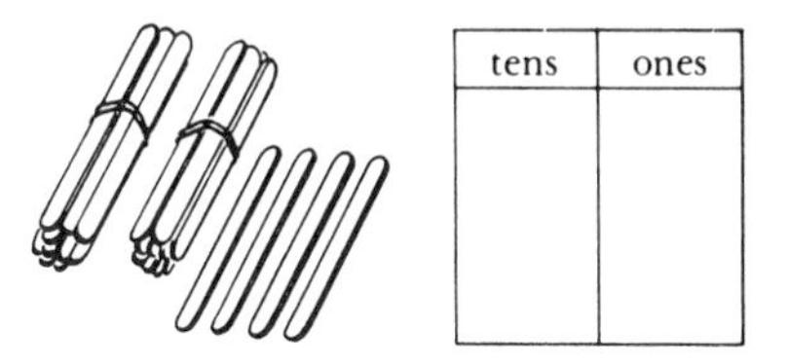

After the pupils filled in the frames, discussion revealed that the two methods shown below were considered to be good representations.

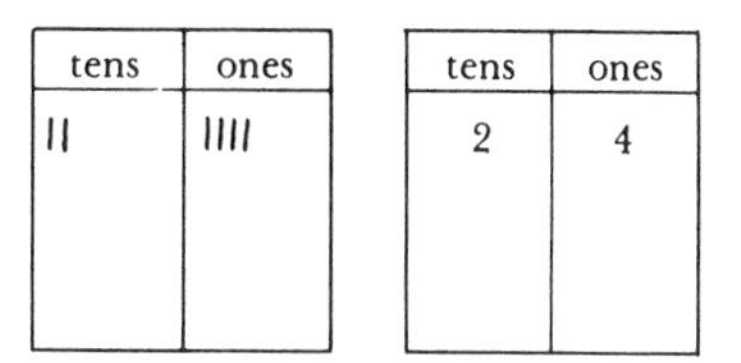

The class agreed that it would not be good form to use the method below because a person looking at the frame could not tell if it meant two tens or two groups of 10 tens.

tens	ones
~~IIIIIIIIII~~ ~~IIIIIIIIII~~	IIII

A laboratory approach can be used with the concept of a chocolate or candy factory, in which bites (ones) of chocolate are thought of as being put in a machine and coming out in the form of bars (tens) and bites (ones).[5] From this, it is possible to move to boxes (hundreds) and later to cartons (thousands). Candy can be represented by duplicated squares $\frac{1}{2}$ inch × $\frac{1}{2}$ inch for bites, $\frac{1}{2}$ inch × 5 inch for bars, and 5 inch × 5 inch for boxes. Marbles, or paper or plastic bags, of two sizes can also be used.

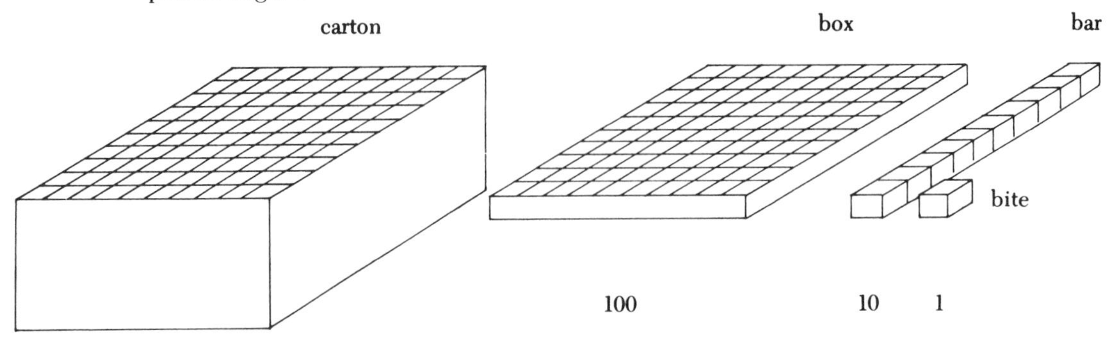

In addition, children can have experience buying and selling at a classroom store in which all the money is dollars, dimes, and pennies.

When some place-value understanding has been developed, duplicated sets of dials can be used with paper-clip spinners to generate numbers to be written or substituted in a place-value frame.

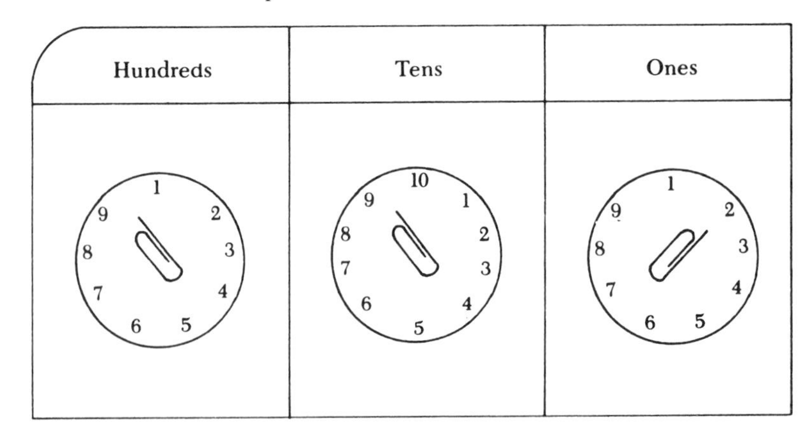

A game can be played by using this device. Two children each spin a number and record the number. The child who has the larger number scores a point if he or she has named the number correctly. The game can be played equally well with three or four children.

Devices such as Cuisenaire rods, tens, and ones blocks, Stern materials (strips of 10 and single squares), counting discs in plastic bags, and a simple abacus may also be used.

[5]See Roald Dahl, *Charlie and the Chocolate Factory* (New York: Knopf, 1964). This book can provide a good motivation lead-in for the chocolate-factory idea.

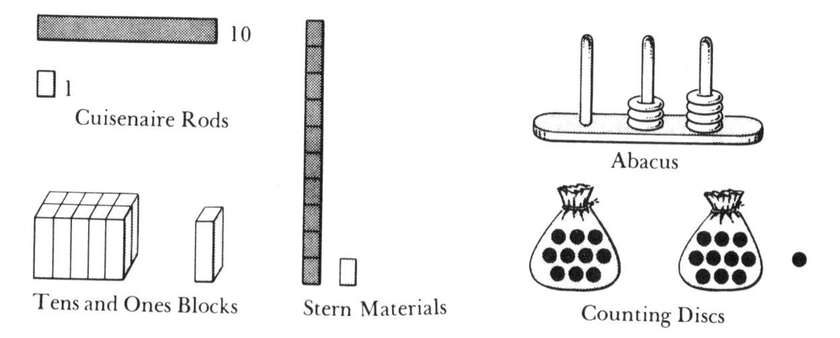

As children in succeeding grades mature in their understanding of the number system, first expanded notation and then exponents are used to continue work on place value. (See Chapter 11).

Place Value and Estimation

If I could use only one instructional technique to develop place value concepts, I would use estimation. The more that children estimate additions, subtractions, multiplication, and divisions, the better they understand place value. The teacher of elementary school mathematics should be constantly alert to situations that naturally lead to estimation. The teacher should stress that a good understanding of place value greatly aids the students' ability to estimate.

One teacher used the accompanying series of pictures (Figure 3–10) and followed the teaching suggestions. The lesson was followed by a series of lessons on place-value estimation and reviewing basic computation. (*Note*: This lesson is at the fourth-grade level.)

KEEPING SHARP

Self-Test: True/False

_______ 1. Children who begin studying mathematics in kindergarten can explore a wide variety of ideas with little emphasis on mastery.

_______ 2. At this level, concept accuracy is of more importance than whether or not the concepts make sense to the pupils.

_______ 3. One of the advantages of programs such as DMP (the material under "A Look at Kindergarten Topics") is that the child is actively involved in the learning experiences.

_______ 4. Classification is relatively unimportant in building early mathematics concepts, so it can safely be ignored.

_______ 5. Developing patterns is basic to mathematical thought and can easily begin in kindergarten activities.

_______ 6. It makes little difference to later computation skills if children count by rote or count rationally.

_______ 7. The reading and writing of numerals are developed with counting and comparison of numbers.

_______ 8. Children rarely experience difficulty in learning to write the numerals 0, 1, and 4.

_______ 9. Verbal problems should not be introduced until pupils can read them.

_______ 10. Verbal problem-solving skills can be developed successfully in early primary grades if the problems are presented orally.

_______ 11. Place value is basic to an understanding of our numeration system and should be begun as soon as children count beyond 10.

_______ 12. Counting by tens and naming larger numbers in tens and ones (e.g., 4 tens and 6 ones) provides useful experience with place-value concepts.

_______ 13. The use of manipulatives (concrete materials) helps your children visualize and understand developing mathematics concepts.

Vocabulary

classification
equalization
rational counting
ordinal numbers
cardinal numbers

1

Counting and Estimating

Student book pages 2, 3, 4, 5

Purposes
The purposes of this lesson are:

1. To refamiliarize the students with numbers.
2. To review counting, base-ten arithmetic, and estimation.

More systematic review of these topics is given in later lessons: counting and base-ten in lessons 2 and 3, and estimation in lessons 17–19. These topics also come up in various contexts throughout the year. A checkpoint for assessing mastery of numerical sequence is included in lesson 3.

Mental arithmetic
Reintroduction to numbers Provide mental drill as described below. (Time is provided in the next 2 lessons for continuing these drills, so it's not necessary to cover each topic in this lesson.) Try to select problems that are easy for the entire class; then gradually work toward slightly more difficult problems.

1. Do unison counting up and down in a number range appropriate to the class's ability. Gradually work toward slightly more difficult problems.
2. Give problems like these: "What number comes after 75?" and "What number comes before 721?" This type of exercise will be especially appropriate if you are using number-wheel response cards. (See page 425 of this guide for a complete discussion of the use of these cards and for suggestions on doing mental arithmetic without the cards.)
3. Give addition and subtraction fact problems, but note that a formal review of the addition and subtraction facts begins in lesson 5. (For more information about the Mental Arithmetic component of lessons, see page 425 of this guide.)

Student pages 2–5
Treat pages 2–5 as a light-hearted activity for the class to work on together. Each student should have a pencil and paper to

4 Lesson 1

Estimating

A BASKET OF APPLES

The Seedtown Apple Orchard gave Richard a basket of apples for his school. Help him estimate how many apples there are in the basket.

- Write down your best guess of how many apples there are. Guesses will vary; they may range between 30 and 5000.
- How did you make your guess? Answers will vary.
- Do you think you have a good chance of being exactly right? no

2

S. Willoughby et al., *Real Math, Grade Four* (LaSalle, Ill.: Open Court Publishing Co., 1985), 1–3. With permission.

Figure 3–10

Richard's friends helped him carry away the apples.

- 10 children took 10 apples each. How many apples did they take? 100

There are still lots of apples left.

- Now do you know exactly how many apples there were altogether? no
- Make a second guess of how many apples there were.
  Reasonable guesses may range from 300 to about 2000.

Then 10 more children came. They each took 10 apples. Now the basket looks like this.

- So far, 20 children have taken 10 apples each. How many apples is that? 200
- Make a third guess of how many apples there were.
  Reasonable guesses should be in the neighborhood of 700.

3

write an answer for each problem. Discuss each set of answers after the students have made guesses. Use the questions at the bottom of page 5 as a guide for discussing how the guesses changed as more and more information became available.

Keep the tone of the discussion light. Neither you nor the students should place too great an emphasis on precise answers. Those who make poor estimates (especially on the first 2 guesses) should be made to realize that not enough information was given at those points for making refined estimates.

Note 1: On pages 2–5 we use the term *guess*. However, somewhere during the lesson, and this will vary from student to student, the answers will become more estimates than guesses because more information will have been used in arriving at the approximate number of apples. The issue, though, is not the terminology but that the students learn to use available information to make reasonable estimates. Generally, after this lesson we use the term *estimate*, unless very little or no information is used to arrive at an approximate quantity.

Note 2: All questions marked with • are discussion questions, intended to be read and discussed with the whole class. Numbered questions and problems can, as a rule, be done by the students on their own or in groups. However, even some of these you will want to do or discuss with the class (this is usually pointed out in the corresponding page comment in the guide).

Student page 3

Discussion questions Before making a second guess, the students should see that many more apples remain than have been taken. However, it isn't clear how many more. Thus, although it's known that 100 apples have been taken, any guess between 300 and about 2000 could be considered a reasonable estimate of the original number of apples.

After the students make their third guesses, let them discuss their reasons for their guesses here (and discourage them from turning the page during the discussion). At this point, there is a systematic way of making an estimate. There are about 100 apples between each 2 consecutive hoops of the basket. That's 6 sections of about 100 apples each. Add to that the 100 apples that were on top, and you have a good estimate of around 700. If the students don't come up with this reasoning on their own, ask how many apples there were between the top 2 hoops. Then ask how many such sections there are in the basket. Finally, remind them of the 100 apples that were on top.

Lesson 1 5

Figure 3–10 *Continued*

THINK ABOUT

Beginning Instruction: Foundation Experiences, Counting, Meaning of Numbers, Reading and Writing of Numerals

1. When should arithmetic instruction begin? Kindergarten, grade 1, grade 2, grade 3?
2. What should be the nature of the first instructional material?
3. Describe some promising procedures for developing the meaning of numbers. This description may consist of only two or three procedures to be representative; it may be extensive, as a source for teaching or supervision; or it may be a critical review of issues, claims, status, etc.
4. Describe a program for the improvement of numeral writing in grades 1 through 6. In your description, answer such questions as "What form of each numeral should be adopted as a standard?" and "Are the arguments for beginning with manuscript numerals as valid as the arguments for beginning with manuscript letters?"

5. Should foundation number experiences in the primary grades deal with any or all of the following points: set language, decimal notation, fractions, algebra, geometry?
6. Describe a series of five or six activities to develop an understanding of place value; use several different materials.
7. Begin building an activity file of verbal problems, including those with minimal vocabulary and cartoon sequences. Exchange ideas with several friends. Try to gather at least 20 such activities to use in the early primary grades. (You might consider tape-recording some problems.)

SUGGESTED REFERENCES

BARATTA-LORTON, MARY, *Workjobs II: Number Activities for Early Childhood,* Menlo Park, Calif.: Addison-Wesley, 1979.

BAROODY, ARTHUR J., *Children's Mathematical Thinking,* New York.: Teachers College Press, 1987.

BITTER, GARY G., MARY M. HATFIELD, AND NANCY T. EDWARDS, "Number Readiness: Early Primary Mathematics." In *Mathematics Methods for the Elementary and Middle School: A Comprehensive Approach,* chap. 5. Boston, Mass.: Allyn and Bacon, 1989.

BURTON, GRACE M., *Towards a Good Beginning: Teaching Early Childhood Mathematics,* Menlo Park, Calif.: Addison-Wesley, 1985.

CLEMENTS, DOUGLAS H., *Computers in Early Primary Education,* Englewood Cliffs, N.J.: Prentice Hall, 1985.

CLEMENTS, DOUGLAS H., *Computers in Elementary Mathematics Education,* chap. 6. Englewood Cliffs, N.J.: Prentice Hall, 1989.

Counting and Early Arithmetic Learning, Washington, D.C.: The National Institute for Education, 1985.

KAMII, CONSTANCE K., *Young Children Reinvent Arithmetic: Implications of Piaget's Theory,* New York: Teacher College Press, 1985.

LOVELL, KENNETH, *The Growth of Understanding in Mathematics: Kindergarten through Grade Three,* New York: Holt, Rinehart and Winston, 1971.

Mathematics Learning in Early Childhood Education, Thirty-Seventh Yearbook. Reston, Va.: National Council of Teachers of Mathematics, 1975.

NELSON, DOYAL, and JOAN WORTH, *How to Choose and Create Good Problems for Primary Children,* Reston, Va.: National Council of Teachers of Mathematics, 1983.

POST, THOMAS R., ed., *Teaching Mathematics in Grades K–8: Research Based Methods,* chaps. 3 and 4, Newton, Mass.: Allyn and Bacon, 1988.

Research on Mathematical Thinking of Young Children, Reston, Va.: National Council of Teachers of Mathematics, 1975.

TAYLOR, JOYCE, *The Foundations of Mathematics in the Infant School,* London, England,: George Allen and Unwin, 1976.

WIEBE, JAMES H., *Teaching Elementary Mathematics in a Technological Age,* chap. 8. Scottsdale, Ariz.: Gorsuch Scarisbrick, 1988.

chapter 4

PROBLEM SOLVING

OVERVIEW

The National Council of Teachers of Mathematics Standards emphasizes problem solving as "the central focus of the mathematics curriculum. Not only is the ability to solve problems a major reason for studying mathematics, but problem solving provides a context in which concepts and skills can be learned. In addition, problem solving is a major vehicle for developing high order thinking skills."

Students in kindergarten through grade four should be able to

1. Use problem-solving approaches to investigate and understand mathematical content.
2. Formulate problems from everyday and mathematical situations.
3. Develop and apply strategies to solve a wide variety of problems.
4. Verify and interpret results with respect to the original problem.
5. Acquire confidence in using mathematics meaningfully.

Students in grades five through eight should be able to

1. Use problem-solving approaches to investigate and understand mathematical content.
2. Formulate problems from situations within and outside of mathematics.

3. Develop and apply a variety of strategies to solve problems, with emphasis on multistep and nonroutine problems.
4. Verify and interpret results with respect to the original problem situation.
5. Generalize solutions and strategies to new problem situations.
6. Acquire confidence in using mathematics meaningfully.

Verbal problem solving has long been an area of elementary school mathematics instruction of great concern to teachers and the cause of much pupil anxiety. This situation has led to the formulation of many proposals for improving the teaching of verbal problem-solving methods, but interest in many of these proposals has waned quickly after teacher tryout or after studies to test their validity have failed to produce clear-cut evidence of their worth. The overall result has been that, during instruction on verbal problems, the major portion of the time is devoted to the solution of the same verbal problems by all the students in a class by means of general or poorly defined procedures.

The majority of textbook problems are simply practice-related to the computational topic of the chapter. If the children simply do what they were last taught, they will be able to correctly answer 60–80 percent of their textbook problems without even reading them.

Thus, the strength or weakness of a problem-solving program rests primarily with the individual teacher. The procedures presented in this chapter are a composite of results of classroom testing and research that have proven successful in improving children's problem-solving ability.

TEACHER LABORATORY

1. You have $1.00. You are going to buy one or more pencils, erasers, and paper clips at a stationery store. Pencils cost 10¢, paper clips are two for 1¢, and erasers cost 5¢ each. How much of each item will you buy if you have 100 items when you leave the store?
2. A farmer did not want his neighbor to know how many cows he had, so when the neighbor asked, he replied, "I have 35 cows and chickens. They have a total of 78 legs." How many cows did the farmer have?
3. Janet baked several cakes to sell in the neighborhood. She went to Mrs. Scott, who said, "I'll take one-half your cakes and one-half a cake more." Janet gave Mrs. Scott a number of cakes (without cutting the cakes). Mrs. Scott went to the phone and called Mrs. Smith and Mrs. Black, telling them each to ask for one-half the cakes and one-half a cake more. Janet made these deliveries, and when she finished with Mrs. Black, she was out of cake. (*Note:* She had not at any time cut a cake.) How many cakes did Janet begin with?
4. Get a mail-order catalog. Use it to make up a set of problems for children at any grade level you choose.

TAKE INVENTORY

Can You:

1. Develop a lesson featuring one of the following problem-solving improvement procedures?
   a. Open-ended problem
   b. Problems without numbers
   c. The mathematical sentence
   d. Orally presented problems
   e. Nonverbal problems
   f. Diagrams and drawings
   g. Pupil formulation of problems
   h. Restatement
   i. Problems with too much or too little data
   j. Estimation as a problem-solving tool
2. Explain how orally presented problems develop skills in other curricular areas, such as reading, social studies, etc.?
3. Describe how children's interest in problem solving can be improved?
4. List five suggestions to eliminate "rote-thinking" in problem solving?
5. Discuss the role of the hand-held calculator in solving problems?
6. Develop a set of problem-solving lessons, using one or more of Polya's heuristics (informal procedures)?
7. Develop a set of nonroutine problems and nonroutine ways of solving problems?
8. Understand the nature of problem solving in elementary school mathematics?

The problem solving referred to in this chapter has two major facets: (1) open-ended, nonroutine situations that require observing, gathering data, making predictions, testing the predictions, arriving at tentative solutions, and testing the solutions (this phase may occur in interdisciplinary-problem-oriented units), and (2) quantitative situations presented in oral or written context in which a question or questions are asked without an accompanying statement concerning the mathematical operation required. Such problems can vary in difficulty from almost disguised computation to situations requiring a great deal of thought.

A problem is not a problem if it can be solved by a set algorithmic procedure. For real problem solving, the child must pull together a number of skills and knowledge from his or her past and put them together in a new way to arrive at a solution. In a NCTM Yearbook, Kantowski proposes that the teacher may aid in developing children's ability in problem solving by changing roles as the children develop maturity, as shown in the accompanying table.

Children vary in the type of experience they require to understand a given situation. Problem situations also vary in the attack required. Therefore, it is essential that the teacher use a variety of procedures and teaching techniques; those described in the following section are effective in improving verbal problem-solving ability. It is hoped that the present trend toward more laboratory-type materials and open-ended teaching will improve the problem-solving program. The suggestions that follow are aimed at that goal.

Characteristics of Students at Each Level of Problem-Solving Development and the Role of the Teacher at Each Level

First Level

Students have little or no understanding of what problem solving is, of the meaning of strategy, or of the mathematical structure of the problem. Most students at this level do not know where to begin to solve a nonroutine problem.
Teacher assumes the role of *model.*

Second Level

Students understand the *meaning of problem solving,* of strategy, and the mathematical structure of a problem. *They are able to follow someone else's solution and can often suggest strategies to be tried for problems similar to those they have seen before.* Although they will participate actively in group problem-solving activities or instructional episodes, many feel insecure about independent problem solving.
Teacher acts as *prosthesis,* or crutch.

Third Level

Students begin to feel comfortable with problems. They suggest strategies different from those they have seen used. They understand and appreciate that problems may have multiple solutions and that "no solution" may be a perfectly good solution.
Teacher becomes *problem provider.*

Fourth Level

Students are able to select appropriate strategies for most problems encountered and are successful in finding solutions much of the time. They show an interest in elegance and efficiency of a solution and in finding alternate solutions to the same problem. *They suggest variations of old problems and are constantly searching for novel problems to challenge themselves and others.*
Teacher serves as *facilitator.*

Mary G. Kantowski, "Some Thoughts on Teaching Problem Solving," in *Problem Solving in School Mathematics, 1980 Yearbook* (Reston, Va.: National Council of Teachers of Mathematics, 1980), pp. 198–99. Used with permission.

The current New York State *Elementary Science Syllabus* provides an excellent model for problem solving. The careful student of elementary mathematics teaching can gain a great deal of insight by a careful study of the model for problem solving.

PROCEDURES FOR IMPROVING PROBLEM SOLVING

Open-Ended Problems

Many of the problems involving mathematics that face both children and adults out in the world involve more than a single answer. Thus, through the elementary school mathematics program, the teacher should provide children with open-ended–type problem situations.

One teacher brought in several toy catalogs and began a problem-solving laboratory lesson by saying, "Many people call catalogs like this one 'wish books.' Have you ever looked through catalogs thinking about what you would like to buy?" The children indicated that they had often done this. Then the teacher said, "Pretend that you have won a gift certificate for $50 to buy things from this catalog. Make out several lists that you might use in making your order. When you've finished, pretend you have won $100 to spend on your family. Decide what you would buy, and make out a list of things for each family member. I don't have enough catalogs for each of you, so we will use them over a period of time. If you have free time during the day, you can take one of the catalogs and work on the project."

During the school year, many situations arise that can be developed into open-ended problem lessons (for example, planning a party, including the cost per person and comparative shopping for the needed items; making presents; going on scavenger hunts; and keeping data on various team sports). Also, the situation games often used in social studies or science can provide rich opportunity for developing problem-solving skills.

USING ORALLY PRESENTED PROBLEMS

One of the most valuable techniques for the improvement of problem solving is the use of orally presented problems. In addition to the improvement of "in-school" problems, the orally presented problem is much nearer the average "out-of-school" problem than is the pencil-and-paper problem. Children and adults are often confronted with mathematical problems that are stated orally rather than presented in writing. Oral presentation is also of value in inducing pupils to listen carefully and to concentrate on the most significant aspects of the problem.

These additional suggestions may aid the teacher in effectively using orally presented problems:

1. Because orally presented problems proceed at a fixed pace, it is suggested that no more than 5 or 10 minutes at a time be spent on this procedure.

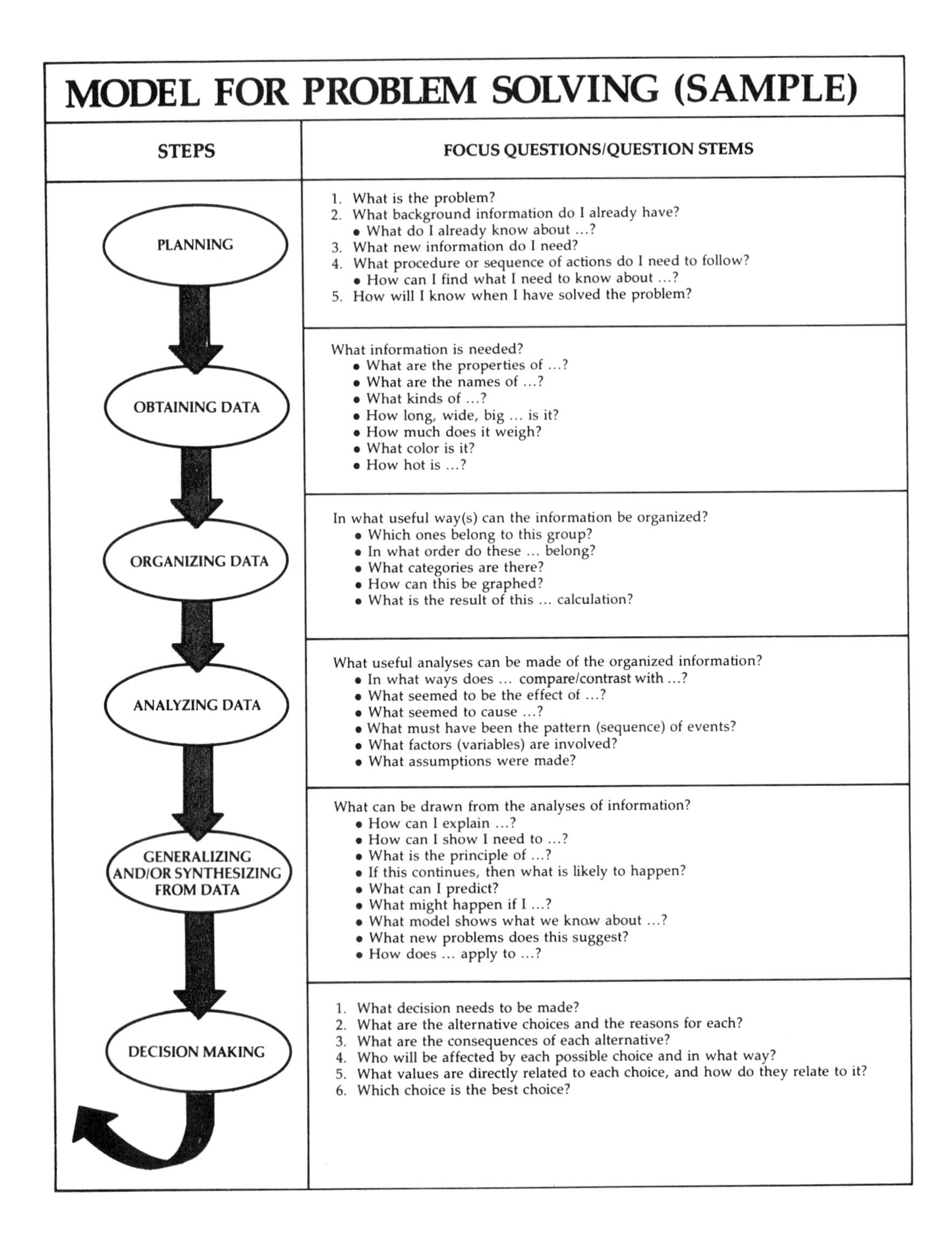

MODEL FOR PROBLEM SOLVING (SAMPLE)

STEPS	FOCUS QUESTIONS/QUESTION STEMS
PLANNING	1. What is the problem? 2. What background information do I already have? • What do I already know about ...? 3. What new information do I need? 4. What procedure or sequence of actions do I need to follow? • How can I find what I need to know about ...? 5. How will I know when I have solved the problem?
OBTAINING DATA	What information is needed? • What are the properties of ...? • What are the names of ...? • What kinds of ...? • How long, wide, big ... is it? • How much does it weigh? • What color is it? • How hot is ...?
ORGANIZING DATA	In what useful way(s) can the information be organized? • Which ones belong to this group? • In what order do these ... belong? • What categories are there? • How can this be graphed? • What is the result of this ... calculation?
ANALYZING DATA	What useful analyses can be made of the organized information? • In what ways does ... compare/contrast with ...? • What seemed to be the effect of ...? • What seemed to cause ...? • What must have been the pattern (sequence) of events? • What factors (variables) are involved? • What assumptions were made?
GENERALIZING AND/OR SYNTHESIZING FROM DATA	What can be drawn from the analyses of information? • How can I explain ...? • How can I show I need to ...? • What is the principle of ...? • If this continues, then what is likely to happen? • What can I predict? • What might happen if I ...? • What model shows what we know about ...? • What new problems does this suggest? • How does ... apply to ...?
DECISION MAKING	1. What decision needs to be made? 2. What are the alternative choices and the reasons for each? 3. What are the consequences of each alternative? 4. Who will be affected by each possible choice and in what way? 5. What values are directly related to each choice, and how do they relate to it? 6. Which choice is the best choice?

SKILLS	PRODUCTS
• Communicating information • Creating models • Formulating hypotheses • Manipulating ideas • Predicting • Questioning • Recording data • Using cues	1. A statement of the problem 2. List of facts (background information) 3. List of questions (related to later steps in the process) 4. Sequential plan (a list of tasks, student assignments, and times for completion) 5. Sketch or description of the expected (predicted) final product
• Acquiring information • Developing vocabulary • Manipulating materials • Measuring • Observing • Recording data • Using cues • Using numbers	• Collections • Counts • Definitions • Lists • Photographs • Sketches • Tape recordings
• Classifying • Communicating information • Creating models • Manipulating ideas • Manipulating materials • Replicating • Using numbers	• Calculations or computations • Charts, tables • Diagrams, scale drawings • Graphs • Groups, categories of information • Outline • Sorted objects
• Identifying variables • Inferring • Interpreting data • Manipulating ideas • Using cues	• Description of a pattern or sequence • List of variables • Statements of cause and effect relationships • Statements of similarities and differences • Summary
• Acquiring information • Communicating information • Creating models • Formulating hypotheses • Generalizing • Manipulating ideas • Predicting • Questioning	• A model or simulation • A new hypothesis • A new prediction, problem, theory • Applications to new situations • Statements of principles • Statements which accept or reject hypotheses • Statements which confirm or do not confirm predictions • Written report
• Acquiring information • Communicating information • Making decisions • Manipulating ideas • Questioning	1. Statement of the decision to be made 2. List of alternative choices, supported by reasons 3. List of consequences of each alternative 4. List of persons directly affected by each choice and the way each is affected 5. List of values related to each choice supported by statements of how the values relate 6. A personal choice, supported by defendable reasons for the choice

Elementary Science Syllabus, Albany, NY: New York State Education Department (1985), pp. 18–19.

2. The use of orally presented problems is an effective means of filling in a minute or two before lunch, recess, or the arrival, for instance, of the music teacher.
3. Problems that involve complicated computation may be handled by directing that the pupils determine only the operations to be used, writing A for addition, S for subtraction, M for multiplication, and D for division. The pupils can also be instructed to write only the mathematical sentence that is necessary to the solution of the problem.
4. A teacher who wishes to determine the pupils' answers as the problem set progresses may use 3×5 cards on which the various numerals have been written. Pupils hold the correct cards in front of them. For example, if the answer were 35, the pupil would hold a card with the numeral 3 in the right hand and a card with the numeral 5 in the left hand.
5. Orally presented computational or "follow-me" exercises are also helpful in problem solving. They form an important part of orally presented materials. A teacher may say, "See if you can follow me. Start with 5, add 7, subtract 2, square this number, subtract 50, divide by 5. What is the answer?" Some children become interested in a "non-number" approach, such as, "Take today's date, subtract the grade you're in, add your age," etc.
6. The tape recorder is a useful device for orally presented problems. Pupils are intrigued by the tape recorder, and they also pay greater attention to it than to a teacher because they realize that the recorder will not repeat the problem.
7. A sequence of pictures can be used with the orally presented problem as a means of helping children visualize the action and remember the problem situation.

NONVERBAL PROBLEMS

The presentation of a problem situation by using a single picture or a multipicture cartoon format with a minimum of words is often referred to as the nonverbal problem-solving approach.

There are several features of nonverbal problems that make their use a valuable part of the problem-solving program: (1) The teacher can present problems that are very much "real life." (2) Nonverbal problems allow children to focus quickly on a problem situation without reliance on reading skills, which is a great help at the prereading level and also for children with reading handicaps. (3) The flexibility of the nonverbal format allows the children to generate several problems from a single situation. This is an aid to realism and helps to provide for individual differences. It also provides an opportunity for the children to exercise some originality and creativity in finding solutions to the problems. (4) Nonverbal problems, much like pupil-formulated problems, aid the children in expressing mathematical situations in their own words. This is an aid to most children, but specifically it can be helpful to children with culturally different or poor language patterns.

USE OF DIAGRAMS, GRAPHS, AND DRAWINGS

In the past, the use of diagrams, graphs, and drawings was often reserved for only the slow student, but these are useful problem-solving aids for pupils of all ability levels. They force the pupil to consider the problem situation, since he or she cannot just manipulate numerals without understanding the basic structure of the problem. Another valuable feature of graphic aids is versatility. Drawings may be used by a slow pupil to solve problems that would otherwise seem unsolvable, and the advanced student finds drawings of value in the solution of problems that involve complex relationships.

The use of diagrams is particularly helpful for solving problems that involve rates, distances, and measures of quantities. It can and should begin at the early primary level and continue in each succeeding grade.

USE OF PROBLEMS WITHOUT NUMBERS

Problems that do not contain numbers can be presented and the pupils required to explain how the problem would be worked if the numbers had been given. Such problems are very helpful in forcing pupils to analyze the problem situation rather than just finding means of computing with the numbers.

Two formats can be used effectively in this method: The problem can be presented and the pupils asked to write a short description of the procedures they would use in solving the problem, or several possible solutions can be presented and the pupils asked to pick the correct or best procedure for solving the problem. An example of each technique follows:

1. Read the problems carefully. You will find that they do not contain numbers. You can, however, decide the procedure that would be used if a solution were attempted. In the space below each problem, write a sentence or two telling how you would solve it.
   Problem: Ann weighs herself and her cat together. Then she weighs herself. How can she find the weight of her cat?
2. Choose the correct solution for the problems.
   Problem: You go to the store and buy a comic book, a package of gum, and a notebook. How can you tell if you have enough money?
   a. Multiply the cost of the items and compare it with the amount of money you have.
   b. Add the cost of the items and then divide it by the number of items.
   c. Add the cost of the items and then compare it with the amount of money you have.
   d. Correct answer is not given.

USE OF RESTATEMENT OR ANALOGIES

Asking students to restate a problem in their own words is helpful in problem comprehension. The teacher may also suggest that students develop a problem that is of the same structure as the original but uses simpler numbers. In some cases, the teacher may wish to present orally a simpler problem of the same type, ask students to solve that problem, and then compare the simpler problem with the original one. For example, "A ranch 2.67 miles long and 1.82 miles wide contains how many square miles?" can be paralleled with "A ranch 4 miles long and 2 miles wide contains how many square miles?"

"Polya Sequences"

Professor George Polya is considered one of the master teachers of problem solving. Over the years he has developed a set of *heuristics* (informal procedures) that are useful in solving a problem. In fact, the procedures listed in Box 4–1 were considered so important by the *Problem Solving in School Mathematics* yearbook committee that they were presented on the front and back inside covers of the 1980 yearbook.

Study each of Polya's steps and substeps carefully. Keep in mind that these heuristics cannot be used as a rote step-by-step procedure. For years the majority of elementary school mathematics textbooks have given a four-step procedure similar to the one shown here with little success. In fact, a study done many years ago indicated that children did not use the procedure. When I began teaching elementary school mathematics in the 1950s, I tried a four-step procedure with little results. Later I learned it was necessary to try out—discuss and break down—the procedures in ways that helped children to discover solutions.

Study Polya's chart carefully. Make notes. With the notes from the chart, go through the remainder of the problem-solving suggestions and determine how you can incorporate Polya's heuristics with the other procedures.

USING PUPILS' FORMULATION OF PROBLEMS

When pupils formulate, criticize, and refine problems that they themselves have composed, they often gain insight into the makeup of verbal mathematics problems. Graham[1] and Keil[2] have found that students profit from making up their own problems.

The teacher may begin this phase of problem solving by stating, "One of the best means of improving your problem-solving ability is to make up your own problems. Today let's use the newspaper article on 'Active Stocks' as the basis for your problems. You may check other sources of information if you wish. Here is an example of a problem I developed: How much more would it cost to buy three shares of Apple Computer stock than three shares of Chrysler if both shares were purchased at the closing cost?" The class may discuss the form of the problems and the importance of clear expression. It is usually wise to suggest that students formulate problems that use various number operations. In addition to this, the development of multistep problems should be emphasized.

Pupils can formulate problems at all levels of the elementary school. First-grade pupils can make up oral problems to present to the class, and higher-grade students can post problem sets on the bulletin board. On occasion, two or

[1]V. G. Graham, "The Effect of Incorporating Sequential Steps and Pupil-constructed Problems on Performance and Attitude in Solving One-step Verbal Problems Involving Whole Numbers" (Ph.D. dissertation, Catholic University of America, 1978).

[2]G. E. Keil, "Writing and Solving Original Problems as a Means of Improving Verbal Arithmetic Problem Solving Abilities" (Ph.D. dissertation, Indiana University, 1964).

Box 4–1 Polya's Heuristics

1. You Must Understand the Problem

What is the unknown? What are the data? What is the condition?
Is it possible to satisfy the condition? Is the condition sufficient to determine the unknown? Or is it insufficient? Or redundant? Or contradictory?

Draw a figure. Introduce suitable notation.

Separate the various parts of the condition. Can you write them down?

2. Devise a Plan

(Find the connection between the data and the unknown. You may be obliged to consider auxiliary problems if an immediate connection cannot be found. You should eventually obtain a *plan* of the solution.)

Have you seen it before? Or have you seen the same problem in a slightly different form?

Do you know a related problem? Do you know a theorem that could be useful?
Look at the unknown! And try to think of a familiar problem having the same or a similar unknown.

Here is a problem related to yours and solved before. Could you use it? Could you use its result? Could you use its method? Should you introduce some auxiliary element in order to make its use possible?

Could you restate the problem? Could you restate it still differently? Go back to definitions.

If you cannot solve the proposed problem, try to solve first some related problem. Could you imagine a more accessible related problem? A more general problem? A more special problem? An analogous problem? Could you solve a part of the problem? Keep only a part of the condition, drop the other part; how far is the unknown then determined, how can it vary? Could you derive something useful from the data? Could you think of other data appropriate to determine the unknown? Could you change the unknown or the data, or both if necessary, so that the new unknown and the new data are nearer to each other?

Did you use all the data? Did you use the whole condition? Have you taken into account all essential notions involved in the problem?

3. Carry Out Your Plan

Carrying out your plan of the solution, *check each step*. Can you see clearly that the step is correct? Can you prove that it is correct?

4. Examine the Solution Obtained

Can you *check the result?* Can you check the argument?

Can you derive the result differently? Can you see it at a glance?

Can you use the result, or the method, for some other problem?

three students will benefit from working together on the formulation of verbal problems. Pupils may also be challenged to develop their own variations of puzzle-type problems.

COMPUTER USAGE

The microcomputer can be a very valuable adjunct to the problem-solving program. Several of the possibilities are suggested in Chapter 5. Probably the most important contributor to the problem-solving program is the work that children do in LOGO or BASIC to develop a solution to a problem. The development of a LOGO program also involves the use of logic, sequential thinking, and true upper-level problem-solving skills. Care should be taken that such teaching does not become "cookbook programming."

Furthermore, teachers and parents who have access to a microcomputer for a long period of time have found that children develop problem-solving skills by using programs such as "Snooper Troopers" and "In Search of the Most Amazing Thing."[3] These programs deal with solving mysteries ("Snooper Troopers") and finding missing treasure or people. These programs do take a great deal of time and thus are more appropriate for after-school or home use.

THE MATHEMATICAL SENTENCE

Early in the elementary grades, children can learn to write mathematical or number sentences to express in numbers the ideas conveyed in a problem and to write the different names for a number ($5 = 3 + 2, 4 + 1$, and so on).

Mathematical sentences can be of several types: (1) $3 + 4 = 7$ is a "true sentence" and also an "equation"; (2) $3 + 4 = 8$ is a "false sentence"; (3) $3 + 4 = \square$ and $5 - 3 = N$ are "open sentences." The symbols $\square$ and N are called *place holders,* or *variables.* The open sentence is made true or false by the numeral used to replace the place holder. (4) In addition, sentences that indicate relationships other than equality or inequality may be written. For example, $5 > 3$; $N + 2 > 7$; and $\square - 5 < 3$. In each mathematical sentence that has a place holder, there may be one or more than one replacement that makes the sentence a true sentence. A replacement or replacements that make a sentence true are called the *solution set* or *roots* of the sentence. The solution set for $\square - 5 < 3$ would be $\{5, 6, 7\}$ if the set of whole numbers is used as the basis for the sentence. Several other sentences and their solution sets are illustrated below. Many problem situations ask questions that can be expressed in this manner:

Sentence	Solution Set
$3 + 4 = \square$	$\{7\}$
$N + 2 > 7$	$\{6, 7, 8, \ldots, N\}$
$6 \times \square < 60$	$\{0, 1, 2, 3, 4, 5, 6, 7, 8, 9\}$

[3]Developed by Spinnaker Software, 215 First St., Cambridge, MA 02142.

Solving Problems by the Use of Mathematical Sentences

At the early primary-grade level, the teacher can begin the use of the mathematical sentence with the following development: "The past few days we've been working problems that I've read to you, asking you to find the answer. Today I'm going to ask you to see if you can make a mathematical sentence about the problem. Let's try one together. 'Bill bought three 4¢ stamps. How much did they cost?' Can you state in words the basic question asked?"

The pupils suggested two possibilities: (1) How many are three 4s, or three 4s equal what number? (2) How many are 4 plus 4 plus 4, or 4 plus 4 plus 4 equals what number? The teacher suggested that the questions be written in mathematical-sentence form by using mathematical symbols rather than words, and quizzed the class about the substitutions in symbols that could be made. The pupils suggested that 3×4 replace the three 4s; that = replace *equals;* and that $\square$ or N replace *what number.* Similarly, in the second case, the pupils wrote the mathematical sentence as $4 + 4 + 4 = \square$.

During the remainder of the problem-solving session, the teacher suggested that the pupils develop a mathematical sentence that could be used in solving the problem, solve the problem, and then verify their answers with a drawing or a diagram. On occasion, for variety, the teacher suggested that the pupils write only the mathematical sentence and not continue to find the number for the place holder.

When the mathematical sentence had been introduced and the pupils seemed to understand all parts of it, the teacher developed by student consideration the terms *open sentence, equation, solution set,* and *place holder.* The use of these terms is optional.

After the children have had a variety of experience in writing mathematical sentences while solving verbal problems, they should have an opportunity to experiment with the development of procedures for solving equations. Using the basic ideas that relate addition to subtraction and multiplication to division, children can arrive at the following generalizations:

1. The same number can be added to both members of an equation.
2. The same number can be subtracted from both members of an equation.
3. Both members of an equation can be multiplied by the same number.
4. Both members of an equation can be divided by the same number, with the exception of zero.

Because of limited space, illustrative lessons on the development of generalizations for the solution of mathematical sentences are not included here. This does not mean that the teacher should present the material above in an explanatory manner. Pupils can effectively discover for themselves, or be led to discover, the generalizations for working with equations and other types of mathematical sentences.

Multistep Problems and Mathematical Sentences

Although pupils quickly develop skill in writing mathematical sentences for problems that require a single mathematical operation, this is not true for problems whose solutions involve several steps. Some can work a multistep problem but cannot develop the mathematical sentence that concisely describes the problem; others manipulate the numbers, hoping to arrive at some solution.

At present, few specific suggestions regarding developing pupil skill in solving multistep problems are available to teachers. Because of this lack, a lesson that was successfully used in fourth-, fifth-, and sixth-grade classrooms is presented. Its purpose was to show a need for setting apart the processes involved in a multistep problem. The pupils had an opportunity to discover the most logical use of parentheses.

The teacher wrote this problem on the chalkboard: Jeff was collecting money to purchase groceries for a camping trip. He was to buy the following items (*Note:* The teacher took this problem from an old text (1950s). Use current prices.):

1 tin of Spam® at 45¢ a can
5 cans of beans at 12¢ a can
1 box of crackers at 24¢ a box
6 cans of juice at 12¢ a can
1 bag of potato chips at 75¢ a bag

How much money did Jeff need? (Tax was not charged on the groceries.)

The pupils were instructed to write a mathematical sentence that could be used to solve the problem.

As the faster workers finished, they were asked to write their mathematical sentences on the chalkboard. When most of the pupils had finished, a discussion was held to determine the best method of writing the mathematical sentence. The teacher had purposely allowed several pupils to write incorrect solutions on the board and began with a discussion of one of these.

Mel said that the problem was an addition and multiplication problem and could be written as $45 + 5 \times 12 + 24 + 6 \times 12 + 75 = 7{,}635¢$. He computed from left to right in the following manner: $45 + 5 = 50$. $50 \times 12 = 600$. $600 + 24 = 624$. $624 + 6 = 630$. $630 \times 12 = 7{,}560$. $7{,}560 + 75 = 7{,}635$. The answer was in cents, so Mel converted his answer to $76.35.

Immediately there was a stir of disapproval from class members. Jane noted that the answer was too large. A scout troop couldn't possibly pay that much for a few groceries. Another pupil noted that the multiplications should have been performed before the addition.

The class then considered other ways of writing the problem. The methods used included:

NANCY: $ .45
.60
.24
.72
.75
$2.76

HOWARD: $45 + 60 = 105 + 24 = 129 + 72 = 201 + 75 = 276¢$[4]
JOHN: $45 + (5 \times 12) + 24 + (6 \times 12) + 75 = 45 + 60 + 24 + 72 + 75$
$= 276¢$ or \$2.76

The class decided that Nancy had the correct answer but had not shown all of her thinking. Howard's answer was also considered correct, but he had not shown all of his thinking either. Furthermore, he had made incorrect use of the equals sign. John was asked to explain why he had used parentheses. He said, "First, I took the 45¢ and put that down. Then, so I could show I didn't mean that 45 + 5 was to be multiplied by 12, I enclosed the computation 5 × 12 in parentheses."

Special comment was then made about the possibility of working a problem incorrectly unless the work to be done was clearly indicated, and the children discussed the value of parentheses as an aid to showing exactly the processes involved.

Practice was then given in writing mathematical sentences for other multistep verbal problems.

Formulas

Another use of the mathematical sentence is in the development of formulas. In the traditional mathematics program, the formula for finding the answer to a particular type of problem was given to the pupil for study, then problems were given in which the formula was used.

With the guided-discovery approach, a problem situation is presented and analyzed. Then several other problems of this type are solved. After acquiring a good understanding of this problem type, the pupil is challenged by the teacher to develop a formula for the solution of similar problems. For example, a pupil who has worked several problems that involve finding the distance when the speed (rate) of a vehicle and the time traveled are known will make the statement, "The distance traveled equals the rate of the car times the time taken for the trip." This is shortened to: Distance traveled = Rate (speed) × Time; and finally, $D = R \times T$.

After the pupil has developed a formula, he or she should be challenged to develop variations of it. For example, What is the formula for the rate? What is the formula for the time? The pupil made use of basic axioms of equations, and, beginning with $D = R \times T$, divided both members by T to find the rate:

$$D = R \times T; \qquad \frac{D}{T} = \frac{R \times T}{T}; \qquad \frac{D}{T} = R$$

This approach to the development of formulas has several advantages: (1) If pupils have developed a formula on their own, they are much less likely to forget it than if they memorize someone else's formula. (2) If pupils do for-

[4]Note the incorrect use of the signs of equality. This is a mistake frequently made by both children and adults. Howard is stating that 45 + 60 = 276¢.

get the formula, they can once again go through the process necessary for its development. (3) The pupils develop confidence in formulas that they have discovered themselves and are less frightened by letters that name numbers. At the present time, adults often have difficulty when any symbols except numerals are used. Thus, whereas almost any adult can easily understand $2 + 3 = 3 + 2$, many are baffled by $a + b = b + a$.

SIDE TRIP

In the section that follows, Sherlock Holmes is discussing his approach to solving a murder. As you read the material, consider the skills involved in "thinking backward."

> In solving a problem of this sort [murder] the grand thing is to be able to reason backward. That is a very useful accomplishment and a very easy one, but people do not practice it much. In the everyday affairs of life it is more useful to reason forward, and so the other comes to be neglected. There are fifty who can reason synthetically for the one who can reason analytically.
>
> ... Most people, if you describe a train of events to them, will tell you what the result would be. They can put those events together in their minds, and argue them from them that something will come to pass. There are few people, however, who, if you told them a result, would be able to evolve from their own inner consciousness what the steps were which led up to that result. This power is what I mean when I talk of reasoning backward, or analytically.*

1. How should this skill be taught?
2. Think up a lesson that would develop this skill. How could you use detective stories?

INTERDISCIPLINARY PROBLEM SOLVING

The mathematics used outside the school classroom comes from a wide variety of situations. There are routine problems, such as keeping track of the budget (not that this can't be very challenging), and very nonroutine problems, such as helping a child discover the real and scale speed of a slot car as it moves about a track. (This is a good lab activity for advanced upper-grade children.)

With the current emphasis on basic mathematical skills and very specific instructional objectives, it is necessary to be sure that sufficient time is allowed for "real-problem solving" that cuts across instructional areas and across mathematical topics. Such problem solving may be fostered by the use of interdisciplinary problems such as those developed by Unified Science and Mathematics for Elementary Schools (USMES). USMES has published teacher guides to such topics as "Finding Your Way," "Play Area Design and Use," "Bicycle

*A. Conan Doyle, *A Study in Scarlet, The Complete Sherlock Holmes*, v. 1 (Garden City, N.Y.: Doubleday, 1930), pp. 83, 84.

Transportation," "Ways to Learn," "Traffic Flow," "Lunch Lines," "Soft Drink Design," "Consumer Research Product Testing," "Weather Predictions," and "Designing for Human Proportions."

Each of the problems or "challenges" is designed to be real in several respects: (1) A solution is needed and not currently known, (2) the students are involved in complete situations with all the accompanying variables and complexities, (3) the problem applies to some aspect of student life in the school or community, and (4) the problem is such that the work done by the students can lead to some improvement in the situation. The expectation of useful accomplishment is the motivation for children to carry out the comprehensive investigations needed for a solution to the challenge.[5]

A challenge may read, "I've brought along five boxes of different breakfast foods and some milk; let's form groups of four and try to find out all we can about the cereals." During the course of the study, the children decide to test the cereals for sogginess, taste, crunchiness, and so on, and to analyze the printed vitamin-mineral content, sugar content, fibrousness, calories, and the like, with various groups comparing notes at appropriate discussion points.

OTHER TECHNIQUES TO IMPROVE PROBLEM-SOLVING ABILITY

In addition to the procedures previously suggested for developing problem-solving skills, several other techniques can be used. A number are briefly described below:

1. Make use of problems with *too much* or *too little* data. Such problems force the pupils to focus on the important aspects of the problem situation and are probably more typical of everyday problems than those with the exact amount of data. Periodically, a teacher can say to pupils, "Today some of your problems may have too much or too little information. If there is too much information, work the problem and then compose a short sentence that contains the information that was not needed. If there is the correct amount of information, just work the problem. If you do not have enough information to work the problem, indicate 'too few data' and then write a short sentence telling what data you need to work the problem."

   In addition to specific lessons that use this type of problem, many of the regular assignments should contain problems with too much data.
2. *Estimating the answer* without working a problem is another device to improve the pupils' analysis of problems. This technique is often difficult to use, since many conscientious pupils will work the problem first and then obtain an estimate from the solution. To avoid this practice, it is suggested that the estimation-of-answer technique be used in conjunction with orally presented verbal problems. After pupils become familiar with the value of estimating the answer, they will often use it on their own. Approximating a logical answer is particularly useful in problems involving fractions and decimals. To estimate an answer in these situations, the pupil rounds off the fraction or decimal to a whole number and computes mentally. The use of this procedure helps to eliminate the incorrect placement of a decimal point, or, in the case of frac-

[5]Unified Science and Mathematics for Elementary Schools, *Orientation* (Newton, Mass.: Education Development Center, 1975), p. 3.

tions, it helps to eliminate the use of the dividend as the divisor in division-of-fractions problems.

3. The use of "How's Your P.Q.?"–type problems is one of the most valuable problem-solving techniques. A portion of the arithmetic bulletin board can contain these difficult problems (usually at two or three levels of difficulty). Several can be presented each week and adequate time given for pupils to think through the problem (at least two days). In addition to the improvement of problem solving, the "How's Your P.Q.?" problems act as a good motivational device.
4. Problem analysis was once a very popular phase of problem solving. Textbooks contained directions to follow these steps in solving problems: (a) decide what is given, (b) think about what is asked, (c) think about what operation should be used, (d) estimate the answer, (e) solve the problem, (f) check the answer. Studies such as the one reported by Burch have indicated that formal analysis is not of great value, since pupils do not normally make use of the six-step procedure.[6] Thus, it is probably better to make use of only one or two of the steps at a time.
5. Use problems from a *variety of sources.* Pupils show great interest in problems taken from old textbooks, foreign textbooks, and scientific books. The alert teacher can find old textbooks in secondhand stores and at used-book dealers. Presently, several publishing companies sell translations of foreign textbooks. The problems that the teacher finds in these sources are not superior in quality to those found in the standard textbooks, but the idea that the problems were used by pupils of the past or are are used by pupils in foreign countries is intriguing students.

Additional suggestions from research. There have been over 500 studies dealing with problem solving in school mathematics. The suggestions that follow, added to the previous suggestions, should provide the basis for superior teaching of problem solving in the elementary school. It is also suggested that you read carefully several of the references given at the end of the chapter. Some findings to remember follow:

1. Children learn better when the problem situations used are within their interest and understanding range.
2. Young children need concrete problems rather than abstract problems.
3. Early childhood learners need direct aid in recognizing the relevant facts in a problem.
4. Trial-and-error methods are often used by young children. They can be helped to develop individual strategies.
5. A set of steps to be used in problem solving works less well with young children than middle-grade children.
6. The actual materials of a problem situation are more important with young children than with middle-grade children.
7. Praising pupil progress improves achievement.
8. Working problems in two- or three-person teams improves problem-solving ability.
9. Helping pupils correct their own problems improves pupil achievement.

[6]Robert L. Burch, "Formal Analysis as a Problem-solving Procedure," *Journal of Education 136* (November 1954): 44–47.

10. Problems using one type of classification (dogs and dogs) are easier for young children than problems using more than one classification (dogs and cookies).
11. Rigid problem-solving procedures should not be used.

See:

GROSSNICKLE, F., and others, *Discovering Meanings in Elementary School Mathematics.* 7th ed. New York: Holt, Rinehart and Winston, 1983, pp. 188–90.
KANTOWSKI, MARY GRACE, "Problem Solving." In *Mathematics Education Research: Implications for the 1980s.* Washington, D.C.: Association for Supervision and Curriculum Development, 1981.
LESTER, F. K., JR., "Research on Mathematical Problem Solving." In *Research in Mathematics Education.* Reston, Va.: National Council of Teachers of Mathematics, 1980, pp. 286–323.
RIEDESEL, C. A., and PAUL C. BURNS, "Research on the Teaching of Elementary School Mathematics." In *Second Handbook of Research on Teaching,* edited by Robert M. W. Travers. Chicago: Rand McNally, 1973, pp. 1160–61.

SUMMARY COMMENTS ON PROBLEM SOLVING

1. No one "best" method of improving problem-solving ability has been developed. Therefore, the teacher should stress several approaches.
2. Some of the programs in elementary school mathematics treat problem solving after a topic has been introduced and developed rather than making use of problems in the introductory phase and again at a later phase of the topic development. This practice is questionable. One of the chief jobs of a mathematician is to "take a physical situation or a portion of the real world and represent it by a mathematical model."[7] Use of verbal problems at the introductory stage of a new topic develops skills in this phase of mathematical thinking.
3. The ability to read is an important adjunct to skill in verbal problem solving. The teacher must take care to develop the reading skills required in verbal problem solving.[8]

KEEPING SHARP

Self-Test: True/False

_______ 1. "Problem solving" in mathematics is concerned *only* with the ability to read and find the answer to story problems related to the computation skills being taught.

_______ 2. Open-ended, nonroutine problems are valid components of the mathematics curriculum.

[7]Robert A. Sebastian, "The New Mathematician," *The Mathematics Student Journal* 12 (November 1964):2.

[8]For helpful treatments of reading in elementary school mathematics, see Peter L. Spencer and David H. Russell, "Reading in Arithmetic," in *Instruction in Arithmetic, Twenty-Fifth Yearbook* (Washington, D.C.: National Council of Teachers of Mathematics, 1960), pp. 202–24; and C. Alan Riedesel and Paul C. Burns, "The Role of Reading in Mathematics and Teaching Problem Solving," in *Handbook for Exploratory and Systematic Teaching of Elementary School Mathematics,* chap. 6 (New York: Harper & Row, 1977).

_______ 3. Primary-grade pupils can begin to use mathematical sentences to represent problem situations.

_______ 4. Among the generalizations useful in solving verbal problems is that the same number can be added to one member of the equation and subtracted from the other member of the equation.

_______ 5. Pupils should not be allowed to record or write on the board incorrect or inefficient solutions to problems.

_______ 6. Not only are pupils who develop a formula for solving a particular type of problem less apt to forget the formula, but, if it is forgotten, they can repeat the process and develop it again.

_______ 7. One of the advantages of the interdisciplinary approach to problem solving is that it gives real problem-solving practice by isolating mathematics from other subject areas.

_______ 8. Orally presented problems of the "follow-me" type provide practice in mental computation.

_______ 9. Both nonverbal and orally presented problems allow poor readers to participate fully in the problem-solving activities without being penalized for their reading performance.

_______ 10. Nonverbal problems are more flexible because several problem situations may be suggested by one picture.

_______ 11. The use of diagrams and graphs should be discouraged, because they do not focus attention on computation practice.

_______ 12. Good teachers avoid problems without numbers because they do not provide computation practice.

_______ 13. Problems without numbers allow pupils to analyze the situation and determine the correct procedure without worrying about getting the answer right.

_______ 14. Problems with too much or too few data force pupils to focus on significant aspects of the problem situation.

_______ 15. Estimating the answer before working a problem is not one of the recommended techniques for improving problem solving.

_______ 16. There is no one best method for improving every pupil's problem-solving ability, but several approaches are useful.

THINK ABOUT

1. The term *problem* in mathematics teaching is often a source of confusion. Why?
2. What are the major roles of word problems in mathematics instruction?
3. Do mathematics books give much space (time) to any one procedure of improving problem-solving ability? Use at least two different series of books to substantiate your answer.
4. What makes the oral presentation of problems and computation exercises that pupils are to solve without use of pencil and paper such a good procedure for improving problem-solving ability?
5. Why must the source of the major part of the program for improving problem-solving ability be nontextbook?
6. What is the role of word problems in the typical first- and second-grade mathematics textbooks? Use at least three series of books to support your answer.

SUGGESTED REFERENCES: (Note that the starred references are "classic" to be compared with more recent references)

*BROWNELL, WILLIMA A., "Problem Solving." In *The Psychology of Learning, Forty-first Yearbook,* pp. 415–43. National Society for the Study of Education, Chicago, Ill.: University of Chicago Press, 1942.

CLEMENTS, DOUGLAS H., *Computers in Elementary Mathematics Education,* chap. 5, Englewood Cliffs, N.J.: Prentice Hall, 1989.

GAROFALO, JOE, (Ed.) *Mathematical Problem Solving, Issues in Research,* Philadelphia, Penn.: The Franklin Institute Press, 1982.

GILBERT-MACMILLIAN, KATHLEEN, and STEVEN J. LIETZ, "Cooperative Small Groups: A Method of Teaching Problem Solving." *The Arithmetic Teacher* 33 (March 1986): 9–12.

*HENDERSON, KENNETH B., and ROBERT E. PINGRY, *"Problem-Solving in Mathematics."* In *Learning of Mathematics, Its Theory and Practice, Twenty-first Yearbook,* chap. 8. Reston, Va.: National Council of Teachers of Mathematics, 1953.

MORRIS, JANET, *How to Develop Problem Solving Using a Calculator,* Reston, Va.: National Council of Teachers of Mathematics, 1986.

OVERHOLT, JANE L., JANE B. RINCON and CONSTANCE A. RYAN, *Math Problem Solving for Grades 4 through 8.* Boston, Mass.: Allyn and Bacon, 1984.

POYLA, GEORGE H., *How to Solve It,* 2nd ed. Princeton, N.J.: Princeton University Press, 1971.

Problem Solving in the Mathematics Classroom, Math Monograph No. 7., Edmonton, Alberta: Alberta Teacher's Association, 1982.

SAUNDERS, HAL, *When Are We Ever Gonna Have to Use This?,* Palo Alto, Calif.: Dale Seymour, 1981.

SCHOENFELD, ALAN H., *Mathematical Problem Solving,* Orlando, Fla.: Academic, 1985.

SILVER, EDWARD A., (Ed.) *Teaching and Learning Mathematical Problem Solving: Multiple Research Perspectives,* Hillsdale, N.J.: Lawrence Erlbaum Associates, 1985.

chapter **5**

COMPUTERS AND CALCULATORS

OVERVIEW

Experimentation with computers and calculators in teaching elementary school mathematics began in the 1960s. With the advent of the microcomputer in the late 1970s, this activity has greatly increased. Calculators have dropped in price from over $50 to under $10. Microcomputers have also dropped greatly in price. In fact, at the present time the great majority of elementary schools have at least one microcomputer and many elementary schools have several.

More work has been done with computers in the field of mathematics teaching than in any other teaching field. This is natural, since the majority of persons in computer science have a strong mathematical background, and many persons working in mathematics education feel comfortable with computers.

The National Council of Teachers of Mathematics Standards suppose that all elementary school students have appropriate calculators available at all times; that a computer is available in every classroom for demonstration purposes; that every student has access to a computer for individual and group work; and that students learn to use the computer as a tool for processing information and performing calculations to investigate and solve problems.

It is important to keep in mind that the same criteria of pupil involvement, discovery, and critical thought used with other tools of instruction should be applied to calculators and computers.

There are four major uses of microcomputers in teaching elementary school mathematics: (1) using the computer in problem solving and logical thinking to program, (2) computer-assisted instruction, (3) using LOGO and other graphic programs to develop geometric concepts, and (4) management of instruction in mathematics via computers.

TEACHER LABORATORY

1. If possible, using LOGO draw a square, a triangle, a hexagon, an octagon, a circle, and a flower.
2. Suggest uses of drill and practice, simulation, and tutorial computer-assisted materials in mathematics?
3. Pick a grade level. What activities would be appropriate for developing computer literacy at that level?
4. Develop a discovery-oriented lesson that involves a computer-mathematics concept.

TAKE INVENTORY

Can You:

1. Explain uses and limitations of calculators and microcomputers in elementary mathematics instruction?
2. Suggest uses of drill and practice, simulation, and tutorial computer-assisted materials in mathematics?
3. Give examples of motivational and recreational uses of calculators?
4. List some criteria for selecting educational courseware in mathematics?
5. Discuss uses and abuses in learning to program in BASIC and LOGO?
6. Understand the computer terms contained in the glossary at the back of the book?
7. Suggest several means of keeping up-to-date in this ever-changing field?

COMPUTER-ASSISTED INSTRUCTION

Computer-assisted instruction (CAI) has been used for over 2 decades. CAI is a teaching process directly involving the computer in the presentation of instructional material in a mode designed to provide active involvement with the student. These modes are usually subdivided into drill and practice, tutorial, simulation, instructional games, and problem solving.

Studies on the effectiveness of CAI strongly indicate that CAI either improves learning or shows no difference when compared to traditional instruction; CAI is effective at all grade levels; CAI is effective in drill-and-practice, tutorial, gaming, problem-solving, and simulation modes; CAI speeds up learning when compared to conventional methods; students have a positive attitude toward CAI; a positive attitude of the teacher increases CAI effectiveness; CAI is helpful in review; and the cost-effectiveness of CAI is improving.[1]

Drill and Practice

There are a number of drill-and-practice programs in elementary school mathematics that make use of games. Drill and practice typically require a one child–one computer ratio. This makes it difficult to widely use drill-and-practice materials in situations without several computers available.

In situations where drill and practice can be used, consider game-format disks. There are several public-domain disks[2] (disks that can be legally reproduced and are thus inexpensive) that have good game formats.

Well-developed drill and practice disks often have most of the following characteristics. (1) They deal with topics that lend themselves well to computer use. (2) They drill on topics that require immediate feedback, such as addition or multiplication facts. (3) They use the computer's ability to randomize and generate a variety of problems. (4) They use the computer's ability to keep track of errors and give further practice on exercises that are answered incorrectly. (5) They use sound where appropriate. (6) They use a game format that increases interest in the math rather than the game aspect.

Tutorial

Programs designated as tutorial are typically associated with learning new material rather than practicing material already studied. Good tutorial programs require that the author provide many branches to aid the learner. They also require a variety of teaching strategies. To date, the quality of tutorial programs in elementary school mathematics has been quite low.

Before purchasing any tutorial CAI mathematics program, it is suggested that you (1) make sure the topic is one you wish to develop by using the computer rather than by teaching it personally, (2) make sure that there are effective means of handling logical errors (for example, if $25 + 18$ is presented for the first time in a program, there should be a remedial branch that provides questions to help the student who answers 313, a common early error: $25 + 18 = 313$, in which the student does not "carry"), and (3) make sure that the tutorial makes use of the unique capabilities of the computer, such as move-

[1]For listing of appropriate CAI mathematics material, see: current issues of the journals listed at the end of the chapter and Clements, Douglas H., *Computers in Elementary Mathematics Education,* (Englewood Cliffs, N.J.: Prentice Hall, 1989.)

[2]*Softswap,* SMERC Library, San Mateo County Office of Education, 330 Main Street, Redwood City, CA 94063.

ment, branching on the basis of error, and keeping track of errors and pacing accordingly, moving a child along rapidly when it is evident that few examples are necessary for understanding.

Simulation

There are a number of instructional programs that can be classified as simulations. Simulations are used by the space program to test out ideas when a given environment is developed. Simulations are used to pretest cars. The game of monopoly is a game simulation.

There are several inexpensive or public-domain simulations that can be effectively used in the elementary school mathematics program. One is the old economic simulation "Hamurabie," in which the players take the role of the ruler of a country in the Middle East. They have to make the right decisions in order for the country to prosper and to prevent revolution. Each of the decisions involves a good deal of mathematical problem solving.

Another good simulation is "Lemonade Stand," in which the children act as operators of a lemonade stand. The simulation can be played in teams, and each team decides the cost of lemonade, advertising, and how much to make. With each round there can be a difference in temperature, precipitation, and/or workers in the area. The children keep track of the profit for each decision and by mathematical problem solving keep improving their profit.

Simulations not only make good use of the computer, they also help in the problem of having one computer for 30 children by using simulations, the class members can be divided into teams to work together to find answers.

Simulations are one of the better modes of CAI. It is suggested that the teacher of elementary school mathematics study catalogs carefully to find simulations that can be used in the elementary mathematics program.

Problem Solving

Microcomputers can be used for a variety of problem-solving activities. Some of these involve the development of computing programs by the children. There are a number of other problem-solving uses of microcomputers. For example, one teacher developed a program in which children were given a display of groceries from each of several supermarkets. Each week a new set of figures was inserted from newspaper ads. The children figured the amount they would save by making a combination of store purchases.

One of the best uses of the computer in problem solving involves a variety of activities with LOGO (suggestions for this use are given later in the chapter).

Game Formats

Any of the previous CAI modes can take the form of a game. Games that are well done can be effectively used for from 50 to 70 percent of the drill-and-practice activities necessary in elementary school mathematics.

COMPUTER-MANAGED INSTRUCTION

Microcomputers can help the teacher by performing a variety of management tasks. For example, (1) keeping attendance, (2) keeping class records, (3) scoring and analyzing tests, (4) administering tests in an individualized, interactive situation, and (5) managing assignments by coordination of quizzes with textbooks and other instructional materials are good tasks.

Several mathematics management systems are available for microcomputer use. Some are tied to a particular textbook and thus are valueless unless one uses that series primarily. Others are keyed to several textbooks or allow one to key one's classroom materials. Typically, these systems provide for testing students, evaluating the tests, prescribing appropriate work, and recording progress within a specific curriculum. One such system designed for students on a fourth- through eighth-grade level manages mathematics work in concepts of operations with whole numbers, fractions and mixed numbers, and decimals and percentages; ratios; proportions; problem solving and estimation; geometry; mathematics; and statistics and probability ("Classroom Management System—Mathematics B"; Science Research Associates, Chicago, 1982). This system, keyed to six major basal mathematics textbook series and several SRA (Science Research Associates) programs, is typical of some Computer-managed instruction (CMI) systems.

Students first take a survey, a wide-ranged test. They do this directly on the computer or, if computer time is at a premium, on reproduced paper copies, entering their responses on the computer at a scheduled time. If the survey indicates a possible area of need, the student is directed to the appropriate probe test to pinpoint specific weaknesses. If questions are missed on the probe test, a prescription is given. If, after completing the material, the student fails the probe test again, the student must see the teacher. The teacher can also customize the system by omitting certain surveys or probes for the class or individuals.

The teacher can call up the following records: class lists; survey or probe status reports; individual student records; class records; grouping reports; graph grouping reports, which provide a bar graph of the number of students working on each probe; and prescription reports.

At first, CMI appears to offer great hope in individualizing the elementary mathematics curriculum. However, it is only as good as the basic developmental teaching preceding its use. Some school districts have relied on CMI as the basis for their mathematics programs, with poor results. It should only be used in connection with discovery-oriented introductions by the teacher.

SELECTION OF SOFTWARE

Locating good software programs is a major undertaking. While there is some similarity between the search for a good textbook series and the search for a good CAI disk, there are a number of other factors that must be considered when selecting software. Teachers and administrators have had high hopes when purchasing a program only to find that it turns out to be an electronic page turner or even that it goes against the teaching philosophy of the district.

The scope of this book does not allow for a complete guide to software selection. However, because so many mathematics materials are individual disks rather than complete programs, all teachers should have a good idea of what to look for in choosing a CAI disk. Here are a few guiding principles of software selection designed to help the teacher of elementary school mathematics:

1. Know the software before you buy.
   a. Preview all software.
   b. Study reviews of software you are considering.
   c. Use an evaluation scale that fits your need.
   d. Know why you want a particular piece of software. Answer the question, "Why am I buying software?"
2. Find the answers to these critical questions concerning software:
   a. How does the program motivate the child to study the mathematics? Is the interest the result of an interesting approach, or does the motivation come from some outside source? In general, it is better to have the motivation come from the mathematics itself or from good uses of mathematics?
   b. What is the teaching method? Does it involve critical thinking by the children? Do children have an opportunity to discover mathematics?
   c. What background is required to use the material? What does the child need to know before starting?
   d. How does the program use the child's time? Is it time-effective?
   e. Is the program user friendly? That is, can a first-time computer user rapidly learn to use the material?
   f. Does it have a good manual (typically called "documentation")?
   g. Is the sequence of topics in keeping with the curriculum of the school?
   h. Do the materials fit into the regular mathematics program or with enrichment or remedial strands you want to add? Is the main thrust supplemental or basic?
   i. Is the material free from mathematical errors and misconceptions?
   j. Are the graphics well done?
   k. Is the program a good use of the computer? Does it make use of the unique strengths of the computer?
   l. Does the material make good use of situations appropriate for the background of the children?
   m. Does the material provide for a variation in mathematical content to meet individual needs?
   n. Are back-up copies provided?
   o. Is the program cost-effective in terms of use? Is the program written so that the material can be varied? For example, if it is a multiplication drill-and-practice program, can the teacher add material to use it for division?
3. Have the disks been developed with the involvement of authors with recognized background in teaching elementary school mathematics? Because the children are often so directly involved with CAI, the philosophy of the authors is tremendously important.
4. Test the material with children who will be using it.
5. Check on the possibility of home use of some of the materials.
6. Be sure to consider materials for teacher use, such as grade books, worksheet developers, and text construction programs.
7. Be open in your selection. Some of the best CAI elementary school mathematics materials have been developed by companies that were not in operation in 1978, (the first year of mass production of microcomputers), while some of the electronic

page-turners are produced by companies with a long reputation of producing elementary school mathematics books.

8. Send suggestions to manufacturers.

PROBLEM SOLVING AND PROGRAMING

The computer can make a major contribution to the child's problem-solving ability. In the elementary grades this occurs primarily through the use of the programing language LOGO and, to a lesser extent, the language BASIC. The material that follows gives a short description of the type of work that can be used. For a more thorough treatment of the topics, check the selected references at the end of the chapter.

LOGO

LOGO is a procedural language. That is, it proceeds step by step under the thoughts of the individual working with it. Because of this feature, a child with no knowledge of LOGO can begin and, by a series of experiments and discoveries, learn a great deal in a very short time. The way that LOGO can be learned is very much in keeping with the philosophy of this book.

LOGO was developed under a National Science Foundation Grant at Bolt, Beranek and Newman, a Cambridge, Massachusetts, high-tech consulting firm. Of the LOGO developers, probably the best known educationally is Seymour Papert (see *Mindstorms* in the selected readings), who combined computer science expertise with his belief in Piaget-type learning. Papert's philosophy of education combines the theories of Piaget with programming approaches developed for research in artificial intelligence.

Seymour Papert, and another developer of LOGO, Harold Abelson, suggest that LOGO is a philosophy of education in which a learning environment is established that gives people personal control over powerful computer resources. This enables them to establish intimate contact with powerful ideas of mathematics and science (such as model building and artificial intelligence).

With this in mind, every attempt should be made to help children control the computer in self-directed ways from their very first exposure to LOGO.

The portion of LOGO most often used in the elementary school is its "turtle graphics." A *turtle* is a computer-controlled triangle that resides on the display screen and responds to LOGO commands such as FORWARD, BACK, LEFT, or RIGHT. As the turtle moves, it traces its path and makes geometric shapes or drawings (see Chapter 13 for an example of a LOGO development of a square). As the child makes drawings and figures out ways of moving the turtle, many mathematical ideas are developed.

LOGO can be taught as soon as the children have developed some sense of direction. One third-grade teacher used LOGO to develop the majority of the geometric concepts in the regular program. When the class considered the topic of shapes, they went through a sequence of pointing out like and different shapes and then set about marking off two-dimensional shapes with masking tape on the classroom and hall floor.

The teacher then asked the children to direct a blindfolded classmate to walk around the shape. This required settling on commands such as forward, right, and left. When the children were able to "walk" a few shapes the teacher introduced them to LOGO. She began by asking, "What does the small triangle in the middle of the computer monitor look like?"

The children gave several responses such as "a dot, a kite, a box." The teacher told them that it was called a turtle and that they could get it to move by asking her to type in commands on the computer keyboard. She suggested that they try several commands.

The children began with FORWARD 40 and observed the results. They were not sure what to do with the right turn. They suggested RIGHT 40 and noted that the turtle turned right at an angle but did not move. After some discussion they realized that the right and left statements moved the turtle in a direction but that it was necessary to move FORWARD or BACK to move the turtle. The accompanying diagram represents the final sequence.

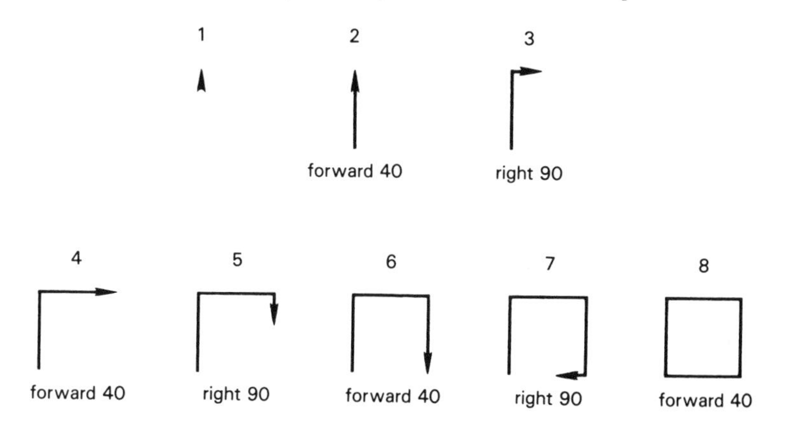

BASIC

BASIC was developed as an easy but powerful language to introduce college students to the computer. It became popular and is now the language that comes in the majority of microcomputers (to use LOGO you have to buy LOGO disks and load them into the computer). While BASIC does not lend itself to early school use, many middle schools have found BASIC to be a good learning experience for children in grades five to eight.

If you are going to use BASIC with your children, it is suggested that you

1. Find things to program that show good uses of the computer.
2. Use some structure, such as a simple flow chart, to develop with the children the steps that are to be taken in programing.

3. Stress that to be able to solve problems by using the computer, a person must be able to think through the problems. The computer helps, but does not do the hard thinking.

The scope of this book does not allow for any in-depth look at programing in LOGO or BASIC. However, it is important to remember that all of the teaching principles suggested in the chapters of this book apply for teaching problem solving and programing with the computer. To deal with the computer in rote and/or show-and-tell fashion is not only deadly, it misses the point of the computer as a key tool.

SIDE TRIP

Beat the Calculator

Tell a friend who uses a calculator that you can add faster. Have your friend write down two multidigit numbers, such as 1,682 and 456. You then write down the third. Say, "Let's see who can add these three numbers faster. You use the calculator and I'll use pencil and paper."

The trick: The number you write is important. It should be a number in which each digit is equal to 9 minus the corresponding digit in your friend's second number. In the case above, (9 − 4), (9 − 5), (9 − 6) makes 543. The sum of the three numbers will be simply the first number plus 1 followed by as many zeros as there are digits in the second number, then subtract 1.

Friend adds	You add
1,682	1,682
456	1,000
543	2,682
2,681	− 1
	2,681

Try to figure out how it works.

CALCULATORS

For many years, some teachers, mathematics educators, and mathematics teaching associations have been advocating the increased use of the hand-held calculator in the elementary school mathematics program. However, little progress has been made. With today's interest in the NCTM Standards, particularly since a number of elementary school mathematics textbooks series are being prepared by using them, we should be experiencing increased usage.

There is also a movement to make the use of calculators a part of the standardized testing program. When this occurs, teachers should feel much more comfortable with the using of calculators.

Over 200 research studies have been conducted on the use of calculators in schools. No ill effects have been reported. If anything, those children who have been using calculators are better in paper and pencil computation.

What follows are among the positive results from using calculators:

1. An increase in student enthusiasm and confidence in problem solving.
2. More positive attitude toward mathematics.
3. Greater persistence in problem solving.
4. Learning when and when not to use a calculator.
5. Increased comfortableness with technology.
6. Improved "number sense."
7. Improved master of basic number facts.
8. Willingness of children to seek alternate solutions of problems.
9. Improved skill in generalization from working many examples.
10. Development of skill in discovering relationships and number properties.
11. Clearer understanding of problems.
12. Improved speed in problem solving.
13. Solution of problems with greater complexity. True Polya-type procedures.
14. Better use of computational algorithms.
15. Increased consumer awareness.
16. Peer cooperation.
17. Increased interest and ability to check answers.
18. Greater individualization, for children are more independent in the development of new concepts.

A study of these 18 findings make clear that calculator use in the elementary grades should be strongly encouraged.

Help in estimation. A teacher reviewed with the children the fact that rounding numbers to the nearest multiple of 10 makes it easier to mentally estimate their products and then gave the children the sheet that is partially reproduced below. When the children had completed their sheets, a discussion was held concerning, "Why were some of your estimates much closer than others? What pattern did you see?"

Rounding, Estimates, and Patterns

USE YOUR HEAD			*USE YOUR CALCULATOR*			
Task	*Rounded Factors*	*Estimate*	*Task*	*Exact Product*	*Difference (Est./Prod.)*	*Pattern*
21 × 79	20 × 80	1600	21 × 79	1659	59	57
22 × 78	20 × 80	1600	22 × 78	1716	116	
23 × 77	20 × 80		23 × 77			
24 × 76			24 × 76			
25 × 75			25 × 75			
25 × 75			25 × 75			
26 × 74			26 × 74			

See: Callahan, Leroy, *Calculator Booklet,* Albany, N.Y.: New York State Education Department, 1990.

Improved thinking on operations. The teacher directed the children to use addition and subtraction signs to make sentences true and suggested the use of calculators for the activity. Here are three of the 10 exercises suggested.

1. Use addition and subtraction to make the sentences true.
   a. 96 ( ) 17 ( ) 19 ( ) 8 = 87
   b. 71 ( ) 23 ( ) 5 ( ) 23 = 66
   c. 91 ( ) 21 ( ) 22 ( ) 8 = 100
2. Developing generalizations.* The teacher gave the children a worksheet containing a number of multiplications and said, "Use your calculator to solve the multiplications. See if you can fill the blanks at the bottom of the page."

56 × 78 =	12 × 24 =	79 × 65 =	11 × 17 =
112 × 23 =	345 × 16 =	567 × 89 =	102 × 21 =

3. Make up more multiplications if you need, and fill in the blanks.

   tens × tens = thousands and hundreds
   tens × hundreds = ____________
   hundreds × hundreds = ____________
   hundreds × thousands = ____________

Emphasis on place value. The teacher used an overhead projector and uncovered one set of number pairs at a time (see below). The children were directed to tell what number operations they were to perform to get to the second number. The children entered the first number in their calculator and then were to show the second number on the calculator.

786	subtract 24	762
578	________	610
532	________	32
867	________	807

A variety of activities of this sort strengthens place-value concepts, addition and subtraction computation, and problem solving.

Problem Solving

The possibility for computer use and problem solving is almost endless. Beginning in kindergarten, children should use calculators to solve problems of appropriate difficulty. One teacher brought in a grocery list and the newspaper with ads from three supermarkets. The children were directed to find the most inexpensive way to buy all of the groceries and also to determine which store was the least expensive if all of the groceries were bought at one store. At another time the children were to estimate the time necessary to go to three stores and the gasoline that it would take.

*See Chapter 8 for further details of these generalizations.

Another teacher made use of Project Impact "Calculator Problem Solving Cards."

B9

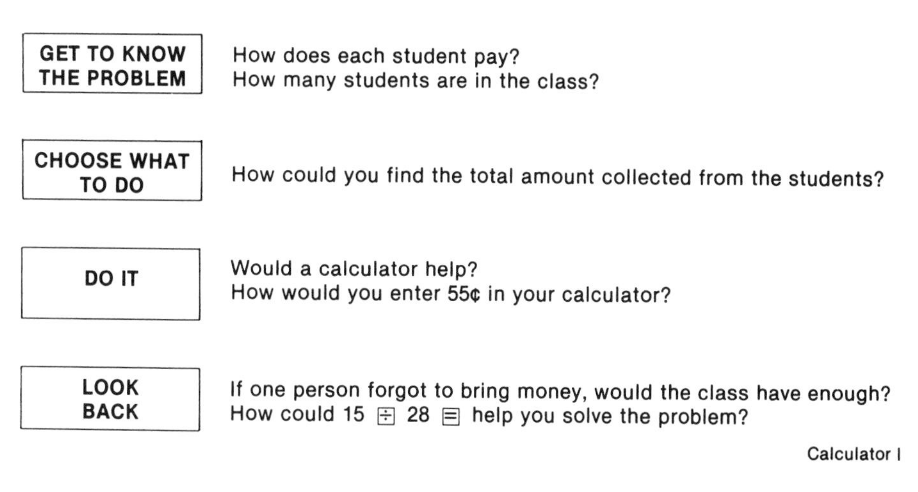

GET TO KNOW THE PROBLEM	How does each student pay? How many students are in the class?
CHOOSE WHAT TO DO	How could you find the total amount collected from the students?
DO IT	Would a calculator help? How would you enter 55¢ in your calculator?
LOOK BACK	If one person forgot to bring money, would the class have enough? How could 15 ÷ 28 = help you solve the problem?

Calculator I

B10

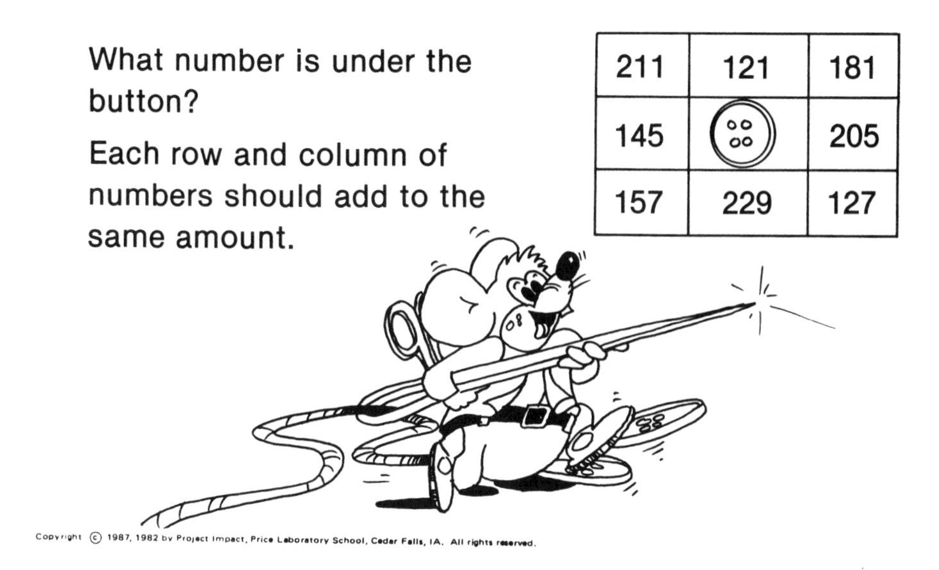

What number is under the button?

Each row and column of numbers should add to the same amount.

211	121	181
145		205
157	229	127

The NCTM Standards suggested the diagramed pattern for calculator use in problem solving:

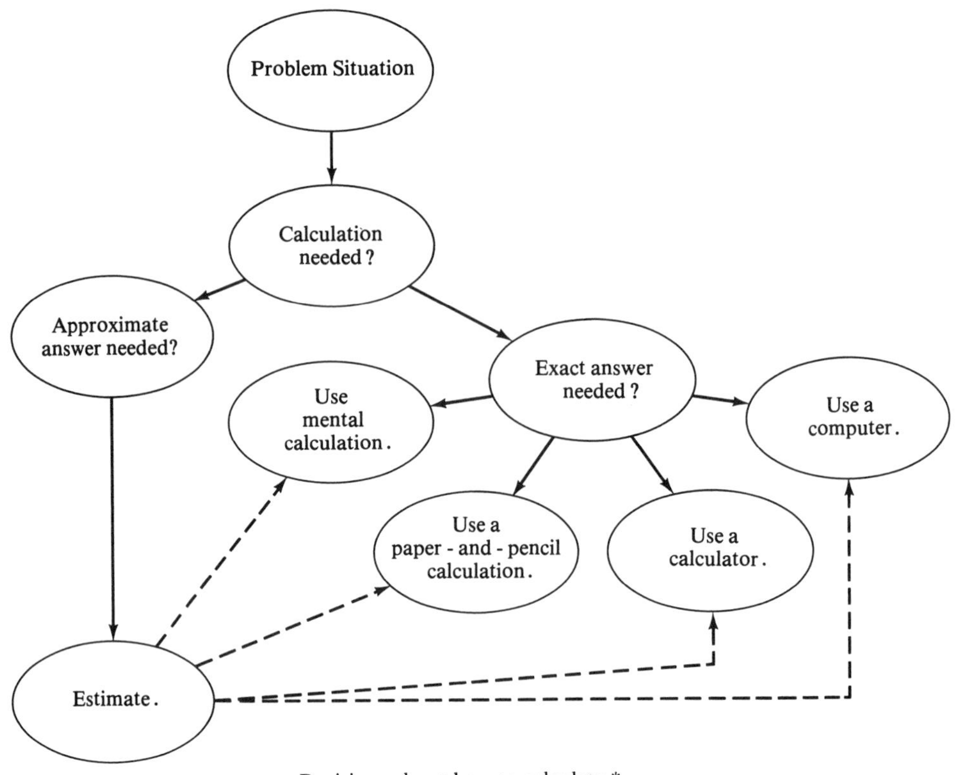

Decisions about how to calculate.*

A NOTE

As a teacher, remember that there are dozens of research studies that show that the calculator helps children in problem solving and computation, and none has found that the calculator is harmful to mathematical development. Therefore, use the calculator as often and as creatively as you can.

Also, there are many good books with good calculator exercises. Get some of them and try them out.

KEEPING SHARP

_______ 1. LOGO should only be taught after BASIC.
_______ 2. Calculators and computers serve the same basic function.
_______ 3. Always buy computer software from a company that has been developing textbooks for many years.
_______ 4. Usually, drill-and-practice materials in mathematics are better written than tutorial materials.

*Commission on Standards for School Mathematics, *Curriculum and Evaluation Standards for School Mathematics,* Reston, Va.: National Council of Teachers of Mathematics, 1989, p. 8.

_______ 5. CMI should be only a part of the elementary school mathematics program.

_______ 6. It is important for individual schools to evaluate software by having the teacher and children review it.

_______ 7. Both computers and calculators can be effectively used to improve the problem-solving skills of elementary children.

_______ 8. Computers and calculators are useful for enrichment, remedial, and basic use.

_______ 9. LOGO lends itself to developmental- (discovery-) oriented teaching.

_______ 10. It is possible for teachers to learn to develop their own CAI material.

_______ 11. The microcomputer can become an effective teacher aid.

_______ 12. Keeping up-to-date with computers in elementary school mathematics takes less time than keeping up-to-date on changes in the elementary school mathematics program.

_______ 13. Helping children learn how to use computers wisely should be a goal of microcomputer use in elementary mathematics programs.

Vocabulary

BASIC
CAI (computer-assisted instruction)
CBI (computer-based instruction)
CMI (computer-managed instruction)
computer
courseware
disk
microcomputer
LOGO
simulation
software
word processing

SUGGESTED REFERENCES

The Arithmetic Teacher, "Calculators, Focus Issue," February 1987.

Calculators, Reading from the Arithmetic Teacher and the Mathematics Teacher, Reston, Va.: The National Council of Teachers of Mathematics, 1979.

CLEMENTS, DOUGLAS H., *Computers in Elementary Mathematics Education,* Englewood Cliffs, N.J.: Prentice Hall, 1989.

CLEMENTS, DOUGLAS H., "Longitudinal Study of the Effect of LOGO Programming on Cognitive Abilities and Achievement," *Journal of Educational Computing Research* 3 (1987): 73–94.

Computers in Mathematics Education, 1984 Yearbook, Reston, Va.: National Council of Teachers of Mathematics, 1984.

HEMBREE, RAY, and DONALD J. DESSART, "Effects of Hand-held Calculators in Precollege Mathematics, a Meta-Analysis," *Journal for Research in Mathematics Education,* (March 1986): 83–99.

MORRIS, J., *How to Develop Problem Solving Using Calculators,* Reston, Va.: National Council of Teachers of Mathematics, 1987.

REYS, ROBERT E., ET. AL. *Keystrokes: Calculator Activities for Young Students,* Palo Alto, Calif.: Creative Publications, 1979.

RIEDESEL, C. ALAN, and DOUGLAS CLEMENTS, *Coping with Computers in the Elementary/Middle School,* Englewood Cliffs, N.J.: Prentice Hall, 1985.

SHUMWAY, RICHARD J., "Calculators and Computers." In *Teaching Mathematics in Grades K-8, Research-Based Methods,* chap. 13, Thomas R. Post (ed.), Newton, Mass.: Allyn and Bacon, 1988.

THIESSEN, DIANE, MARGARET WILD, DONALD D. PAIGE, and DIANE L. BAUM, *Elementary Mathematical Methods,* 3rd ed., New York: Macmillian, 1989.

WEIZENBAUM, JOSEPH, *Computer Power and Human Reason,* San Francisco, Calif.: W. H. Freeman, 1976.

WIEBE, JAMES H., *Teaching Elementary Mathematics in a Technological Age,* chap. 3 and 4, Scottsdale, Ariz.: Gorsuch Scarisbrick, 1988.

Computer Journals for Educators

The journals below devote the majority of their space to computers in education. Each has its own particular approach. Get a copy of each journal, and see what type of material it contains. Which are most helpful to you? We suggest that after careful study you subscribe to one and find ways to review the others each month.

Aeds Monitor, 1201 16th Street, N.W. Washington, DC 20036
Classroom Computer Learning, 19 Davis Dr., Belmont, CA 49002
The Computing Teacher, Department of Computer and Information Science, University of Oregon, Eugene, OR 97403
Educational Computer Magazine, Educational Computer, P. O. Box 535, Cupertino, CA 95015
Educational Technology, 140 Sylvan Avenue, Englewood Cliffs, NJ 07632
Electronic Education, Suite 220, 1311 Executive Center Drive, Tallahassee, FL 32301
Electronic Learning, Scholastic, Inc., 730 Broadway, New York, NY 10003
School Courseware Journal, Suite C, 1341 Bulldog Lane, Fresno, CA 93710
Teaching and Computers, Scholastic, Inc., 730 Broadway, New York, NY 10003
T.H.E. Journal, P. O. Box 992, Acton, MA 01720

Software Directories

The software directories contain brief descriptions of software available for use with particular computers.

The Apple Software Directory, Vol. 3, "Education", WDL Video, 5245 West Diversey Avenue, Chicago, IL 60639
The Blue Book for the Apple Computer, Visual Materials, Inc., 4170 Grove Avenue, Gurnee, IL 60031
The Book of Apple Software, The Book Company, 11223 South Hindryb Avenue, Los Angeles, CA 90045
The Commodore Software Encyclopedia, Commodore Corporate Offices, Education Dept., 487 Devon Park Drive, Wayne, VA 19087
Educator's Handbook and Software Dictionary, Vital Information, Inc., 350 Union Station, Kansas City, MO 64108
Index to Computer Based Learning, Educational Communications Department, University of Wisconsin, P. O. Box 413, Milwaukee, WI 53201
International Microcomputer Software Directory, Imprint Software, 420 South Howes Street, Fort Collins, CO 80521
Reference Manual for Instructional Use of Microcomputers, JEM Research, Discovery Park, University of Victoria, P. O. Box 1700, Victoria, BC V8W 2Y2, Canada
School Microwave Directory, Dresden Associates, P. O. Box 246, Dresden, ME 04342
The Software Directory, 11990 Dorsett Road, St. Louis, MO 63043
Swift's Directory of Educational Software, Apple II Edition, Sterling Swift Publishing Co., 1600 Fortview Road, Austin, TX 78804
The TRS-80 Sourcebook and Software Directory (available at local Radio Shack stores)
Vanloves Apple Software Directory, Vital Information, Inc., 350 Union Station, Kansas City, MO 64108

Software Clearing Houses

Public domain software (software that can be legally used and copied without fee) is available from some of these clearing houses. They are worth contacting to find out the function each performs.

Conduit, P.O. Box 388, Iowa City, IA 52244
Microcomputer Educations Application Network (MEAN), 256 North Washington Street, Falls Church, VA 22046
Softswap, San Mateo County Office of Education, 333 Main Street, Redwood City, CA 94063

chapter 6

ADDITION OF WHOLE NUMBERS

OVERVIEW

Addition is the first mathematical operation taught to elementary school children. The readiness for addition begins as soon as the child begins to count with meaning and to group objects together. The basic properties and principles of addition are used through college mathematics and in the everyday life of all of us. Because of the social and mathematical importance of addition, we should be very certain that children develop a depth of understanding as well as computational proficiency.

The National Council of Teachers of Mathematics (NCTM) Standards suggests that children should be able to

1. Model, explain, and develop reasonable proficiency with basic facts and algorithms.
2. Invent and use a variety of paper-and-pencil algorithms and mental arithmetic techniques.
3. Estimate results.
4. Use calculators as a computational tool.
5. Select and use an appropriate way to compute a problem situation and to determine whether the result is reasonable.
6. Develop operation sense.

For mastery of basic facts and for developing multidigit paper-and-pencil algorithms, the NCTM Standards suggest the sequences shown in the table. Note that there is no hard and fast distinction between easy and hard facts. However, easy facts can be solved efficiently by informal counting strategies, while hard facts require strategies such as using doubles and near doubles. These strategies are developed later in the chapter.

Reasonable Expectations for Learning Basic Facts

	Grade Level				
Topic	*K*	*1*	*2*	*3*	*4*
Addition/Subtraction					
Easy facts	explore	strategy	recall	recall	recall
Hard facts	—	explore	strategy	recall	recall

Reasonable Expectations for Learning Paper-and-Pencil Algorithms

	Grade Level			
Topic	*1*	*2*	*3*	*4*
Addition/Subtraction				
2-digit	—	explore	develop	maintain
3-digit	—	—	explore/ develop	maintain

TEACHER LABORATORY

Materials

Straws, Popsicle® sticks, toothpicks, or coffee stirrers (about 50).

Experiment

1. Bundle 30 of the sticks into tens. Pretend that you know how to add all the single-digit numbers (up to 9 + 9) but do not know how to add numbers such as 23 + 49 with pencil and paper. However (fortunately), you do have an idea of adding tens to tens. Now, find the sum of 18 + 27 by using the sticks. Make up a problem for this addition that would make real sense to a third grader.
2. Make several drawings similar to this one:

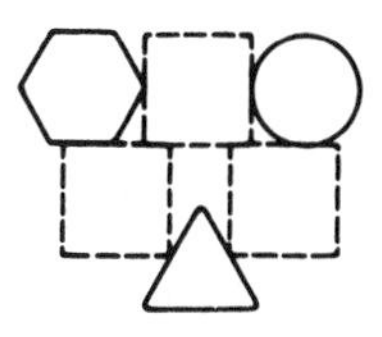

Think of a number for the hexagon, the circle, and the triangle. Write the sums of each set of numbers in the dotted squares. Give the completed drawing to your partner. It is his or her job to determine the original numbers. For example,

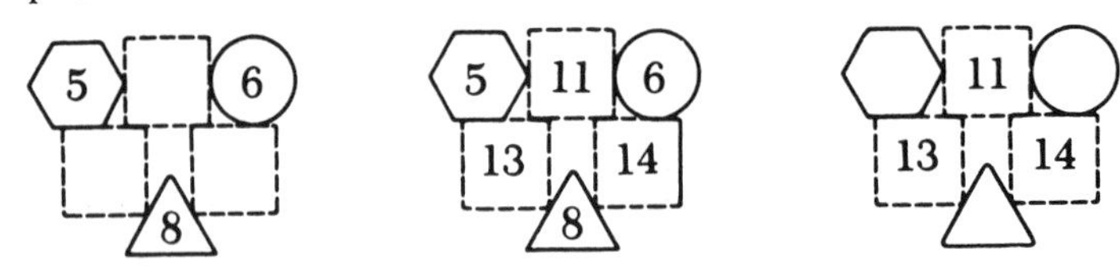

3. Try the same idea by using multidigits.
4. Try using these designs:

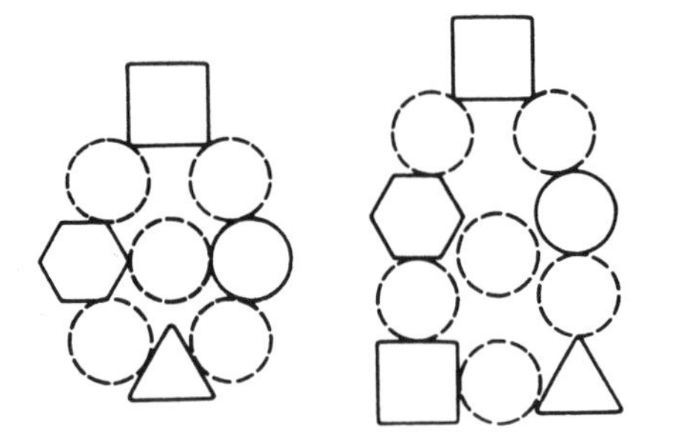

TAKE INVENTORY

Can You:

1. Describe three addition readiness experiences that you could use with kindergarten children?
2. Develop a laboratory lesson to introduce addition of whole numbers?
3. Suggest four ways to develop early addition ideas?
4. List four practice activities for addition?
5. Suggest three manipulative materials that could be used to develop the idea of renaming in multidigit addition? Suggest methods for using them?
6. List several ideas for reintroducing addition facts, multidigit addition?
7. Develop a guided discovery lesson for the following topics?
   a. The commutative property of addition
   b. The associative property of addition
   c. The role of zero in addition
   d. Multidigit addition

Reprinted by permission of UFS, Inc.

WHAT IS ADDITION?

If you were to stop almost any person on the street and ask, "Do you know what addition is?" the answer would come without hesitation: "Certainly I know what addition is." But if you were to ask for a precise definition, the person might have some difficulty. Often we are able to make use of mathematical ideas without being able to explain them concisely.

Before addition can be explained, certain terms must be developed. First, addition is a binary operation. That is, we can only add two numbers at a time. This is true both mathematically and psychologically. For example, $5 + 7 + 3$ becomes $(5 + 7) + 3$. We add $5 + 7$ and then add $12 + 3$. The numbers to be added, 5 and 7, are called addends, and the result of the addition is called the sum. For the majority of real-life situations, addition can be considered a rapid way to count.

In mathematical terminology, addition is defined from the set operation of a union performed on disjoint sets (sets that have no members in common). Recall that the union of two sets is the set composed of all the elements that belong to either of the sets. An element that belongs to either set also belongs to their union. Example: Suppose the Smith children—Mary, Bob, and Ken—go over to play with the Jones children, Harry and Nancy. This can be thought of as forming the union of two sets. With concrete objects this simply involves moving two piles of objects together and then finding the total.

$$\{\text{Mary, Bob, Ken}\} \cup \{\text{Harry, Nancy}\} = \{\text{Mary, Bob, Ken, Harry, Nancy}\}$$

The sum of two cardinal numbers[1] can be defined as the cardinal number of the union of the sets.

$$n\{\text{Mary, Bob, Ken}\} + n\{\text{Harry, Nancy}\} =$$

$$3 \quad + \quad 2 \quad =$$

$$n\{\text{Mary, Bob, Ken, Harry, Nancy}\}$$

$$5$$

If the two sets are not disjoint (if they have members in common), a difficulty arises.

$$\{\text{Tom, Fred}\} \cup \{\text{Mary, Fred, Alice}\} = \{\text{Tom, Fred, Mary, Alice}\}$$

$$n\{\text{Tom, Fred}\} + n\{\text{Mary, Fred, Alice}\} \neq \{\text{Tom, Fred, Mary, Alice}\}$$

$$2 \quad + \quad 3 \quad \neq \quad 4$$

The sets must be disjoint if addition is to be defined in terms of their union.

To avoid a common difficulty, the operation "union of sets" should not be considered the same as "addition of numbers." Children have often been told, "You cannot add unlike things." One does not add things; one adds numbers.

[1]When numbers are used to indicate how many, they are said to be used in the cardinal sense.

It is possible to form a union of a set containing three pears and a set containing six apples. The set formed is a set of fruit containing three pears and six apples. We may also think of the union of three pears and six apples as a set containing nine pieces of fruit.

Also, three sticks 1 foot long are not the same thing as one stick 3 feet long, although $1 + 1 + 1 = 3$. In length, however, the three 1-foot sticks laid end to end and the 3-foot stick are the same length.

Also, addition is the function that maps every ordered pair of numbers on their sum.

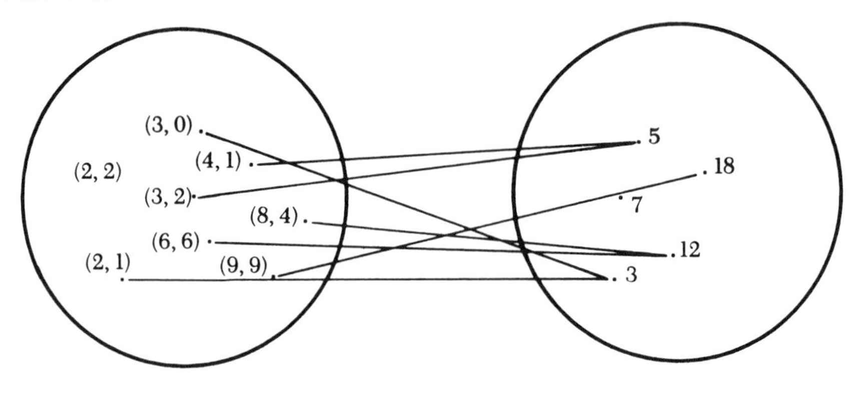

Addition: (a, b) (a + b)

FOUNDATION EXPERIENCES

The beginning of the addition idea starts in the earliest days of school life when children combine groups of objects and count or match the objects with a set of known number to find "how many." Soon, many children recognize without counting or matching with a known set that the result of combining a set of three objects with a set of two objects is a set of five objects. At this time, the children are adding.

Orally presented problem situations in which children have a reason to find "how many" form the framework for early addition readiness. These situations provide a physical-world setting for later abstractions concerning addition and also aid in developing mathematics as a useful and worthwhile activity.

The following materials are helpful in the exploration of pre-addition problems and activities:

1. *The actual objects described in the problem:* "Ken brought us three pictures, and Mary brought us four pictures. How many pictures are there? How can we find out? Lloyd, do you want to count the pictures you see? Don't count aloud: We'll count along with you silently." "Yesterday I gave each of you four crayons. I have four more for each of you." (The teacher passes out the crayons.) "How many do you have now?"
2. *Number strips or rods:* Strips of 1-inch squared paper can be cut to various lengths and the numeral written on the back of each strip. Wooden rods can also serve this purpose. For a problem situation such as, "Jack had three match-

box cars. His mother gave him two more. How many did he have then?" A 3 strip can be selected, then a 2 strip. Next, the two strips can be compared with the longer strips until they are matched with the 5 strip.

3. *A simple beam balance:* Children can experiment to find "how many" by using a simple beam balance made with a piece of pegboard and a wooden block. For counters, use plastic rods (wooden rods are too light to make a measurable difference on the balance) or washers. The washers can be tied with thin wire. Depending on whether or not the children need counting experience, the rods and the bundled washers can be numbered by using a strip of masking tape and a felt pen.
4. *The number line:* "John said he has three new books, Jill said she has five new books. Use the number line to find out how many new books the two children have. Would it make any difference whether we started with five books or started with three books?"

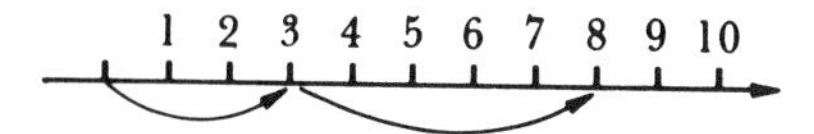

5. *Representative objects to be counted in place of the actual objects:* "Nancy has four dolls and Jill has five dolls. How many dolls do they have in all? If you need help, use Popsicle® sticks to represent the dolls. You may also use blocks if you wish."

   Other representative objects that may be used include beads, buttons, pictures, bottle caps, beans, paper strips, washers, grains of corn, building blocks, and play money.
6. *Children's drawings:* The children can make simple drawings of the objects described in a problem and count them to find the answer to an addition situation.
7. *Computer exploratory programs:* There are a few microcomputer programs for children that allow them to explore the addition facts. Check the catalogues of software publishers. Be sure that the programs are exploratory and not drill and practice.

It is often suggested that pupils should follow a three-step program in developing mathematical abstractions: first, an *enactive* (concrete) stage in which pupils work only with the actual objects; second, an *iconic* (semiconcrete) stage in which counters and pictures are used to replace the actual objects; and, third, the *symbolic* (abstract) stage in which numerals are used to represent the number ideas. In actual practice, children sometimes do not need all three stages. If orally presented verbal problems are used, some pupils can think from the concrete problem to semiconcrete material or to the actual number idea. Because children vary greatly in their maturity in abstract thinking, opportunities should be provided for the pupil to choose appropriate methods of attacking early addition situations.

The Basic Addition Facts

The basic addition facts are formed by the combination of all one-digit addends. When the zero facts are included, there are 100 basic addition facts. The addition facts are often grouped into combinations having sums of 10 or less and facts having sums of more than 10. The first are often called the "easy" addition facts, and the latter the "hard" addition facts. The words *easy* and *hard* actually have little meaning in terms of actual difficulty. For example, many pupils find 9 + 2 easier than 5 + 4.

Discovering the Addition Facts

After the pupils had had many experiences with joining sets, counting to find "how many," and matching sets in one-to-one correspondence (sometime about the middle of grade 1), the teacher asked, "How could you find an answer to this problem? Mary has five arrowheads in her collection. How many will she have if her grandfather gives her three more? See if you can think of the exact question the problem asks, and then solve the problem. Try to show in as many ways as you can that your answer is correct."

The children made use of paper and pencil in working the problem. The teacher moved about the class, noting methods of attacking the problem. When the majority were finished, several pupils drew on the chalkboard their solutions to the problem and explained their reasoning.

MIKE: I drew a set of five dots to represent the arrowheads. Then I drew a set of three dots to represent the arrowheads that Mary's grandfather will give to her. Then I drew a picture of the union of these two sets. I counted the number of dots in the union of the sets.

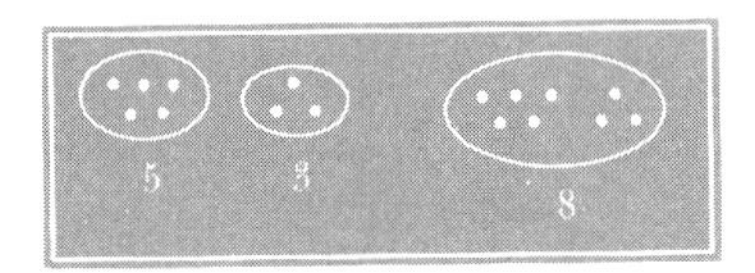

SALLY: I drew a picture of the five arrowheads Mary has. Then I drew a picture of the three arrowheads her grandfather will give her. Then I counted the total number of arrowheads.

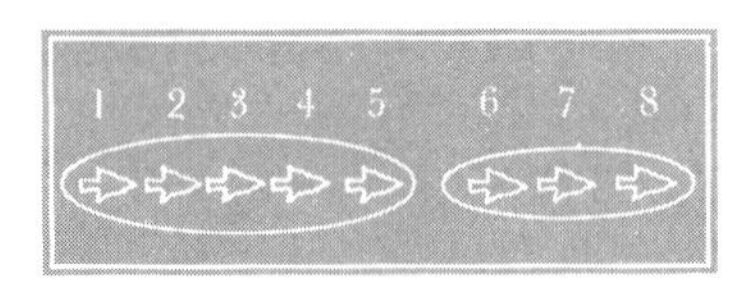

GENE: I used the number line. I moved to five to represent Mary's arrowheads. Then I counted three more to represent the arrowheads her grandfather will give to her.

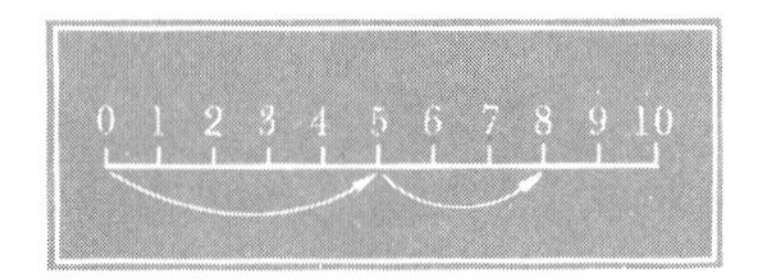

JOYCE: I used a different type of number line. I made a number line using squared paper, then I cut out a strip of five squares and a strip of three squares. When I put them on the number line, I could see my answer was 8.[2]

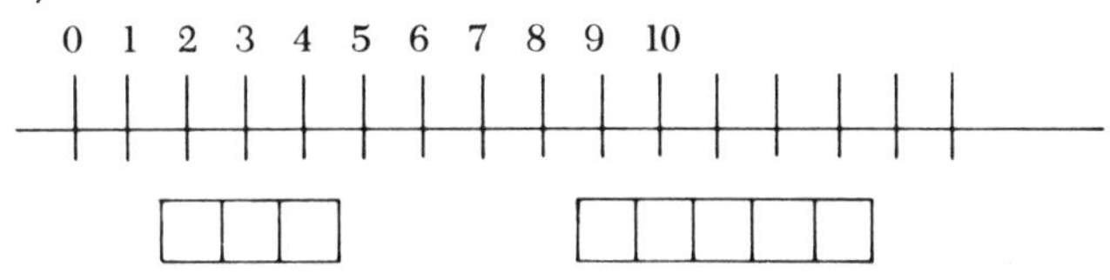

BILL: I used the balance. I put on washers marked 5 and washers marked 3. I guessed that they would "make" 7. I was a little off and found that it was right with 8.

CLAUDIA: I know that Mary has five arrowheads. Then I thought, "How many arrowheads would three more be?" The answer 8 popped into my head.

After some discussion, the teacher asked the pupils if they had been able to identify the exact question the problem asked. After some argument, the pupils agreed that the question was "What is 5 and 3 more?" The teacher restated the question: "5 with 3 equals what number?"

Several other addition problems were worked, and then the teacher said, "When we combine two numbers such as 6 and 3, we usually say, 'Six plus three equals what number?' " (The teacher wrote this on the board.) "In mathematics, we have a shorter way of writing a sentence that uses numbers. Do any of you know how to write a number sentence in a short way?"

Several of the pupils raised their hands and said, "We could replace the six and the three with numerals." "I think my older brother uses '+' to mean plus. He also uses '=' to mean equals." "We could use a question mark to stand for the *what number?*"

Following the suggestions of the pupils, the teacher wrote "6 + 3 = ?" on the board. Then she said, "You may also want to use a box [□] to indicate the number you are seeking. Then, when you have an answer, you may fill in the box." (She wrote "6 + 3 = □".) "We call what I have written on the board a 'number sentence' or a 'mathematical sentence.' Which name would you like to use for the sentence?"

On the second day, the teacher began the lesson with several verbal problems. The pupils were asked to write the number or mathematical sentence that described the problem and then to solve the mathematical sentence. The pupils were also directed to show in as many ways as they could that their answer was correct. Pupils who were experiencing difficulty writing the mathematical sentence that described the problem were asked to solve the problem in another manner and then to raise their hands. The teacher gave individual help to those having trouble.

Later in the period the pupils were given a worksheet containing several mathematical sentences, were asked to write in the box the numeral that

[2]*Note:* Joyce's approach is the beginning of the idea of associating two numbers and mapping these on to a third number. Her number-line usage is addition rather than counting.

made the sentence correct, and were told to show why their answer was correct.

$$2 + 3 = \square \qquad 6 + 3 = \square$$
$$5 + 1 = \square \qquad 2 + 2 = \square$$
$$4 + 2 = \square \qquad 4 + 4 = \square$$

Pupils counted to themselves and made use of the number line, rods, squared paper, a balance, sticks, and so on, to arrive at the correct answer to the addition combination.

On the third through fifth days, time was spent in solving problems involving addition and developing further the use of the mathematical sentence.

By now, the pupils had had many opportunities to develop by inductive discovery the answer to addition questions. So, for the sixth day's work, the teacher felt that it would be worthwhile to make use of some deductive methods of discovery. This took the form of "if–then" statements. The teacher began by developing with the class the pattern for addition involving 6: "What is $6 + 1$? If $6 + 1 = 7$, then what will $6 + 2$ equal? What will $6 + 3$ equal?" Pupils were then given a photocopied sheet containing the material shown below. The teacher suggested that sticks could be used to help with the material. Also, at the early stages of addition work, children find that their fingers are very useful counters.

If $6 + 1 = 7$	If $5 + 1 = 6$
and	and
$6 + 2 = 8$	$5 + 2 = 7$
then	then
$6 + 3 = \square$	$5 + 3 = \square$
$6 + 4 = \square$	$5 + 4 = \square$
$6 + 5 = \square$	$5 + 5 = \square$
$6 + 6 = \square$	$5 + 6 = \square$
$6 + 7 = \square$	$5 + 7 = \square$
$6 + 8 = \square$	$5 + 8 = \square$
$6 + 9 = \square$	$5 + 9 = \square$

It should be noted that at no time during the addition work did the teacher suggest that the pupils memorize the combinations. Rather, each was to make use of the method that seemed most appropriate for arriving at an answer.

During the remainder of the addition work in grade 1, the teacher continued to make use of verbal problems and mathematical sentences, with no stress on memorization of addition facts. The teacher did not use only the addition combinations involving smaller numbers, but made use of situations using all the addition facts.

Techniques for Developing Ideas on Addition

It was previously suggested that all single-digit addition combinations be studied at grade 1 but practice for mastery be delayed until grade 2. Teachers have often asked questions such as, "Harry understands addition very well but just doesn't seem to learn the addition combinations; he always counts. What kind of material can I give him to help him to learn the addition facts?" "Nancy gets bored with the addition practice I give to the class. However, she still doesn't know the combinations as well as she should. What kind of practice that will maintain her interest can I give her?" "When I begin practice on the addition combinations, my third graders say, 'We had all of that in second grade.' How can I vary the treatment?"

One of the best procedures for developing mastery of basic number combinations is to present to children a situation in which they must quickly give the sums of a number of addition combinations. A game or a timed exercise can be used. This presentation helps pupils to see that they need further practice in basic number combinations.

The practice material suggested below is designed with two purposes in mind. One is to develop facility in working with basic addition facts. The other is to further the understanding of addition. Some of the suggestions do both; others develop only facility with addition.

1. Finding different names for numbers by using sticks. A good deal of the mathematics of the elementary school involves finding the standard name for a pair of numbers, for example, 8 for 5 plus 3. Laboratory experiences using Popsicle® sticks or stirrers for counters can be an effective means of developing this idea. They also provide experience with the addition combinations.

One teacher gave each child a set of 20 sticks and said, "Pick up four sticks with your left hand and three sticks with your right hand. How many sticks do you have in all? (seven) What would be a way of writing this? (3 + 4 = 7) Now move one stick from your right hand to your left hand; how many sticks do you have? (seven) What would be a way of writing this? (2 + 5 = 7)"

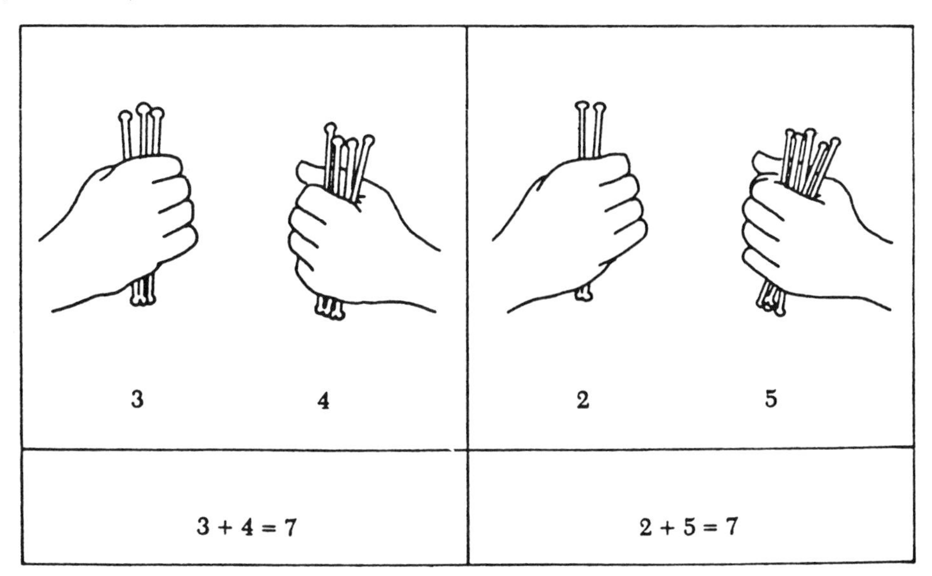

When the children had the idea, each was given the lab sheet shown below and directed to use the sticks to complete the tables.

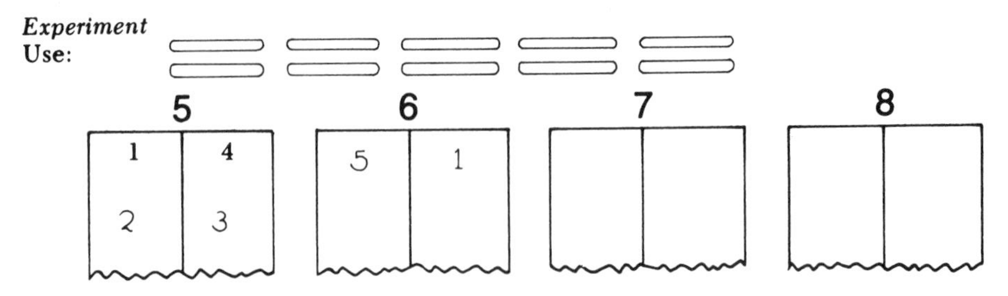

2. Use of fact finders. A simple addition fact finder can be made from a piece of coat hanger and 18 beads (see illustration). The beads may be all of one color, or the first 10 may be of one color and the remaining 8 of another.

The fact finder allows the pupils to count the answer to a forgotten addition combination or to discover the answer to a new combination by counting.

3. Function machines. A variety of "putting-together" function machines can be used. For example, early in the study of addition, a chocolate-bar machine that puts squares of chocolate together to form a bar interests children.

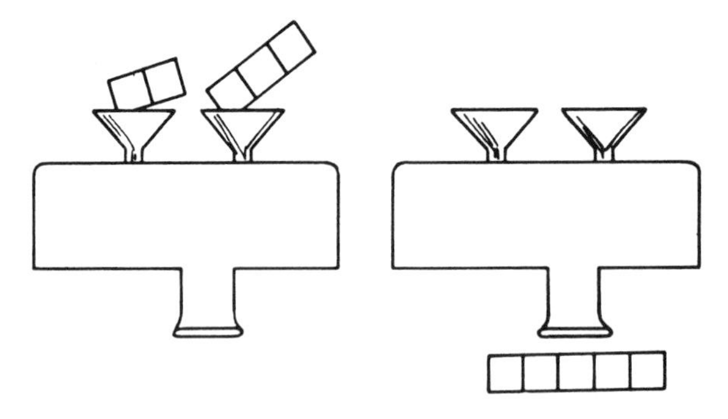

Also, the teacher can make a "machine" from a cardboard box. One child sits in the box and is the programer, and the other children in turn drop two numeral cards through the slot. The programer then slides a card with the sum of the two numbers through the slot at the bottom of the box. Three or four children can engage in this activity together, taking turns being the programer.

4. Function mappings. Using squared paper, the teacher can develop the idea with children that addition can be shown in the two ways diagramed below: (*Note:* There are many more number pairs that could be used.).

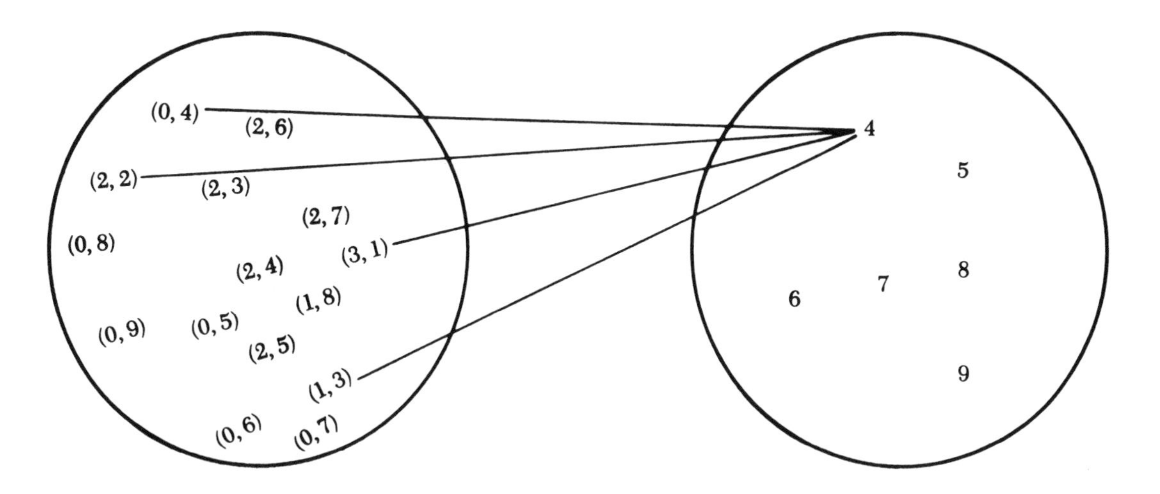

5. Use of the addition table. The addition table is useful in studying the structural properties of addition. The teacher may give each pupil a partially completed addition table and make a statement such as, "I found this table in an old arithmetic textbook. See if you can fill in the remainder of the table."

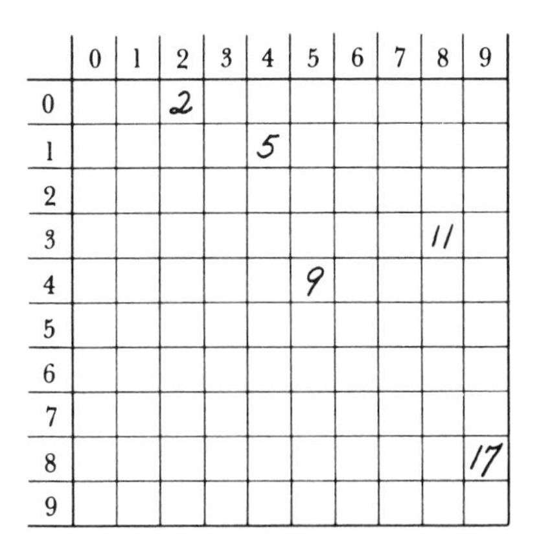

	0	1	2	3	4	5	6	7	8	9
0			2							
1					5					
2										
3									11	
4						9				
5										
6										
7										
8										17
9										

When the pupils have finished the table, the teacher may ask questions such as, "What do you notice about the zeros? Try folding the table in various ways. What patterns can you notice? Why do they work? How many rows with values two larger than the preceding number can you find?"

6. Semiprogramed study sheets. Teachers have found that a sheet of scrambled addition combinations with a slider serves as a valuable study aid. Children are interested in the self-correction feature and in the similarity of the study aid to programed or teaching-machine materials.

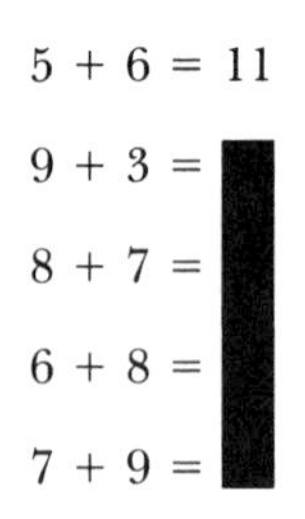

7. Frame arithmetic. The use of many names for the same number and the use of frames provide many worthwhile addition exercises. Questions such as those that follow can be effectively used:

a. What are all the replacements we could use for the box and the triangle?

□ + △ = 9

b. What replacements can we use in each of the following? How many addition facts fit each model?

(△ = △ □ = □; ⬡ = ⬡)

⬡ + ⬡ = □	□ + △ = □
Example: $2 + 2 = 4$	Example: $5 + 0 = 5$
□ + △ = ⬡	□ + □ = □
Example: $5 + 6 = 11$	Example: $0 + 0 = 0$

c. What are the pairs of one-digit addends that will fit each of these models?

□ + 5 = △	Example: $3 + 5 = 8$
6 + △ = □	Example: $6 + 3 = 9$

d. Review of the commutative and associative principles may be accomplished with frames.

5 + □ = □ + △	What is the value of △ if □ = □? (Answer: 5)
(□ + 6) + △ = □ + (⬡ + △)	What is the value of ⬡, if □ = □ and △ = △? (Answer: 6)

What principles are involved in the frames above?

e. Children can plot each of the above ideas on graph paper. They will soon see that when they have found two or three points, they connect them in a straight line to find the other combinations.
(*Note:* The examples are for teacher use. In working with children, the examples should be developed by discussion.)

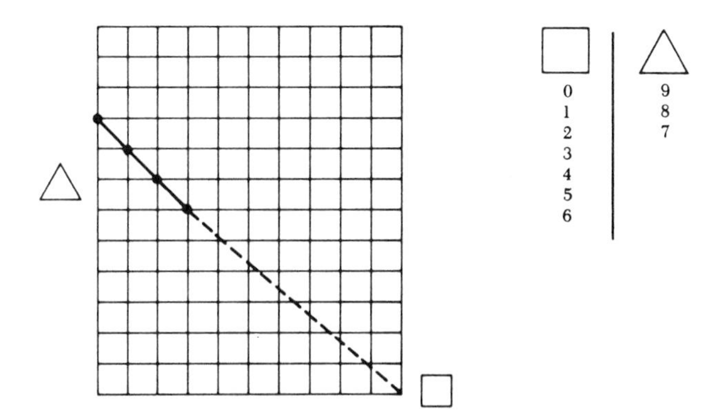

8. Study exercises. Materials such as those illustrated below can be used to give variety in addition practice materials.

a. Find the sum of the number named in the center ring and a number named in the second ring.

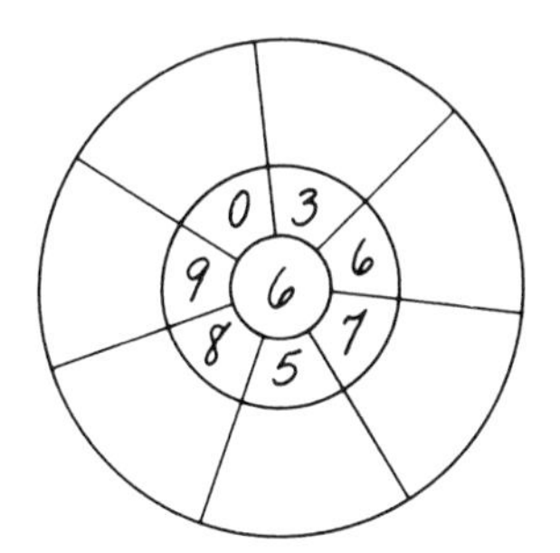

b. Fill in each box with the appropriate numeral.

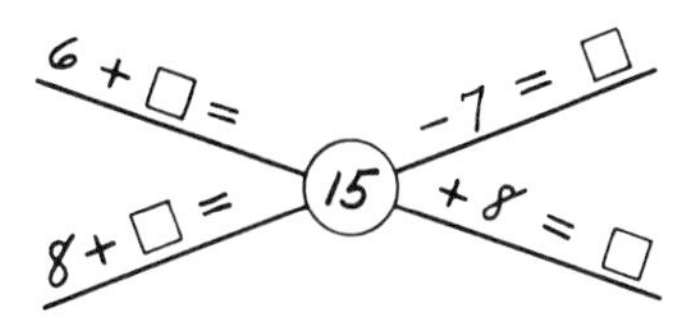

c. Fill in the boxes on the right with the numeral that represents the number that can be added to the number on the left to equal the top number.

14	
9	
8	
6	

9. Games. A variety of addition-oriented games can be used. In fact, the majority of drill and practice on addition facts could easily be accomplished with regular and computer games. (See, *Games and Puzzles for Elementary and Middle School Mathematics,* Reston, Va.: National Council of Teachers of Mathematics, 1975, and the most recent *Creative Publications Catalogue.*)

10. Calculator Activities. Make use of a "guess and test" strategy. Have the children guess the sum of three numbers and then test the correctness with the calculator. If you use this activity often, the children will make great progress in their "addition sense." You may wish to give them several alternative guesses. For example, guess: 196 + 184 + 209. Is the sum less than 500, about 600, more than 600?

11. Computer-assisted drill and practice. There are a number of good drill-and-practice games that allow students to rapidly improve their master of addition facts (about 60 percent faster than regular paper-and-pencil methods). Usually, these programs give the student a record of performance and an increase in speed as the child becomes more proficient. A number of such games are in the public domain (disks that can be legally copied without payment). For a number of addition-oriented computer programs, see the reference to Clements (1989) at the end of the chapter.

12. Other materials. Various commercially prepared colored rods, the number line, Popsicle® sticks, electric addition games, and dice provide materials that lend themselves to practice with addition.

USING THE BASIC PROPERTIES OF ADDITION

The Commutative Property

Recognition of the fact that the order of the addends does not affect the sum (4 + 5 = 5 + 4) is a valuable aid in working with the addition combinations. This property of addition continues to be of great value throughout the study of mathematics.

The commutative property can be developed by pupil analysis of addition situations. At the beginning of a mathematics period, the teacher commented, "At the end of class yesterday, one of you told me, 'I think I can save a lot of time in addition because if I know what 4 + 5 is, I know that 5 + 4 will give me the same amount.' Do you think this person was right? If so, will this be true of all the addition combinations? Try this on several addition combinations, and, if it works, try to develop a means of showing that it is true."

The children tried several combinations and then developed their method of "proof." It should be noted that while the proofs developed by the pupils were not formal proofs, they served a very useful purpose in stressing the need to carefully check number properties.

Several of the proofs given by the pupils were these:

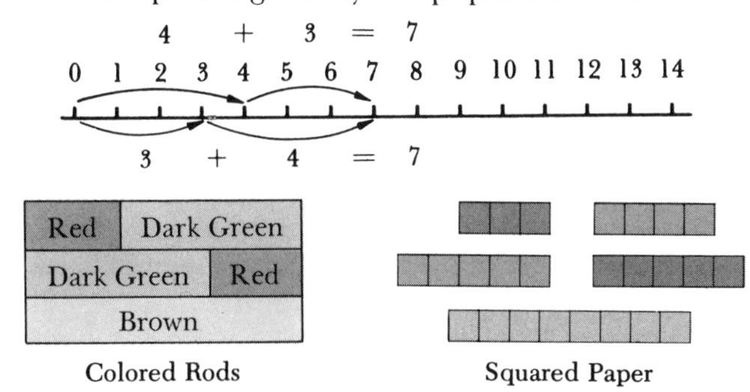

Colored Rods   Squared Paper

First-grade pupils can generalize the commutative property by making statements such as, "It doesn't matter whether you say 3 + 4 or 4 + 3; the answer is the same," or, "The numbers

$$\begin{array}{r} 3 \\ +4 \\ \hline \end{array} \qquad \begin{array}{r} 4 \\ +3 \\ \hline \end{array}$$

can be added up or down." After the students thoroughly understand this number property (usually not until a later grade), the teacher may ask them to state the number property in their own words, have pupils check their definition with a standard definition, and then say, "The number property we have been discussing is called the commutative property for addition." From this time on, the property may be referred to by its name. The teacher should be sure that the stress is placed on understanding the commutative property and not on learning the words *commutative property*. It is better not to introduce the term before meaning has been developed.

The Associative Property

Rather early in problem-solving work, the need to add more than two numbers develops. Since it is not possible to add three or more numbers at the same time, it is often valuable to regroup numbers for addition. Children could discover that the numbers can be regrouped in computation. For example, in adding 7 + 8 + 2, most pupils will find that adding 7 + (8 + 2) = 7 + 10 = 17 is easier than adding (7 + 8) + 2 = 15 + 2 = 17.

Another valuable use of the associative property of addition is the idea that "a number has many names." Since 3 + 2, 4 + 1, 2 + 3, and 1 + 4 all represent the number 5, they may be considered to be other names for 5. This understanding, coupled with the associative property, allows primary-level pupils to solve many addition situations that would otherwise be beyond their grasp. To illustrate: Early in the year, one second-grade pupil was asked to find the sum of 97 + 8. He thought of 3 + 5 as another name for 8. Then he added

$97 + 3 = 100$, and he knew that $100 + 5 = 105$. In written form, this thinking could be represented as follows:

$$97 + 8 =$$
$$97 + (3 + 5) =$$
$$(97 + 3) + 5 =$$
$$100 + 5 = 105$$

The use of another name for a number and the associative property may be introduced in a situation such as the following: The teacher said, "A boy I had last year had trouble with addition combinations such as $8 + 7 = \square$. He changed the name of one of the numbers and added twice. Can you figure out what he did? Look at the way he handled the addition."

$$8 + 7 = \square$$
$$8 + (2 + 5) = \square$$
$$(8 + 2) + 5 = \square$$
$$10 + 5 = 15$$

After studying the procedure, the pupils explained that the boy had renamed 7 as $2 + 5$ and had made use of the enclosures (parentheses) to show that he had added $8 + 2$ first. He knew that $10 + 5 = 15$.

Since the majority of mathematical work is based on the system of place value, the teacher should emphasize often the regrouping of two numbers into tens and ones. Rods, sticks, and strips can be used to help gain understanding of these ideas.

Zero, the Identity Element for Addition

Zero is the additive identity, and thus, $5 + 0 = 5$. Some authorities in the field of elementary school mathematics have advocated that addition combinations such as $5 + 0 = \square$ should not be taught, since normally such situations do not arise in physical-world settings.

Pupils should have experience with number combinations involving zero, since many pupils have a tendency to arrive at the answer 6 when they add $5 + 0$. Also, they may believe that when zero is added to a number, no addition occurs. This is not true. There are a number of physical-world situations in which it is necessary to know whether the addition of zero has occurred or not. For example,

1. Each player gets two turns in a game. We need to know if a player has a score of $2 + 0$ or hasn't had a second turn.
2. Tom and Alan go fishing. They have a contest with Nancy and Elaine to see whether the girls or the boys have caught more fish. Tom catches 3 fish; Alan catches 0 fish. Nancy catches 2 fish; Elaine catches 2 fish. It is necessary to know that the boys have caught $3 + 0$ fish and the girls have caught $2 + 2$ fish.

Other Generalizations Concerning Basic Addition Facts

The following generalizations can be developed through exploratory exercises. They are helpful in developing patterns in mathematics and are useful for many phases of mathematics.

1. Adding ones is equivalent to counting by ones.
2. The sum of a number and the next number larger is one more than the first number doubled: $8 + 9 =$ one more than $8 + 8$.
3. The sum of a number and the next number smaller is one less than the first number doubled: $8 + 7 =$ one less than $8 + 8$.
4. The double of any number is two more than the double of the number that is the next smaller number, and two less than the next larger number.

   $8 + 8 = 16,\qquad$ thus $7 + 7 = 14$

   $8 + 8 = 16,\qquad$ thus $9 + 9 = 18$

   "If-then" exercises can be used to develop this idea.
5. The addition of 9 to any number can be found by adding 10 and subtracting 1.

   $$8 + 9 = (8 + 10) - 1$$

6. Sharing numbers. Both addends may be changed—one increased and one decreased—to make a double. Thus, $7 + 9$ becomes $8 + 8$.[4]

Analysis of the Procedures for Developing the Basic Addition Ideas

1. The use of multiple solutions (solving an addition problem or addition combination in several ways) provides for individual differences and also develops student confidence. A student who has realized that often there are several ways of finding an answer is not as hesitant to attempt new or difficult material.
2. The introductory addition situations make use of laboratory activities and verbal problems. A number of new programs in elementary school mathematics develop the addition facts without verbal problems and then use problems after addition has been developed. The use of verbal problems to introduce addition has two advantages:
   a. It provides a setting in which the pupil can abstract the mathematical idea from the physical world.
   b. It shows a reason for the study of addition.
3. The program stresses the relationship between addition combinations, thus encouraging pupils to form generalizations rather than memorizing each combination as a separate entity.
4. Formal addition is first introduced with a situation requiring the addition of larger numbers, such as $5 + 3$. This is at variance with programs that use combinations such as $1 + 1$ or $2 + 1$ for the first work in addition. Since foundation work has given pupils an opportunity to solve many addition situations by counting, the first situation used in the study of addition should make use of a combination in which addition has a definite advantage over counting. Also,

[4]C. A. Thornton, "Emphasizing Thinking Strategies in Basic Fact Instruction," *Journal for Research in Mathematics Education* (May 1978): 27, 214–27.

1 + 1 and 2 + 1 are not very challenging combinations, even for the slower-than-average pupil.

5. The approaches make use of two early ideas of addition: combining two sets and counting to find a total.
6. Developing the mathematical ideas of commutativity, associativity, and the identity element for addition lays a strong foundation for a structural approach to the study of mathematics. Pupils develop understanding of essential number properties, which are important to all phases of mathematics.
7. Stress is placed on writing the mathematical sentence that could represent the problem situation. The use of the mathematical sentence is an important tool in problem solving at all mathematical levels.

MULTIDIGIT ADDITION

Adding Multiples of 10

Rather early in the addition experiences of children, situations that require the addition of numbers such as $40 + 30 = \square$ can be introduced.

The teacher may begin with a verbal problem, such as, "Ken's mother has 30 trading stamps on one page and 50 trading stamps on another page. How many stamps does she have in all? Is this enough to get a serving spoon that requires 90 stamps?"

When given an opportunity to explore various means of answering the question, pupils can find a solution in several ways. They may find 30 + 50 by (1) counting by tens: 30, 40, 50, 60, 70, 80; (2) using tens and ones blocks and counting them; (3) counting the tens rod of colored rods; (4) adding 3 tens + 5 tens = 8 tens; (5) using sticks (see below); (6) using a place-value chart (below); (7) using strips of squared paper.

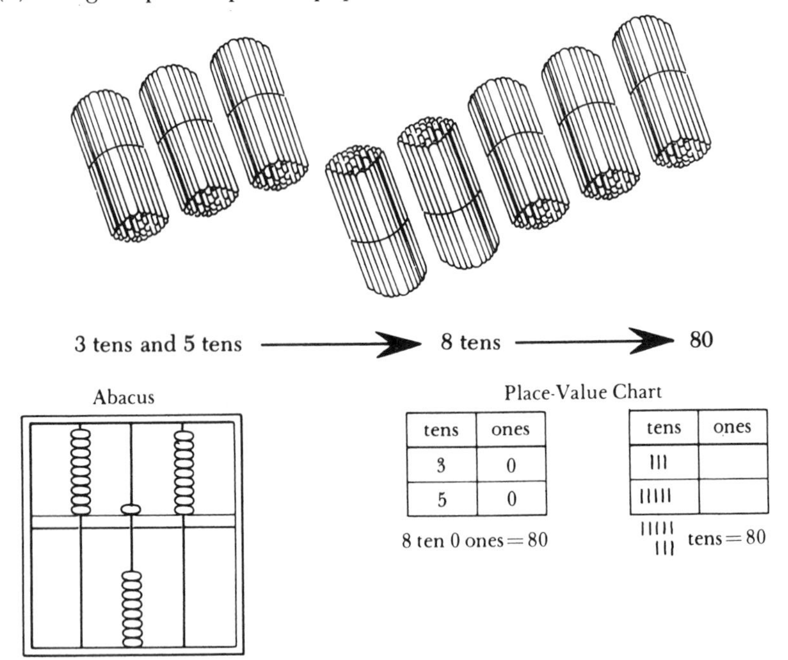

Pupils quickly move to combinations such as $53 + 30 = \square$ and $23 + 34 = \square$ by using the same procedures. Multidigit addition provides a good reason for extensive study of place value, further insight into the addition process, and new settings for basic addition facts.

Higher-Decade Addition

Addition situations in which a one-digit addend is combined with a two-digit addend are often called higher-decade addition. Combinations such as $35 + 3 = 38$ offer little difficulty to most pupils. Situations such as $35 + 8 = 43$ can on occasion cause difficulty. Such situations are often found in column addition and in adding the partial products in multiplication.

Two approaches can be taken to adding $35 + 8$. One method is "adding by endings"; that is, add $5 + 8$ and "bridge" to the next decade. The thinking is, "What is the sum of $35 + 8$? $5 + 8 = 13$; this is greater than 10. My answer will be in the next decade, the 40s. The answer is 43." This method is most commonly taught. The other method is often called "adding to make tens." The thinking of this method is, "What is the sum of $35 + 8$? Rename 8 so that I can make use of renaming and the associative property of addition." This is in keeping with an emphasis on structure in the teaching of elementary school mathematics. Also, most adults and children often use this method quite naturally. For example, in adding a column such as $8 + 9 + 7 + 6$, the thinking might be 8, 17, 20, 24, 30. In this case, the 7 was thought of as $3 + 4$ to form even tens.

Since both methods are used in column addition, there is reason to make use of both approaches in the elementary school.[5] Pupil explanations and guided discussions should be used to develop both. Also, since most addition situations occurring in life situations involve column addition, the teacher should take care to emphasize these two procedures.

Estimating and the Calculator

After the children have experience with addition involving exact tens (for example, $60 + 80$), it is possible to present problems involving renaming and have the children estimate the answers by using their background with adding even tens. The calculator could be used to check out their answers. This procedure has merit for several reasons:

1. It is often difficult to develop the child's skill at estimating the answer to a computational situation. If the child is taught to make estimates *after* learning to compute the addition, there is a tendency for the more conscientious child to work the addition and then round off the answer from the exact computation. The procedure of estimating before learning the exact procedure may aid the child in using estimates at other times.
2. With the almost universal availability of the calculator, a ready check is available for estimates involving computations that the child has not yet mastered.

[5]Grayson H. Wheatley and Charlotte L. Wheatley, "How Shall We Teach Column Addition" in *The Arithmetic Teacher* 25 (January 1978): 18–19.

At the first and early second grade levels the calculator can be used to aid the children in solving important real-life problems that require more computational depth than they possess. For example, one teacher said, "The two first grades are going to the park. There are 26 in our class and 25 in the other class. How many sets of materials will we need for both classes? Guess and answer and then we will use the calculator." This use of the calculator helps to develop number sense and helps the children intuitively think about higher-decade addition.

One teacher said, "You seem to be pretty sharp with adding tens. Let's see how close you can guess the answers to some harder additions. Try this: 45 plus 56. Quick. Hold up your tens and ones cards. Now, let's check with the calculator. Some of you were really close. What did you do?"

TOM: I thought 100. We were adding a little less than 50 to a little more than 50.
KIM: I thought 40 and 50 = 90 and 10 more (about 5 + 6) = 100.
JILL: I guess I didn't estimate. I worked the problem. I thought 40 and 50 equals ninety plus 11 (5 + 6) = 90 + 10 = 100 + 1 = 101.

Over a period of a few days, many such exercises were presented.

Renaming in Addition

Renaming or "carrying" in addition is the most difficult form of the addition algorithms. It is based on the commutative property, the associative property, and place value.

Introductory work in renaming can be started with a laboratory sheet such as the one shown below. The teacher should read the instruction, allow the children to work by using the materials they choose, and hold a group discussion concerning the findings. On some occasions it is quite effective to have one of the class members conduct the discussion dealing with the approaches they have taken.

Experiment

Find the answer to the problems. Work them in several ways.

Materials: Popsicle® sticks, squared paper, number lines, rods, lima beans, abacus, paper and pencil.

1. Mary sold 37 tickets to the school play; Jill sold 25 tickets. How many tickets did the two girls sell?

One group of children responded to the problem in the following manner:

PAULA: I used tens and ones squared paper. I laid out 3 tens and 7 ones, and 2 tens and 5 ones. Then I grouped them together. This gave me 5 tens and 12 ones. I traded 10 ones for another tens strip. This gave me 6 tens and 2 ones, or a total of 62.

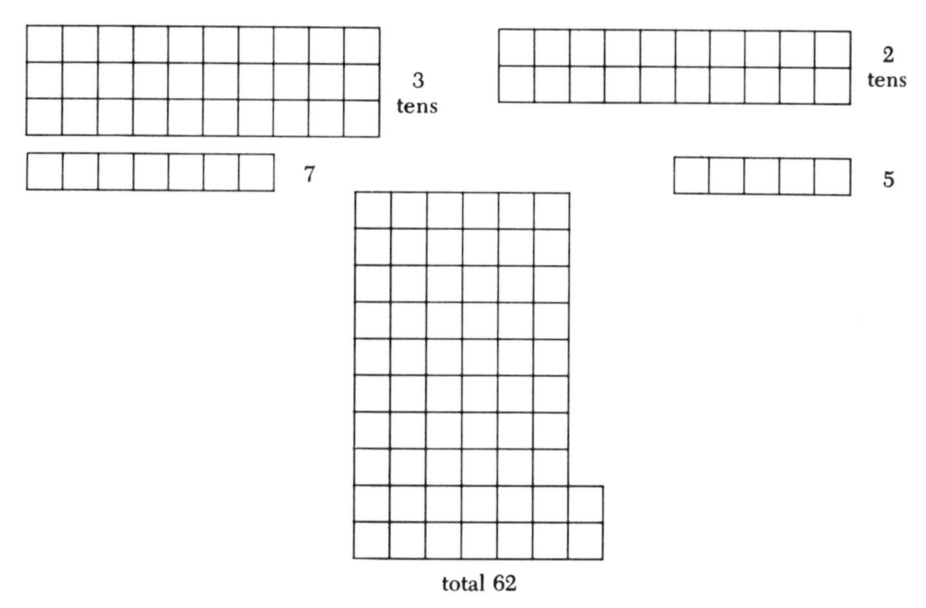

JIM: I used the number line. I marked off 30, 20, 7, and 5. I don't believe that this is the best way to solve the problem, because it takes quite a while.

0 10 20 30 40 50 60 70

CRAIG: I used Popsicle® sticks for tickets. I laid out 3 tens and 7 ones, and 2 tens and 5 ones. Then I grouped them together. This gave me 5 tens and 12 ones. I traded 10 ones for another ten. This gave me 6 tens and 2 ones, or a total of 62.

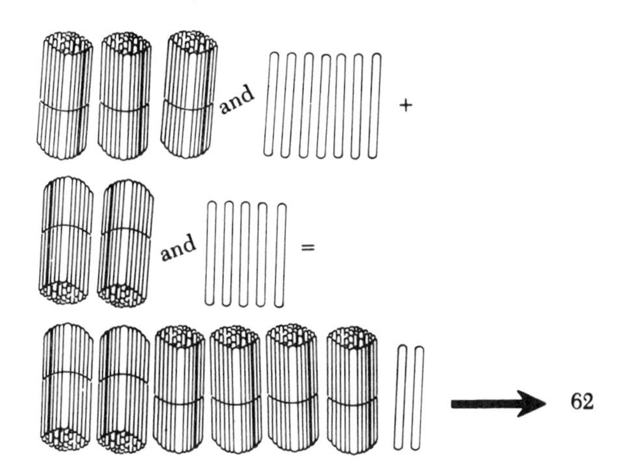

JEAN: I used the place-value frame.

tens	ones
3	7
2	5
5	12

62

or

tens	ones
111	11111 11
11	11111
11111	11111 11111 11

111111 11 =62

JOHN: I used rods. I combined the 5 tens rods and changed the 7 and 5 to 10 and 2. This gave me 62.

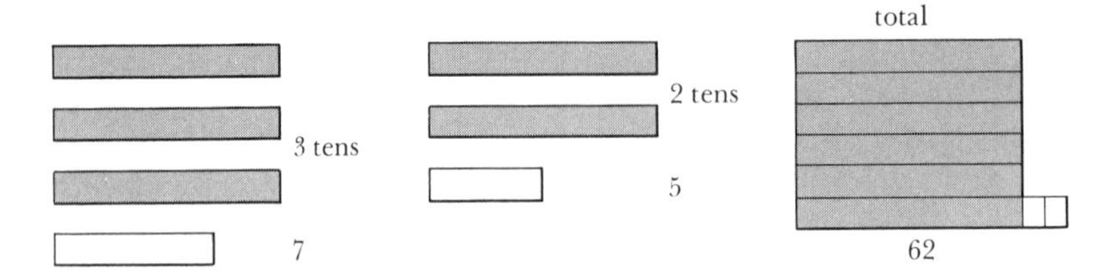

JANE: I used expanded form. I then renamed 12 as 10 + 2 and added again.

$$\begin{array}{rcl} 37 & = & 30 + \ \ 7 \\ +25 & = & 20 + \ \ 5 \\ \hline & & 50 + 12 = (50 + 10) + 2 = 62 \end{array}$$

GEORGE: I added twice or used the associative property.

$$37 + 25 = 37 + (20 + 5) = 57 + 5 = 62$$

KEVIN: I added in the regular way. I first added 7 + 5. I recorded the 12. Then I added 30 + 20. I recorded the 50. Then I added the 12 and 50.

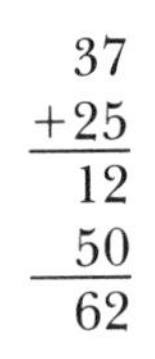

After the completion of several problems, the class members were given a set of addition exercises, a number of which involved renaming. They were instructed to find answers to the exercises in a manner that made sense to them and then to show that their answers were correct by working the exercises in another manner.

During discussions on the following days, other methods of solution were discussed. Several of the students had found a more efficient method of solving addition exercises involving renaming. One pupil explained, "Add

7 ones and 6 ones; this is 13, or 1 ten and 3 ones. Write down the 3 ones and place the 1 ten in the tens place. Then add the tens."

$$\begin{array}{r} {}^{1}\\ 57 \\ +36 \\ \hline 93 \end{array}$$

Another pupil suggested, "To add

$$\begin{array}{r} 57 \\ +36 \\ \hline 93 \end{array}$$

I add 7 + 6, which is 13. I write the 3 and remember the 1 ten. Then I add the 5 tens and the 3 tens. This gives me 8 tens plus the 1 ten I remember, which equals 9 tens. I write 9 in the tens place. My answer is 93."

As the study of multidigit addition progressed, the teacher stressed understanding of the basic idea of renaming and the continued development of mature forms of the algorithm. Those who were not ready for the standard algorithm were not forced to use this form. Periodically, review study questions similar to the following ones were used: (1) What does the 2 that is circled mean? (2) Why do some parents make statements such as "Remember to carry the 1"? What do they mean by "carrying the 1"? (3) What basic properties of addition allow us to compute situations such as 345 + 468?

Children who were experiencing difficulty with the concepts worked in groups using the "chocolate-bar computer." The task was to change all the pieces into tens and ones. Thus, if they were given 4 tens and 6 ones and then 5 tens and 7 ones, they had to trade in the ones for tens and ones. For tens and ones strips, the children used tens and ones made from 1-inch squared paper.

CHECKING ADDITION

Teachers often say, "My pupils hate to check their work in arithmetic." This statement is probably true. It is often difficult for children to realize that in most instances when adults do a computation, they check the work. This reluctance to check addition (or any other) computation can be partially reduced by making use of these suggestions: (1) Pupils often feel that "checking their work" means that they will have twice as much work to perform. This is usually true. The teacher can reduce this feeling by giving fewer exercises when checking is expected. It is usually of greater value to work fewer exercises with a check than a greater number without checking. (2) In some cases, pupils see little reason to check. It is suggested that pupils compare the accuracy of papers they have checked with those that are not checked. If they see a real value in terms of increased accuracy, the use of checking will be better accepted. (3) Often the use of varied and imaginative checks increases pupil interest. Several such checks follow.

Checking Basic Facts

The idea of solving an exercise in several different ways provides a check. In addition exercises, pupils can use counting, the number line, adding twice [$9 + 5 = 9 + (1 + 4) = (9 + 1) + 4 = 10 + 4 = 14$], the abacus, and the commutative property to reverse the addends. All are acceptable checks.

Checking Multidigit Addition

CASTING OUT NINES

Although casting out nines is not a particularly efficient check, it is usually of interest to intermediate-grade pupils and is useful in developing further insight into the number system. The teacher may illustrate the check on the chalkboard and ask questions such as, "This check is often called casting out nines; is that what we're doing? Try dividing each addend by nine. What is the remainder? Why does this check work?"

562	$5 + 6 + 2 = 13 \longleftrightarrow 1 + 3 =$	4
291	$2 + 9 + 1 = 12 \longleftrightarrow 1 + 2 =$	3
243	$2 + 4 + 3 = 9 \longleftrightarrow$	0
+ 122	$1 + 2 + 2 = 5$	5
		$12 \longleftrightarrow 1 + 2 = 3$
1,218	$1 + 2 + 1 + 8 = 12 \rightarrow 1 + 2 = 3$	$12 \longleftrightarrow 1 + 2 = 3$

Study of the nines multiplication table reveals that in every multiple of 9, the sum of the digits is equal to 9. A dot drawing such as the one shown below is also helpful in understanding the procedure.

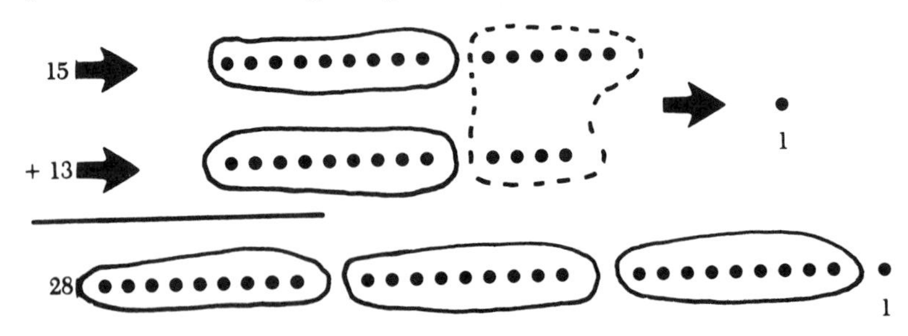

Also, a study of the nines multiplication table is helpful in understanding this interesting characteristic of the number that is one less than the base (10).

ASSOCIATIVE CHECKS

Changing the order of the addends through use of the associative property provides a number of ways of checking.

	(a)	(b)		(c)
3,821	4,653	3,821	7,982	3,821
4,653	7,982	4,653	6,051	4,653
7,982	6,051	8,474	14,033	6,051
6,051	18,686	14,033		14,525
22,507	3,821	22,507		7,982
	22,507			22,507

APPROXIMATION OF SUMS

Children should be given many opportunities to approximate sums. A good teaching strategy to use is a "guess-and-test" procedure, in which the children are to guess the answer to an orally presented addition and then test their answer with paper and pencil.

The guessing puts the children at ease and helps develop skill in estimation. Often, if children are asked to estimate their answer, they work the exercise first and then give an estimate from the completed sum. The "guess and test" pushes them to estimate.

CALCULATOR CHECKS

The calculator can be used at any grade level to check addition exercises. In fact, it may be the best methods for routine checking. There are at least two reasons for suggested wide use of the calculator in checking. First, it is extremely fast and, second, it is so easy that children get the habit of checking.

REINTRODUCTION, DIAGNOSIS, OR EXTENSION

There are many situations in elementary school mathematics that call for a review or further practice on a topic. At times, there has not been enough review on basics. Often a high school mathematics teacher states, "The pupils do not understand the basic principle of renaming in addition. They use 'carry the 1.' I wish the elementary teachers would teach them the basic ideas of mathematics." Such indictments probably arise not so much from a lack of teaching basic principles at an early grade as from a lack of reviewing them at an upper-grade level. Too often, the computational procedure is reviewed but the mathematical meaning is not.

To make reintroduction or review an active learning experience for the pupils, the material should be introduced in a new setting. Pupils who need further study on addition often attack reintroduction work with a lack of fervor. They feel, and justly so, "We had that in the same way last year; we're doing the same old thing." The situations that follow are illustrative of the type of reintroduction materials designed to vary the setting so that pupils may get further insights into addition and also be able to study "something new." Such material is also a valuable aid in diagnosing the level of the child.

Patterns (Upper primary)

The teacher said, "I noticed a girl doodling with addition combinations. Can you tell what she did?" The teacher wrote on the board:

1	2	3	4
1 + 1	1 + 2	1 + 3	1 + 4
2	3	4	5
1 + 2	2 + 3	3 + 4	4 + 5
3	5	7	9
?	?	?	?
?	?	?	?
?	?	?	?

"See if you can complete her chart." The teacher then moved about the class and helped those pupils who had not discovered the pattern of computation. Other pattern-searching exercises can be used at higher-grade levels.

Follow-Me Exercises (Intermediate)

The teacher began the year by saying, "Classes I've had usually enjoy 'follow-me' exercises in mathematics. See how good you are at 'following me.' Without using pencil and paper, work the addition and subtraction I give you orally. Keep up. See if you have the correct answer at the end. Let's begin: 7 plus 2 plus 5 minus 4 plus 50; the answer is—right—60. Try again, 3 plus 5 plus 8 plus 4 minus 10; the answer is—10." The teacher gave several other "follow-me" exercises and suggested that class members having trouble keeping up might benefit from the use of the flash cards, rods, or semiprogramed sheets on the arithmetic table. Pupils using the material at the arithmetic table often worked in pairs, with one child presenting flash cards for the other. In the days that followed, the teacher often used "follow-me" exercises while waiting for the music teacher or for the lunch bell or directly after recess or whenever there were a few moments of uncommitted time available.

Continuous Addition (Intermediate)

The teacher reintroduced column addition by asking, "Have any of you ever heard of someone who could add the numbers on the side of box cars as a train moved by at a railroad crossing?" Most of the pupils had heard of such people or had actually witnessed a performance of rapid addition. "I recently talked with such a person," continued the teacher, "and asked him to explain it to me so that I could show it to you. He said he would give you an idea of his method on a sheet of paper and see if you could figure out how he does the addition. This is what he wrote. Can you find out what he did?"

```
   23       2  3
   46
   34       4→6
 + 52        ↙
 ----
  155       3→4
             ↙
            5→2
          -----
           155
```

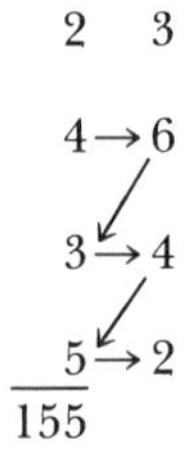

After a little thought, the pupils found the idea was to total as they went along. Think 23, add 40, add 6, add 30, add 4, add 50, add 2, total 155. Many of the pupils felt that this method was no faster than their normal method, and a short argument followed. The teacher suggested that it would be necessary to practice using the new method in order to compare. The class was broken into two teams. One team was to study the new method by working a set of addition exercises using it. They checked their solutions by using the conventional algorithm. The other team worked the exercises in the conventional way and checked by using the new form. Several days later, a contest was held.

This reintroduction accomplished at least two things. First, the pupils had a reason for working on column addition, to find out which procedure was more effective. Second, it introduced a computational procedure that is often handy and is the basis of some forms of rapid addition.

Front-End Addition (Intermediate)

Adding the most important digits first is a method often used by people in business and can be a means of estimation or checking. It also provides a good review of place-value concepts.

An introductory statement that may be used is, "When I bought some clothing the other day, the clerk added my purchases on a sales slip. I've drawn the sales slip on the board. At first I wasn't sure how the clerk added the amounts. Can you find out what procedure was used?"

Sport coat	$95.75
Tie	6.25
Top coat	98.50
Shirt	15.95
Step 1	19
Step 2	24
Step 3	2.3
	.15
Step 4	21
6.45	
216.45	

After a discussion, the teacher asked, "Would this method always work? What do you like about the method? What don't you like about it? Try using this 'front-end' arithmetic to solve the addition exercise in your text. Check using the regular method."

Sand-Board Addition (Intermediate)

The teacher introduced sand-board addition by saying, "Before the development of pencil and paper, the Hindus in India often performed their calculations in the sand or dirt. Often they had a board covered with sand on which they did their work. When they had finished a portion of an addition, they smoothed out the sand. I'll show you the way the sand board looked at various stages. How did they add 596 + 875?"

596 +875	96 +75 13	6 +5 1 136
6 5 146	 1 1461	1471

The Hindu Method (Intermediate)

The pupils were asked to see if they could find the reason for a method of addition used by the Hindus in the twelfth century. The following problem and computational algorithm were given to the pupils: "...if thou be skilled in addition...tell me the sum of 2, 5, 32, 193, 18, 3, and 100 added together."[6]

Sum of the units	2, 5, 2, 3, 8, 0, 0	20
Sum of the tens	3, 9, 1, 1, 0	14
Sum of the hundreds	1, 0, 0, 1	2
Sum of the sums		360

Written as we would:

```
   2
   5
  32
 193
  18
  10
+100
```

Pupils were asked the following questions concerning the procedure: (1) Why does it work? (2) Is it really very different from the form we use? (3) Why can the addition exercise be solved without regrouping? Or is there regrouping? (4) Solve 234 + 12 + 92 + 7 using the Hindu method.

KEEPING SHARP

Self-Test: True/False

_______ 1. Only disjoint sets can be used when defining addition as the union of sets.

_______ 2. Early addition activities should not include the actual objects described in the problem.

_______ 3. One frequently suggested three-step program for the development of abstractions includes a concrete stage, semiconcrete stage, and abstract stage.

[6]From D. E. Smith, *History of Mathematics,* vol. 2, p. 91. Copyright 1953 by Eva May Luce Smith. Published by Dover Publications, Inc., New York 10014, and reprinted by permission of the publisher.

_______ 4. Matching objects with a criterion set is not a good way of finding out "how many."

_______ 5. It is good practice to allow children to use a variety of methods to solve the same problem.

_______ 6. Children need to memorize all the addition combinations before they can successfully solve addition problems.

_______ 7. Changing the order of addends in addition does not change the answer.

_______ 8. It is possible to add more than two numbers at the same time.

_______ 9. Writing the mathematical sentence that represents the problem situation is an important aspect of a good addition program.

_______ 10. Primary children cannot successfully work with higher-decade addition combinations.

_______ 11. Renaming in addition should be introduced by a discussion on how it is done before pupils try to solve problems requiring its use.

_______ 12. There are many ways to check an addition problem.

_______ 13. Patterns and pattern-searching exercises are good ways to reintroduce addition concepts and extend pupil understanding of the concepts.

Vocabulary

associative	identity	disjoint sets	addend
commutative	union of sets	sum	mapping

THINK ABOUT

1. What foundation experiences or learnings should precede initial instruction for systematic study of the addition process?
2. At what grade level should first instruction involving the addition algorithms designed for mastery of basic facts be introduced (for average children)?
3. Should the algorithms be presented first in vertical or horizontal form? Should both forms be used:

$$3 + 2 = \square \quad \text{and} \quad \begin{array}{r} 3 \\ +2 \\ \hline \end{array} ?$$

4. Should early experiences designed for mastery of the basic elements of addition be confined to the addition of one-digit numbers?
5. Is a definition of addition of value to the elementary schoolchild? What are some of the limitations of definitions that are used or that have been suggested?
6. How important is checking, or showing that answers are correct, in the study of addition? If used, when and how should such procedures be introduced?
7. Some authors recommend the simultaneous teaching of the related addition and subtraction facts. Others recommend the teaching of the related addition, subtraction, multiplication, and division facts. Still others recommend the teaching of addition, followed by subtraction, multiplication, and division. What are the merits and demerits of each of these proposals? Use pupil textbooks, professional books, and magazine articles in your presentation.

8. Describe the chief features of the program of instruction in the addition of whole numbers in grades 3, 4, 5, and 6. Cite professional books and pupil textbooks to support the proposals you make.

SUGGESTED REFERENCES

ASHLOCK, ROBERT B., *Error Patterns in Computation*, Columbus, Ohio: Charles Merrill (1984).

BAROODY, ARTHUR J., *Children's Mathematical Thinking*, New York: Teachers College Press, 1987.

BUCKINGHAM, B. R., *Elementary Arithmetic: Its Meaning and Practice*, chap. 5. Boston: Ginn, 1953 (an excellent source of historical algorithms).

CLEMENTS, DOUGLAS H., *Computers in Elementary Mathematics Education*, chap. 7. Englewood Cliffs, N.J.: Prentice Hall, 1989.

Developing Computational Skills, 1978 Yearbook, chaps. 1–5 and 14, Reston, Va.: National Council of Teachers of Mathematics, 1978.

Estimation and Mental Computation, 1986 Yearbook, chap. 19, Reston, Va.: National Council of Teachers of Mathematics, 1986.

KAMII, CONSTANCE K., *Young Children Reinvent Arithmetic: Implications from Piaget's Theory*, New York: Teachers College Press, 1985.

Mathematics for the Middle Grades (5–9), 1982 Yearbook, Reston, Va.: National Council of Teachers of Mathematics, 1982.

MOSER, JAMES M., "Arithmetic Operations with Whole Numbers: Addition and Subtraction." In *Teaching Mathematics in Grades K–8*, Thomas R. Post (ed.), Boston, Mass.: Allyn and Bacon, 1988.

THIESSEN, DIANE, MARGRET WILD, DONALD D. PAIGE, and DIANE L. BAUM, *Elementary Mathematical Methods*, 3rd ed., New York: Macmillan, 1989.

WIEBE, JAMES H., *Teaching Elementary Mathematics in a Technological Age*, chap. 10, Scottsdale, Arizona: Gorsuch Scarisbrick, 1988.

chapter 7

SUBTRACTION OF WHOLE NUMBERS

OVERVIEW

Subtraction readiness begins in the preschool with problem solving by using real physical world objects. It is an important topic in elementary school mathematics because of its social usefulness in problem solving, in its continued mathematical usefulness throughout life, and in the opportunity it provides to compare "undoing thinking" with "combining thinking." The National Council of Teachers of Mathematics Standards concerning subtraction are reported along with those for addition on page 122.

TEACHER LABORATORY

Using Popsicle® sticks or coffee stirrers, solve these subtractions.

$$\begin{array}{r} 33 \\ -12 \\ \hline \end{array} \qquad \begin{array}{r} 45 \\ -21 \\ \hline \end{array}$$

$$\begin{array}{r} 32 \\ -18 \\ \hline \end{array} \qquad \begin{array}{r} 53 \\ -19 \\ \hline \end{array}$$

Make up verbal problems to fit each problem. Develop a set of guided questions you could use to help the children discover renaming (borrowing).

TAKE INVENTORY

Can You:

1. Write a verbal problem to illustrate each of the three "types" of subtraction situations?
2. Illustrate the use of a beam balance in solving a problem such as this? "Bill has 8 cents. How much does he need to buy a 25-cent candy bar?"
3. List five subtraction generalizations?
4. Solve a subtraction by using a cross-number puzzle?
5. Check a multidigit subtraction by using two different checks?
6. Develop a lesson to teach the idea of "compensation" ?
7. Use problems to help children at the concrete, semiconcrete, and abstract levels?
8. Illustrate checks of subtraction?

WHAT IS SUBTRACTION?

Basically, subtraction can be thought of as the inverse or "undoing" of addition. It is standard for a mathematician to consider that there are two basic operations on whole numbers: addition and multiplication. Thus, $8 - 5 = 3$ if and only if $5 + 3 = 8$. In the general case, $a - b = c$ if and only if $b + c = a$.

When we dealt with addition, the two numbers to be combined were called addends and the result the sum. In subtraction, we start with the sum and take away one of the addends to find the other addend. By using the terminology of subtraction, the original sum is called the *minuend;* the addend that is being subtracted, the *subtrahend;* and the remaining addend, the *remainder* or the *difference.* For example,

3	+ □	= 7
addend	missing addend	sum
7	− 3	= □
sum	known addend	missing addend
minuend	subtrahend	difference

The joining of two sets offers an approach to addition. In the same manner, the use of sets leads to subtraction. There are, however, three types of subtraction problems, each of which suggests a different use of sets:

1. *Take-Away Situation.* "Claudia counted eight cardinals at the bird feeder. Three flew away. How many were left?" This situation involves the removal of a subset from a set. We start with eight objects, remove a subset of three objects, and note that a subset of five objects remains.
2. *How-Many-More-Are-Needed Situation.* "Doug wants to buy a 20-cent candy bar. He has 12 cents. How much more does he need to buy the candy bar?" This

situation involves the question of how many set members must be joined to a given set to obtain another set.

3. *Comparison.* "Alice brought eight acorns to school. Bill brought five acorns. How many more did Alice bring than Bill?" This interpretation involves the comparison of two sets by one-to-one correspondence. A set of eight may be compared with a set of five by matching set members.

Subtraction can also be thought of as mapping (function). An ordered pair of numbers can be placed in a function machine, and the difference becomes the output. This idea is shown in the two diagrams below.

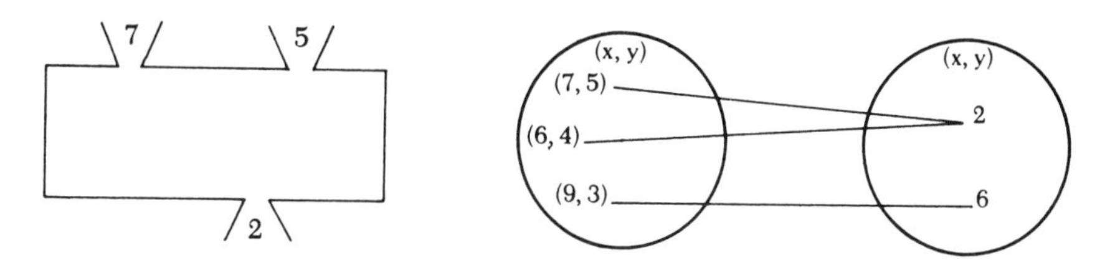

Note that since we are dealing only with a set of positive whole numbers, the first number of the ordered pair must be greater than or equal to the second number. When negative numbers are introduced, it is possible to use any whole number for either number of the ordered pair. Thus, subtraction becomes *closed* (any two whole numbers used as inputs give us an output in the set).

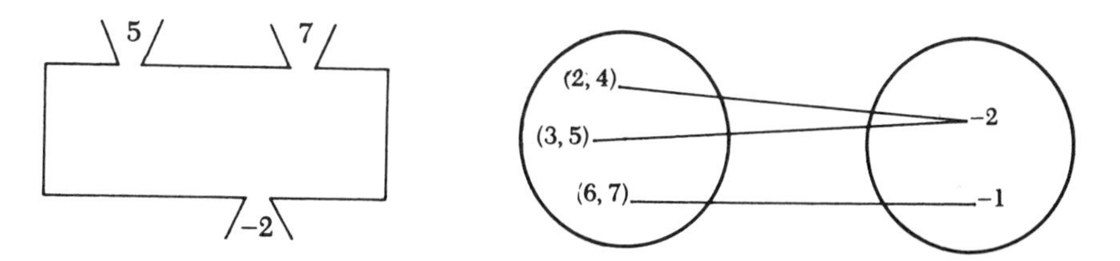

The following type of function thinking is also valuable:

input	rule	output
5	−3	2
6		

input	rule	output
(5,3)	subtract	2
	second	
(6,2)	number	
	from	
(8,)	first	3

FOUNDATION EXPERIENCES

Teachers of elementary school mathematics are faced with several decisions concerning the introduction of subtraction. First, they must decide whether subtraction and addition should be introduced simultaneously. Some writers in the field of elementary school mathematics suggest that subtraction be introduced on the same day as addition in order to highlight the inverse relationship between the two. Others suggest that subtraction be introduced several days after addition. Although there is little research evidence for either procedure, it is my belief that the inverse operation can be developed more effectively if the pupils possess a good understanding of addition. According to the writers of the Nuffield Project,

> This operation [subtraction] has been delayed as long as possible, which may surprise many teachers, but it has to be thought about very seriously indeed because it presents difficulties to children which may be easily overlooked... Literal "taking-away" is of course a very early activity, but there is a long way to go from this process of physical removal to the understanding of the mathematical operation of subtraction.[1]

Major difficulties arise from the nonassociativity; for example,

$$12 - (6 - 4) \neq (12 - 6) - 4$$

and from the noncommutativity; for example,

$$7 - 5 \neq 5 - 7$$

Second, teachers must decide which of the three types of subtraction situations should be used for introductory work. Few authorities advise using the "comparison" type of problem for introductory work. There are, however, proponents of both the "take-away" and the "additive" situations. Additive situations in which pupils search for the missing addend lend themselves to a good development of the inverse relationship of subtraction to addition. Also, some writers believe that a pupil who knows $3 + 4 = \square$ will know $3 + \square = 7$ and $7 - 3 = \square$. There is little evidence to either justify or repudiate this belief.

Take-away situations are usually easier for pupils to understand[2] and can be developed as the removal of members from a given set. For these reasons I suggest that the first formal subtraction situations involve the take-away idea, although informal problems involving all three types of subtraction situations should be used prior to the formal introduction.

INTRODUCING SUBTRACTION

Following foundation work, initial subtraction was begun with this orally presented problem: "Marcia baked eight cupcakes. She gave three of the cupcakes

[1]Nuffield Mathematics Project, *Computation and Structure* (New York: John Wiley, 1968), p. 3.

[2]Glenadine E. Gibb, "Children's Thinking in the Process of Subtraction," *Journal of Experimental Education* 25 (September 1956): 71–78.

to friends. How many cupcakes did she have left?" The pupils gave answers to the problem, then the teacher asked for the methods they had used in solving it. Pupils suggested counting backward from eight, taking eight objects and removing three, and using a number line to help count to find the answer.

The class members were directed to solve several other problems printed on a duplicated sheet, which were first read orally by the teacher, and then to show their answers to be correct by using a different method. All the problems involved take-away situations. During the work time, the teacher noted the methods used by pupils and gave assistance and encouragement to those who were having difficulty.

When the majority of the pupils had worked several problems, the teacher said, "Let's stop for a minute and discuss the process we are using. When we have worked addition problems, we have written the mathematical sentence that can be used to solve the problem. Can you state in words the mathematical question asked in this problem? Grandmother gave Ken nine pencils. Ken's mother said, 'You'll have to share these with your brothers and sisters. I'll take four to give to them.' How many of the pencils will Ken have?"

Pupils suggested, "9 take away 4 leaves what number?" "4 from 9 equals what number?" and "9 remove 4 equals what number?" The teacher then wrote on the board, "9 take away 4 equals what number?" and asked, "How could we shorten this mathematical question?" Pupils suggested that the equals sign (=) replace the word *equals* and that a box (□) replace *what number.* The teacher replaced the words with the symbols and then replaced the words *take away* with the minus sign (−). The teacher said, "We can read this [and pointed to $9 - 4 = \square$] as '9 minus 4 equals what number?' We call this sign [−] the minus sign. Do you know what type of mathematical sentence this is? Remember, we've had addition sentences. Is this addition?" The class agreed that the mathematical sentence was not an addition sentence. Several who had older brothers or sisters said it was a subtraction sentence. The teacher verified that the mathematical sentence $9 - 4 = \square$ was indeed a subtraction sentence.

During the next few days, the pupils worked problems and found solutions to mathematical sentences that involved both addition and subtraction. Dot drawings, the number line, and objects continued to play a vital role in finding answers to the exercises.

After the pupils had a good understanding of take-away subtraction, situations involving "additive subtraction," or "how many more are needed" situations were introduced. The following problems were used: (1) Mary wants to give every person attending her birthday party a paper hat. Mary now has six paper hats. How many more will she need if eight people attend her party? (2) Claudia's mother has finished three dresses. If, in all, she plans to make nine, how many does she have left to make? (3) Jim has 3¢. How many more cents does he need to buy a pencil that costs 7¢?

The pupils were asked to write the mathematical sentence that could be used to solve the problem, to work the problem, and to prove their answer to be correct by solving the problem in a different manner. Several solutions are illustrated below:

JAN: I used a balance. I put six washers on one side for the six hats that Mary had. Then I put eight on the other side for the hats she needed. I guessed that she would need three hats. I tried a set of three washers on the balance and that was too much. Two worked.

TIM: This is the mathematical sentence that represents the problem: $6 + \square = 8$. I thought, 6 plus what number equals 8? The answer is 2.

MARCIA: I knew I could find the answer by subtracting 6 from 8. It isn't the same type of question as some of the other subtraction questions, but I can get the correct answer this way: $8 - 6 = \square$.

PHIL: The number line can be used to count the answer.

0 1 2 3 4 5 6 7 8 9 10

2

KIM: I used this dot diagram to help me. I need 8. I have 6, so I circle six dots. This leaves two dots, which represent the two I need.

The difference between Tim's addition sentence and Marcia's subtraction sentence caused a great deal of class discussion, after which the class decided that either sentence could be used to solve the problem, that the sentence $6 + \square = 8$ represented the actions on the sets of paper hats, and that the sentence $8 - 6 = \square$ represented an effective method of quickly arriving at the answer. It was noted that the balance, rods, and number lines were most helpful for children who needed to experiment to find the answer.

The "comparison" type of subtraction situation was introduced through the following problems: (1) Larry has six coins in his collection. Bob has four coins in his collection. How many more coins does Larry have than Bob? (2) Greta is 7 years old. Her younger sister Sue is 4 years old. What is the difference in their ages?

Again pupils identified the mathematical sentence and showed their answers to be correct by drawings. The majority of the class identified the mathematical sentence as a subtraction sentence. However, they noted that the drawings they needed to show their answer to be correct were different from the type used in their earlier work with subtraction. Examples of the drawings used to solve problem 1 appear below.

WILL: I compared 4 to 6 and found a difference of 2.

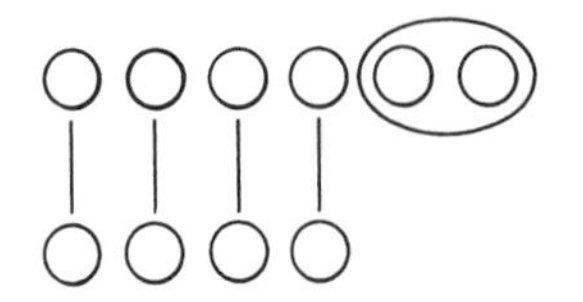

FAY: I marked off 6 on the number line and then marked off 4. The difference between them is 2.

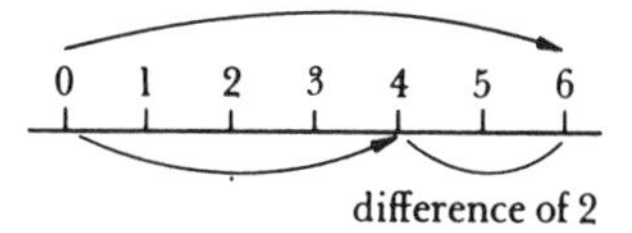

JEFF: I used unifix blocks.

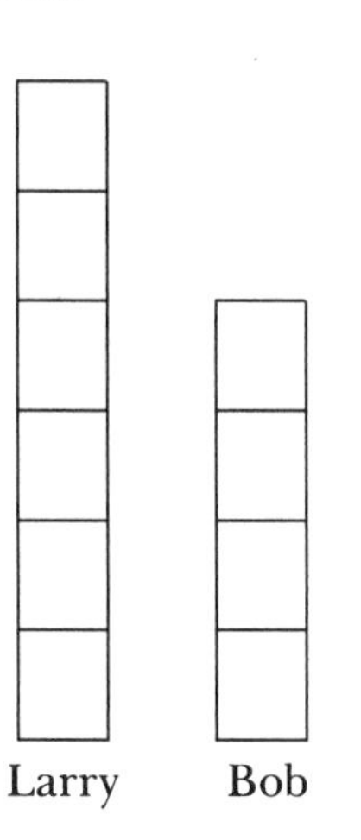

THE TABLE

When most pupils appeared to be progressing well in subtraction, the class studied the addition table to note similarities between addition and subtraction. Several pupils quickly saw that it was possible to use the addition table to solve any subtraction situation. One pupil said, "To find the combination 16 − 9 = □, I look for a 16 in a sum box that is directly across or directly down from a 9; then I follow the column up (or over) to find the answer."

See Chapter 6 for an example of the addition table.

SUBTRACTION GENERALIZATIONS

Once the basic subtraction facts have been discovered, much attention should be given to developing the generalizations or principles that apply to subtraction. In addition to the mathematical understanding involved, such study is helpful in learning the basic subtraction facts. Basically, the most important generalization to remember is that subtraction is the inverse of addition.

Other, related generalizations follow:

1. When zero is subtracted from a number, the number is unchanged.
2. Subtracting a number from itself leaves zero.
3. Subtracting ones from a number is equivalent to counting backward by ones.

4. In situations such as $(16 - 5) - 3 = \square$, the subtrahends may be added and the sum subtracted from the minuend: $16 - (5 + 3) = 16 - 8$.
5. The subtrahend may be renamed (broken into parts) and each part subtracted. For example, $17 - 8 = 17 - (7 + 1) = (17 - 7) - 1 = 9$. This is particularly helpful for reviewing subtraction facts in which a number is subtracted from a multidigit number.
6. The same number can be added to both the sum (minuend) and the known addend (subtrahend) without changing the difference. For example, $17 - 9 = \square$, by adding 1 $[(17 + 1) - (9 + 1)]$, can be simplified to $18 - 10 = \square$. This procedure can be developed by using sticks or counters. The teacher has the children hold a number of sticks in each hand and asks them to find the difference; for example, 13 and 8. Then she or he has them pick up the same number of additional sticks with each hand; for example, two. She or he then asks them to find the difference (15 and 10). Several such experiments will prove helpful in developing this idea, which is often referred to as *compensation* and is explained later in this chapter.

RENAMING IN SUBTRACTION

Subtraction situations such as $34 - 16 = \square$ involve a need to rename one set of tens into ones. Thus,

$$\begin{array}{r} (30 + 4) \\ \underline{-(10 + 6)} \end{array} \longrightarrow \begin{array}{r} (20 + 14) \\ \underline{-(10 + \ \ 6)} \end{array}$$

Before children are introduced to experimentation with paper-and-pencil algorithms, a number of experiments and problems should be conducted by using objects. Bundles of stirrers or Popsicle® sticks and/or squared paper are perhaps the most useful objects. One teacher used the laboratory sheet reproduced below, and got the typical pupil solutions illustrated.

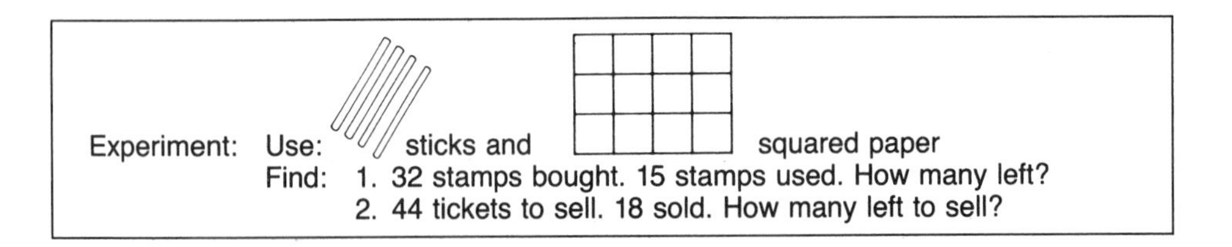

NAN: I cut out 32 squares: I marked out a ten and five ones. There were 17 left.

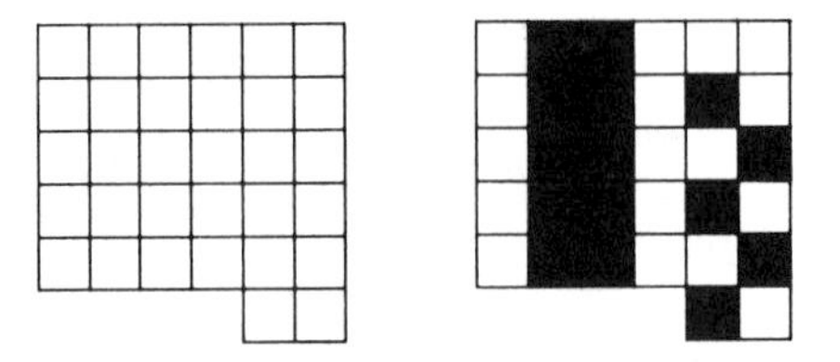

JOE: I laid out three groups of tens and two ones sticks to stand for the stamps. Then I needed to remove 15. I had to undo one of the tens to make it ones before I could remove all 15. There were 17 left.

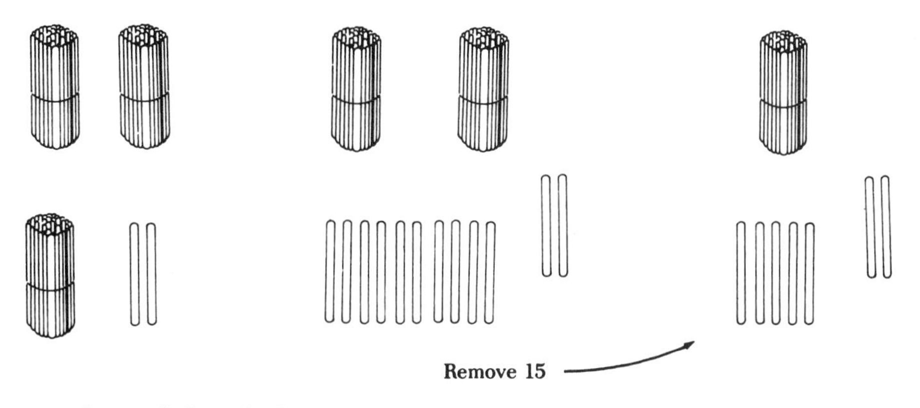

Several days before the introduction of renaming, the teacher may use some exercises in which pupils discuss different ways of naming a number such as 34. This discussion should lead pupils to naming 34 as 30 + 4 and 20 + 14. Naming 34 as 3 tens and 4 ones and as 2 tens and 14 ones may also be developed.

A patterning sheet such as the one shown can be used.

Look:

28 ⟶ 10 + 18   50 ⟶ 40 + 10
39 ⟶ 20 + 19   92 ⟶ 80 + 12

Find: Use the same pattern.

70 ⟶   53 ⟶
67 ⟶   45 ⟶
30 ⟶

The teacher introduced renaming by saying, "Look at problem 1 on the duplicated sheet; it says, 'Betty has a box containing 34 Christmas tree ornaments [see illustration]. If her mother uses 16 of the ornaments, how many will Betty have for use on her own tree?' Write the mathematical sentence that can be used to solve the problem, solve the problem, and then show that your answer is correct in as many ways as you can."

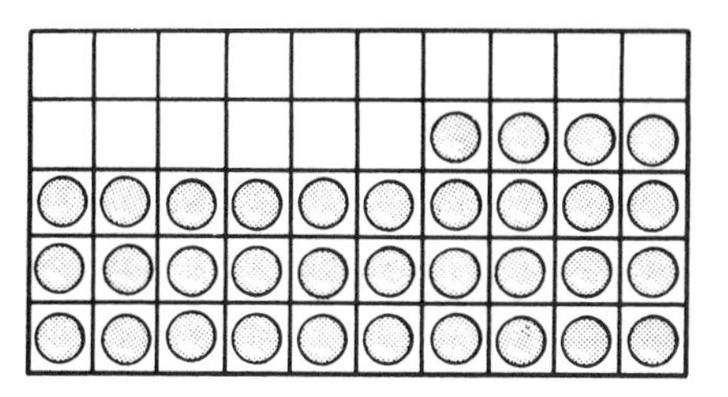

While the pupils worked on the problem, the teacher observed individual class members at work and suggested to those who were having difficulty that they make use of bundles of sticks or the diagram. Several pupils were asked to show their method of attack at the chalkboard. These procedures and pupil statements follow.

The pupils first stated the mathematical sentence as $34 - 16 = \square$,

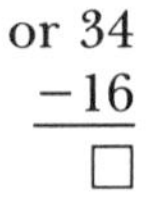

JEFF: I used bundles of tens. I had to remove 16. First I removed 10. I couldn't remove six more without unbundling one of the tens. I unbundled the ten and removed six. I had 18 remaining.

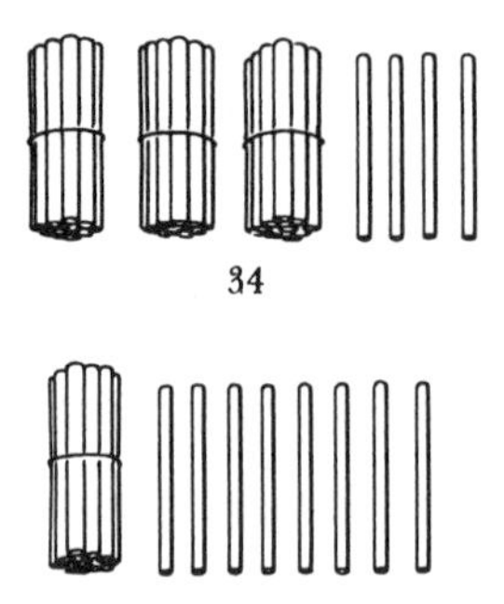

KIM: I used place-value frames. I marked 3 tens and 4 ones. I removed 1 ten. I couldn't remove 6 ones, so I changed one of the tens to ones. Then I removed 6 ones. My answer was 18.

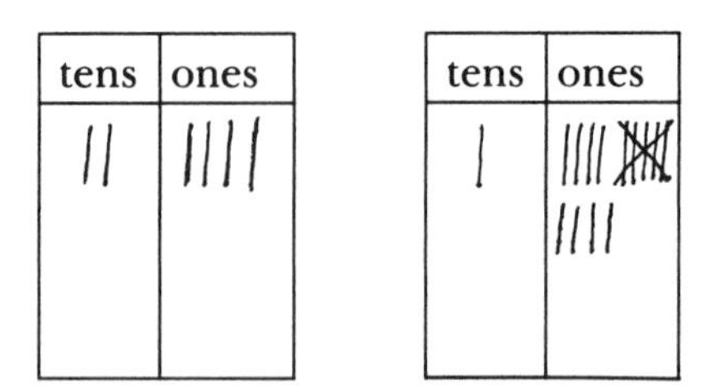

GEORGE: I used the place-value frame using numerals. I couldn't subtract 6 from 4 so I changed the 3 tens to 2 tens and 10 ones. Then I subtracted 6 from 14 and 1 ten from 2 tens. My answer was 18.

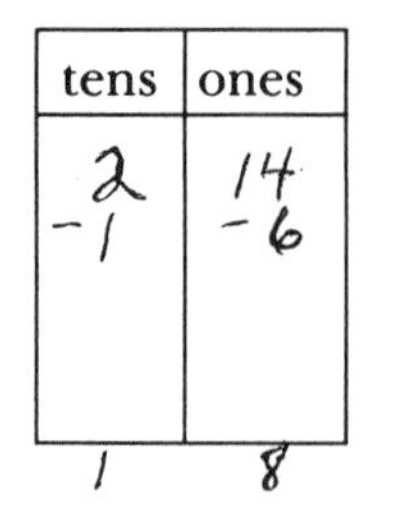

PAUL: I wrote the computation in expanded form. Then I remembered that there are many names for the same number. I renamed 30 + 4 into 20 + 14. Then I subtracted. The answer was 18.

$$\begin{array}{rl} 30 + 4 = & 20 + 14 \\ \underline{-10 + 6} & \underline{10 + 6} \\ & 10 + 8 = 18 \end{array}$$

NED: I used squared paper and marked over the 16 that I needed to remove.

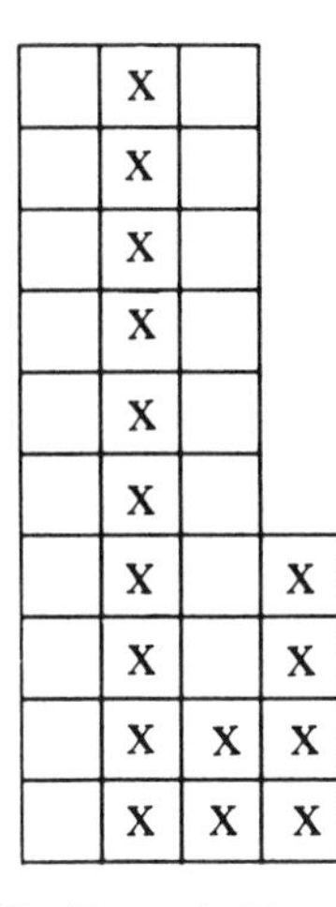

PHYLLIS: I used the abacus. I indicated 34. Then I removed 1 ten. I needed to remove 6 ones. I removed 4 ones. I still needed to remove 2 ones. I exchanged 1 ten for 10 ones and then removed 2 ones.

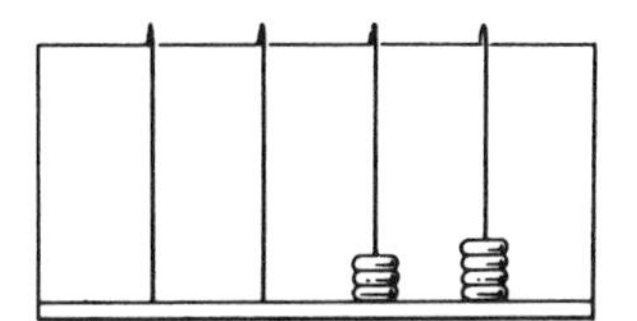

NAN: I had a hard time figuring out what to do, so I counted backward from 34. I got the right answer, but it took a long time.

The pupils discussed the various solutions. They felt that the solution by Paul, using renaming, and the tens-and-ones frame solution by George were probably the best. During this discussion, the teacher emphasized the importance of renaming.

If pupils are not able to "discover" the renaming procedure, the teacher can use leading questions, such as, "What does 74 mean?" "How could we rename 74?" "How could we rename 74 in a way that would help us subtract 38?"

$$\begin{array}{r} 74 \\ \underline{-38} \end{array} \longrightarrow \begin{array}{r} 70 + 4 \\ \underline{30 + 8} \end{array} \longrightarrow \begin{array}{r} 60 + 14 \\ \underline{30 + 8} \end{array}$$

During the next few days, pupils were given their choice of method in solving subtraction exercises involving renaming. The slower pupils often used Popsicle® sticks bundled into tens. As they gained understanding of the process, the teacher helped them to develop the expanded form. The more rapid pupils were directed to try to devise means of shortening the writing process.

When the class members had developed a good understanding of the process, the teacher conducted a discussion on shortening it. The group moved from illustration (a) to illustration (b), and several of the children found that they could think through the problem and remember the renaming without writing it out.

$$\begin{array}{rr} \text{tens} & \text{ones} \\ 5\not{6} & \not{2}12 \\ -4 & 8 \\ \hline 1 & 4 \end{array} \qquad \begin{array}{rr} 5 & 12 \\ \not{6} & \not{2} \\ -4 & 8 \\ \hline 1 & 4 \end{array} \qquad \begin{array}{rr} 6 & 2 \\ -4 & 8 \\ \hline 1 & 4 \end{array}$$

(a) (b) (c)

Subtraction with three or more digits can be developed in the same manner. However, some pupils have difficulty with situations such as $503 - 346 = \square$. It is usually helpful to show that this problem can be renamed in several ways:

$$\begin{array}{r} 503 \\ -346 \\ \hline \end{array} \longrightarrow \begin{array}{r} 500 + 3 \\ 340 + 6 \\ \hline \end{array} \longrightarrow \begin{array}{r} 490 + 13 \\ 340 + \ \ 6 \\ \hline \end{array} \quad \text{or} \quad \begin{array}{r} 400 + 100 + 3 \\ 300 + \ \ 40 + 6 \\ \hline \end{array} \longrightarrow \begin{array}{r} 400 + 90 + 13 \\ 300 + 40 + \ \ 6 \\ \hline \end{array}$$

In the present program, renaming in subtraction is usually introduced late in second grade or in third grade. Since pupils will be using this process over a period, the teacher should take care to develop understanding before developing the shortcut algorithm.

OTHER METHODS

The renaming form as just presented is not the only possible means of developing compound subtraction. Four different (or partially different) methods can be used. They are (1) take-away renaming, (2) take-away equal additions, (3) additive renaming, and (4) additive equal additions. Take-away renaming is the method used for presentation in the teaching situation above. The other methods are explained below.

Take-Away Equal Additions

$$\begin{array}{r} 84 \\ -56 \\ \hline \end{array} \longrightarrow \begin{array}{r} 80 + \ \ 4 \\ 50 + \ \ 6 \\ \hline 80 + 14 \\ 60 + \ \ 6 \\ \hline 20 + \ \ 8 = 28 \end{array} \qquad \begin{array}{l} \text{6 cannot be subtracted from 4.} \\ \\ \text{Add 10 ones to 4.} \\ \text{Add 1 ten to 50.} \\ \text{Subtract.} \end{array} \qquad \begin{array}{r} {}^{7\,1} \\ \not{8}4 \\ 56 \\ \hline 28 \end{array}$$

This procedure is based on the principle that if both terms are increased by the same amount, the difference (remainder) is unchanged. This property is referred to as compensation.

$$\begin{array}{r} 6 \\ -3 \\ \hline 3 \end{array} \longrightarrow \begin{array}{r} 6 + 2 \\ 3 + 2 \\ \hline \end{array} \longrightarrow \begin{array}{r} 8 \\ -5 \\ \hline 3 \end{array}$$

Additive Renaming

$$\begin{array}{r} 84 \\ -56 \\ \hline \end{array} \longrightarrow \begin{array}{r} 80 + \ \ 4 \\ 50 + \ \ 6 \\ \hline 70 + 14 \\ 50 + \ \ 6 \\ \hline \end{array}$$

Rename.
Think 6 plus what number = 14?
Think 50 plus what number = 70?

$$\begin{array}{r} {}^{7\,1} \\ \not{8}4 \\ -56 \\ \hline 26 \end{array}$$

Note that renaming is done in the same manner as in the classroom situation described above. The difference is in using "additive thinking" rather than "take-away thinking."

Additive Equal Additions

$$\begin{array}{r} 84 \\ -56 \\ \hline \end{array} \longrightarrow \begin{array}{r} 80 + \ \ 4 \\ 50 + \ \ 6 \\ \hline 80 + 14 \\ 60 + \ \ 6 \\ \hline \end{array}$$

6 cannot be subtracted from 4.
Add 10 ones to 4.
Add 1 ten to 50.
Think 6 plus what number = 14?
Think 60 plus what number = 80?

$$\begin{array}{r} {}^{1} \\ 84 \\ {}^{6} \\ -\not{5}6 \\ \hline 26 \end{array}$$

Students of the teaching of elementary school mathematics have often debated as to which approach (renaming or equal additions) should be used. One of the most thorough studies on the relative merits of the two approaches was conducted by Brownell and Moser.[3] In this study, four groups of pupils were used. Two groups learned the decomposition (renaming) method, one with understanding and the other mechanically. Two groups learned the equal-additions approach, one group with understanding and the other mechanically.

When the four groups were compared, equal additions were found to be superior to decomposition when both were taught by rote, and decomposition was found to be superior to equal additions when both were taught with understanding. Since a basic tenet of modern programs of mathematics is understanding of the number system, the decomposition method (renaming) is used for initial instruction. A means of making use of the equal-additions procedure, also called the Austrian method, will be discussed in the section on reintroducing subtraction.

[3]William A. Brownell and Harold E. Moser, *Meaningful vs. Mechanical Learning: Study in Grade III Subtraction* (Durham, N.C.: Duke University Press, 1949).

USES OF COMPENSATION WITH MULTIDIGIT SUBTRACTION

It was briefly noted earlier that *compensation* could be used in subtraction. For example, if a child is asked to find $96 - 38 = N$ without using a pencil or paper, the child can think $(96 + 2) - (38 + 2) = N$; or $98 - 40 = 58$. This procedure is easier to handle without pencil and paper than is the conventional thinking: $96 - 38 = N$; change 96 to $80 + 16$; subtract 8 from 16; subtract 30 from 80; answer is 58. This property can be generalized: if $a - b = c$, then $(a + k) - (b + k) = c$, where k is considered to be any whole number. It can also be used in solving problems such as $70 - 32 = N$ by subtracting 2 from each term; $68 - 30 = N$. This generalizes into the following form: if $a - b = c$, then $(a - k) - (b - k) = c$.

The compensation method may be introduced by presenting a number of non–pencil-and-paper subtraction exercises to pupils and asking them to find the answer as quickly as possible. A discussion of the methods they use will often reveal that one or more pupils have discovered the idea of compensation. If this is the case, the teacher should develop a mathematical understanding of the procedure. The teacher may say, "I worked the exercise in this manner," and illustrate on the board. "Why does this work? What mathematical principle allows us to perform subtraction in this manner?"

$$\begin{array}{r} 96 \\ -38 \\ \hline 58 \end{array} \longrightarrow \begin{array}{r} 96 + 2 \\ 38 + 2 \\ \hline \end{array} \longrightarrow \begin{array}{r} 98 \\ 40 \\ \hline 58 \end{array}$$

Note that the additive equal additions can be taught with meaning by using the compensation idea. It is suggested that this form be used as reintroduction in fifth or sixth grade. In fact, some fifth and sixth graders may prefer to make it their normal method of subtraction. Many Europeans do.

OTHER TECHNIQUES FOR STUDYING SUBTRACTION

The suggestions that follow can be used in practice situations to further the understanding of subtraction and to develop an understanding of the relationship between addition and subtraction.

Mathematical sentences (primary level). Work with various types of mathematical sentences involving addition and subtraction, distinguishing between mathematical sentences that are true and those that are false and making use of *greater than, less than,* and *not equal to* in situations using addition and subtraction.

Children can use balances, squared paper, sticks, or pencil and paper alone to explore answers to questions such as those that follow. It is often helpful to use a laboratory format, with two or three children working together and finding the solutions in several different ways.

1. Make a true mathematical sentence of the following.

$5 + \square = 9 \qquad \square + 8 = 15 \qquad 9 - 3 = \square$

$18 - 7 = \square \qquad \square + \square = 18$

2. Insert the correct symbol: =, <, >

$$5 - 3 \bigcirc 2 \qquad 17 - 9 \bigcirc 7$$

$$6 - 4 \bigcirc 3 \qquad 9 - 5 \bigcirc 4$$

3. Decide whether each of the following mathematical sentences is true, is false, or could be either true or false:

$16 - 7 = 8$	true, false, either
$\square + \square = 17$	true, false, either
$\triangle - \triangle = 9$	true, false, either

4. List the possible replacements for the frames. Use only $\{0, 1, 2, 3, \ldots, 17, 18\}$:

$\triangle - \square = 9$	⬡ $- \triangle = \triangle$
$\square + 5 = \triangle$	$9 + \triangle < 16$
⬡ $- \triangle =$ ⬡	$17 - \triangle < 9$

Making change (late primary–early intermediate level). "When I bought several items at the drugstore yesterday, the cost was $3.57. I gave the clerk a $5 bill. The clerk didn't subtract to find how much change I had coming but counted it in the following manner: $3.57, $3.58, $3.59, $3.60, $3.70, $3.75, $4.00, $5.00. When I asked, for fun, 'How much change have you given me?' the clerk didn't know. Yet he said that he was sure that it was the correct change. What did he do? Why was he sure he was right? Is this counting addition or subtraction? Try a few examples with making change in this manner."

The situation above is helpful in additive thinking in subtraction. Most pupils are familiar with it even if they have not analyzed it.

Games. The game that follows is illustrative of the type of games included in the better elementary school mathematics textbook series.

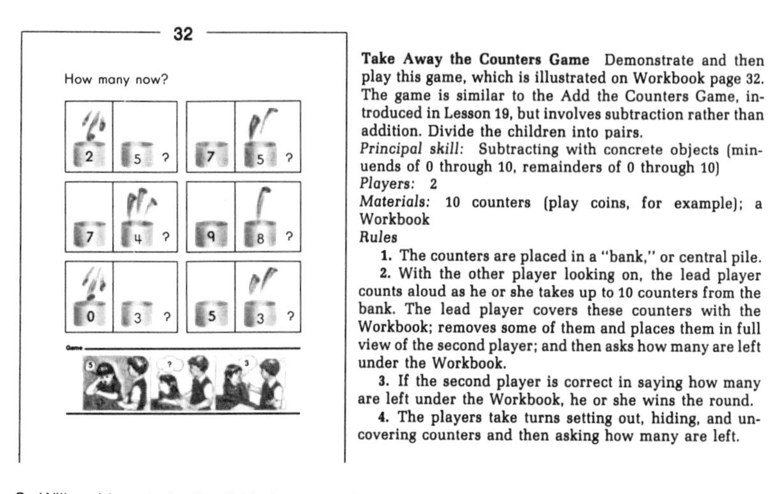
32

How many now?

Take Away the Counters Game Demonstrate and then play this game, which is illustrated on Workbook page 32. The game is similar to the Add the Counters Game, introduced in Lesson 19, but involves subtraction rather than addition. Divide the children into pairs.

Principal skill: Subtracting with concrete objects (minuends of 0 through 10, remainders of 0 through 10)

Players: 2

Materials: 10 counters (play coins, for example); a Workbook

Rules

1. The counters are placed in a "bank," or central pile.
2. With the other player looking on, the lead player counts aloud as he or she takes up to 10 counters from the bank. The lead player covers these counters with the Workbook; removes some of them and places them in full view of the second player; and then asks how many are left under the Workbook.
3. If the second player is correct in saying how many are left under the Workbook, he or she wins the round.
4. The players take turns setting out, hiding, and uncovering counters and then asking how many are left.

S. Willoughby et al., *Real Math, Level One* (LaSalle, Ill.: Open Court Publishing Co., 1985), p. 32. Used with permission.

Estimating. Counting backward by hundreds, tens, fives, and ones in non–paper-and-pencil situations can provide a base for estimating in multidigit subtraction. Bundles of coffee stirrers can be used for those children who need concrete help. Here are a few of the oral exercises that can be used.

1. Start with 80 and count backward by tens. How many tens to 50? (3 tens).
2. Start with 90 and count backward by twenties (70, 50, 30, 10).
3. What is 70 minus 30 (you can count back 3 tens)?
4. What is 800 minus 300?

Another type of exercise that helps with estimation and subtractions concepts is shown below:

Pick out the answers that do not make sense

$$\begin{array}{r}58\\-29\\\hline 72\end{array}\quad\begin{array}{r}31\\-19\\\hline 12\end{array}\quad\begin{array}{r}386\\-243\\\hline 543\end{array}\quad\begin{array}{r}9425\\-6321\\\hline 3104\end{array}\quad\begin{array}{r}899\\-537\\\hline 362\end{array}\quad\begin{array}{r}576\\-189\\\hline 613\end{array}$$

Cross-number puzzles and magic squares (all grades). "I've placed some cross-number puzzles and magic squares on the mathematics table. You may want to spend a portion of the mathematics period working on them. They should be of help to you in subtraction and addition."

The procedure for use with the cross-number puzzle for subtraction is to

1. Write the minuend (17) in the top left box.
2. Write the subtrahend (8) in the bottom right box.
3. Rename the minuend, and write the parts in the spaces to the right of the minuend.
4. Rename the subtrahend, and write the parts above the subtrahend. (For work with pupils before negative numbers have been introduced, the pupil should always rename the minuend so that each part is larger or equal to the parts of the renamed subtrahend.)
5. Subtract each part of the subtrahend from each part of the minuend (10 − 6; 7 − 6; 10 − 2; 7 − 2). The difference is found by finding the sum of either diagonal (4 + 5, or 8 + 1).

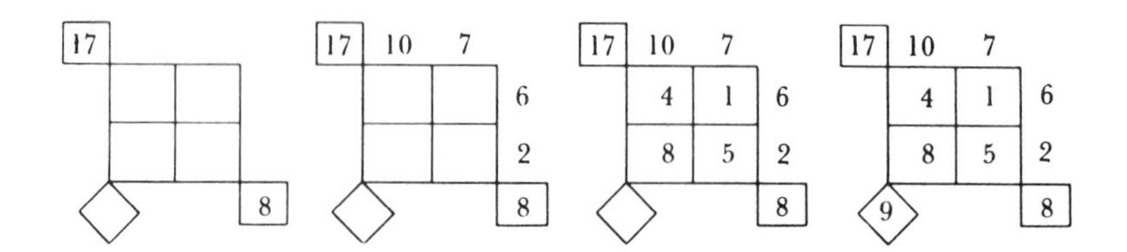

When work with negative numbers has been introduced, the pupil no longer needs to rename the parts of the minuend so that each is larger than the parts of the subtrahend. See illustration below.

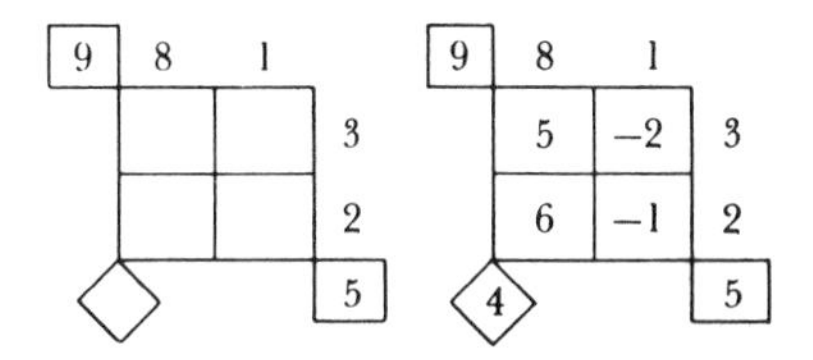

Puzzles of this type are based on the principle that the subtrahend may be renamed and each part subtracted from the minuend. Pupils should be challenged to try to figure out this principle.

A magic square is an array in which the sum of each horizontal column, the sum of each vertical column, and the sum of each diagonal are equal. Study magic square A, and then supply the missing numerals in magic square B. When you become proficient, try to develop your own magic squares.

A

10	3	8
5	7	9
6	11	4

B

7	0	
2		6
	8	1

CHECKING SUBTRACTION

Checking basic subtraction combinations may be done in several ways. The most common is an application of the inverse relationships with addition, thus using addition as the check. Dot drawings, counting backward, the number line, the abacus, and subtracting twice [$16 - 9 = (16 - 6) - 3$] are also appropriate checks.

For multidigit subtraction, one can effectively use addition, subtraction of the remainder from the minuend, casting out nines, and any of the procedures suggested for reintroduction.

Graphs. The teacher identifies points on the grid with numbers. Children find names for the direct paths to the point. Children are also given laboratory lessons in which they identify many names for numbers on the grid.

Number families (primary grades). The relationship between addition and subtraction is often furthered by the study of number families. For example, in the family of 7, how many addition and subtraction facts can you find that have 7 as a sum (addition) or 7 as a minuend (subtraction)?

$4 + 3 = 7$	$7 - 0 = 7$
$3 + 4 = 7$	$7 - 1 = 6$
$5 + 2 = 7$	$7 - 2 = 5$
$2 + 5 = 7$	$7 - 3 = 4$
$6 + 1 = 7$	$7 - 4 = 3$
$1 + 6 = 7$	$7 - 5 = 2$
$0 + 7 = 7$	$7 - 6 = 1$
$7 + 0 = 7$	$7 - 7 = 0$

Mapping. Use function mappings in the same manner as those used in Chapter 6 and the beginning of this chapter.

Function machines (all grades). Several types of function machines can be used. Those most often used are the single-input model and the number-pair-input model, both referred to earlier in this chapter.

"Follow-me" oral exercises (all grades). As in addition, valuable use can be made of oral "follow-me" exercises. They can be used in spare minutes throughout the day. Example: 23 minus 10 plus 7 minus 15 plus 5 plus 8 minus 9 equals what number?

Identification of subtraction types (late primary–early intermediate). "The problem sheet contains problems of all three subtraction varieties. Identify the type of subtraction situation and illustrate the situation with the number line."

$$\begin{array}{r} 654 \\ -386 \\ \hline 268 \end{array} \qquad \begin{array}{r} \text{check} \\ 386 \\ +268 \\ \hline 654 \end{array} \qquad \text{addition check}$$

$$\begin{array}{r} 654 \\ -386 \\ \hline 268 \end{array} \qquad \begin{array}{r} 654 \\ -268 \\ \hline 386 \end{array} \qquad \text{subtraction check}$$

$$\begin{array}{rll} 654 & 6 + 5 + 4 = 15 \rightarrow 6 \rightarrow & 15 \\ -386 & 3 + 8 + 6 = 17 \rightarrow & \underline{-\ 8} \\ \hline & & \\ 268 \rightarrow & 16 \rightarrow \qquad 7 \longleftrightarrow & 7 \end{array}$$

casting out nines
Note: Increase the excess in the minuend by 9 if it is less than the excess of the subtrahend.

Basic facts in different settings (grade 4). "Yesterday you told me that you were subtraction whizzes. Let's see. Put away your paper and pencils. Now start at 78 and subtract 5s until you reach a number in the 50s. Raise you hand when you know. Yes, the answer is 58. Now start at 156 and subtract 7s until you reach a number in the 130s. What is that number?"

This type of subtraction makes use of a "bridging" situation and a series of subtractions. It also provides a good foundation experience for division.

Additive thinking, multidigit (grade 5). "We haven't always renamed in the manner we now use. See if you can figure out the procedure that I've duplicated, the explanation given in *White's Arithmetic,* which was written in 1883. See if you can work the examples using this procedure. Also, see if you can give an explanation for the procedure."

> Since 6 units cannot be taken from 4 units, add 10 units to the 4 units, making 14 units, and take 6 units from the 14 units, and write 8 units (the difference) below. To balance the 10 units added to the minuend, add 1 ten (equal to 10 units) to the 2 tens, making 3 tens, and take 3 tens from 3 tens, and write 0 (the difference) below.
>
> PROCESS
>
> | | |
> |---|---|
> | Minuend | 5334 |
> | Subtrahend | 2726 |
> | Difference | 2608 |
>
> Since 7 hundreds cannot be taken from 3 hundreds, add 10 hundreds to the 3 hundreds, making 13 hundreds, and take 7 hundreds from 13 hundreds, and write 6 hundreds (the difference) below. To balance 10 hundreds added to the minuend, add 1 thousand (equal to 10 hundreds) to the 2 thousands, making 3 thousands, and take 3 thousands from 5 thousands, and write 2 thousands (the difference) below. The difference is 2608.

Pupils commented on the writing of the explanation. They were surprised at the number of times *and* was used. Several subtraction combinations were solved as a total class. Discussion brought out the idea that the procedure made use of the principle of compensation. One gave an explanation as follows:

MARK: Use tens and ones. I added 10 ones to the upper term and 1 ten to the lower term. This makes use of the principle we learned earlier, called compensation. If both terms are increased by the same amount, the difference is not changed.

$$\begin{array}{r} 56 \\ -38 \\ \hline \end{array}$$

tens	ones
5	6 + 10
3 + 1 ten	8
1	8

Complementary method, multidigit (grade 4, 5, or 6). The question, "Do you know what a complement is?" was asked by the teacher. Pupils thought it meant praise or flattery. The teacher wrote *complement* on the board and said, "That would be a homonym, which is spelled c-o-m-p-l-i-m-e-n-t. *Complement* is spelled in the same way as in *complementary colors.* Let's see if you can figure out what a mathematical complement is. The tens complement of 6 is 4; the complement of 3 is 7; what is the complement of 5? Right, 5. What would be the thousands complement of 350? Right, 650. How can you state a generalization that would cover all complements?"

The teacher suggested, "Let's agree that I mean the next largest power of ten." Then the teacher said, "During the 1800s, Europeans often used the 'complementary method' of subtraction. Look at the example on the board, and see if you can work the other examples."

32	The complement of 18 is 82.	32
−18	Add the complement to the sum (minuend).	+82
	Cross out the numeral in the hundreds place.	~~1~~14

After working several of the exercises, the class attacked the problem, "Why does the complementary method work?" Herb suggested that they might get an idea of the workings of the complementary method by recording the entire operation as a mathematical sentence. He worked $74 - 36 = \square$ in the manner shown below:

$$74 - 36 = 74 + (100 - 36) - 100$$

He said, "If we look closely, we can see that 100 has been added to the sentence and then subtracted. This makes use of the principle that the value of a number is not changed if we add and take away the same amount from it. It's just like $(5 + 7) - 7 = 5$." After further clarification of the complementary method, the teacher assigned several exercises to be worked by the complementary method and checked by the regular subtraction algorithm.

Estimating (grade 5 or 6). One teacher used the material that follows to begin a lesson estimating multidigit subtraction. After asking the book question, the teacher gave out a sheet on which the areas in square miles (to give some English system practice) were given. The class played a game in which two state names were drawn from a hat. The players wrote down an approximation of the difference. Calculators were used to check to see which player's estimate was closest to the correct difference.

Using Multidigit Numbers: Area of States

State	Area (square kilometers)
Montana	381,087
Massachusetts	21,386
Missouri	180,487
Ohio	106,765

Use the map and the chart to estimate the area in square kilometers of the following states. In each problem, 4 areas are given in square kilometers but only 1 is correct. Choose the correct area.

1. New Mexico	a. 150,289	(b.) 315,115	c. 105,321	d. 567,402
2. Georgia	(a.) 152,489	b. 472,809	c. 210,450	d. 102,946
3. Pennsylvania	a. 46,315	b. 256,947	(c.) 117,412	d. 437,231
4. Oklahoma	a. 392,025	b. 53,702	c. 108,257	(d.) 181,090
5. Texas	a. 45,327	(b.) 692,535	c. 392,621	d. 192,849
6. Connecticut	(a.) 12,973	b. 194,327	c. 56,240	d. 376,294

19

Student page 19

Go over the first problem with all the students before they begin work on the page. Ask them to compare the area of New Mexico with that of each of the 4 states whose area is given in the chart. Use a sequence of questions such as: "Is New Mexico bigger than Massachusetts?" (yes) "Is it bigger than Ohio?" (yes) "Is it bigger than Missouri?" (yes) "Is it bigger than Montana?" (no) "So New Mexico is bigger than Missouri and smaller than Montana. Which of the choices for area given in problem 1 is bigger than 180,487 and smaller than 381,087?" (315,115) "Then the answer for the area of New Mexico is 315,115 square kilometers."

If the students understand how to estimate area in this way, let them finish the page, working independently or in small groups. If they have difficulty understanding this method of estimation, finish the page as a whole-class activity.

S. Willoughby et al:, *Real Math, Level Six* (LaSalle, Ill.: Open Court Publishing Co., 1985).

KEEPING SHARP

Self-Test: True/False

_______ 1. Subtraction is associative.

_______ 2. Subtraction can be viewed as a taking-away situation or as a how-many-more-are-needed situation.

_______ 3. Since subtraction is the inverse of addition, it is preferable to introduce it after children have a good understanding of addition.

_______ 4. Children can experiment with concrete objects to develop the need for renaming in subtraction.

________ 5. Casting out nines is one way to check subtraction.

________ 6. Subtraction is closed for whole numbers.

________ 7. Problems involving the comparison type of subtraction are the easiest and are good introductory situations.

Vocabulary

equal additions	complementary subtraction
decomposition	compensation method

THINK ABOUT

Subtraction of Whole Numbers

1. With what type of subtraction situation (how many left, how many more, or comparison) should systematic study of subtraction be introduced?
2. Some programs of instruction begin subtraction study with objects and algorithms (8|○ or ○⊗ as representing $2 - 1 = 1$). What advantages and disadvantages do you see in such a beginning procedure?
3. Would the term *change* be a good substitute for *borrow*? Why or why not?
4. By the time able pupils have finished sixth grade, they should have had the opportunity to use how many ways of checking answers for subtraction questions? Illustrate each.

SELECTED REFERENCES

BAROODY, ARTHUR J., "Children's Difficulties in Subtraction: Some Causes and Cures," *The Arithmetic Teacher* 32 (November 1984), 14–19.

BROWNELL, WILLIAM A., and H. E. MOSER, *Meaningful versus Mechanical Learning*, Durham, N.C.: Duke University Press, 1949. (A classic)

BURTON, GRACE M., *Toward a Good Beginning: Teaching Early Childhood Mathematics*, chap. 14., Menlo Park, Calif.: Addison-Wesley, 1985.

CLEMENTS, DOUGLAS H., *Computers in Elementary Mathematics Education*, chap. 7, Englewood Cliffs, N.J.: Prentice Hall, 1989.

Experiences in Mathematical Ideas, vol. 1, Chap. 1 and 2.: Reston, Va.: National Council of Teachers of Mathematics, 1970. (Old but excellent)

Learning to Add and Subtract, Washington, D.C.: National Institute of Education, 1985.

Mathematics in the Middle Grades (5–9), 1982 Yearbook, chap. 5, Reston, Va.: The National Council of Teachers of Mathematics, 1982.

MOSER, JAMES M., "Arithmetic Operations on Whole Numbers: Addition and Subtraction." In *Teaching Mathematics in Grades K–8*, chap. 5, Thomas R. Post (ed.), Boston, Mass.: Allyn Bacon, 1988.

WIEBE, JAMES H., *Teaching Elementary Mathematics in a Technological Age*, chap. 9 and 10. Scottsdale, Arizona: Gorsuch Scarisbrick, 1988.

chapter 8

MULTIPLICATION OF WHOLE NUMBERS

OVERVIEW

Whole number multiplication is one of the most used portions of mathematics. Hardly a day goes by that a person does not use multiplication. Also, the history of multiplication and the various algorithms associated with it make it a topic for exploration, discovery, and problem solving.

The National Council of Teachers of Mathematics (NCTM) Standards suggest frequent use of mental computation, problem-solving strategies, estimation, and calculators in exploring and mastering multiplication.

TEACHER LABORATORY

Materials

Lima beans or countable objects (about 50), squared paper, scissors.

Experiment

1. Count out 36 beans. Arrange the beans in rectangular patterns, and record the number of rows and columns in each pattern. What do you notice about the num-

ber pairs for the rows and columns for each rectangle? (They are factors of 36.) How many different rectangles can you form?

2. Use squared paper and scissors. Cut out a 23 × 27 rectangle of squared paper. Now find the total number of squares. Follow these rules:
   a. You must know how to multiply single-digit factors up through 9 × 9.
   b. You must know how to multiply 10 by a single-digit factor and by 10. Now find the answer in at least three different ways. Which way was most efficient? Compare your methods with those of several other people. What is the smallest number of multiplications you can make?
3. Discuss with two or three others the following question: How could activities 1 and 2 be used to develop multiplication concepts with children?
4. Complete and study the multiplication table shown on page 188. Record all the patterns you can find.

Reprinted by permission of UFS, Inc.

TAKE INVENTORY

Can You:

1. Write a word problem that would lead children to develop a drawing for the following multiplication ideas: cross-product of sets, series of additions, a function mapping?
2. Develop a teaching plan for a pattern-searching strategy for one of the following: learning basic multiplication combinations, the commutative property of multiplication, the associative property of multiplication, lightning multiplication?
3. Illustrate three visual aids that could be used to help children develop the idea of the distributive property?
4. Invent a game that could be used to help children with multiplication?
5. Domonstrate "finger multiplication" ?
6. Tell the advantages of using the "lattice method" for multidigit multiplication?
7. Give three suggestions for reintroduction of multiplication?

WHAT IS MULTIPLICATION?

Multiplication, like addition, is an operation that maps (assigns) a pair of whole numbers (factors) to a unique whole number called its product. Thus, the pair of whole numbers 5, 4 is mapped to the product 20. We may write $(5,4) \rightarrow 20$.

This idea may seem to be pure mathematics; however, there are many physical-world situations for which multiplication is useful. An analysis of books concerned with mathematics for children and mathematics teaching reveals several different means of viewing the operation of multiplication. Here are a few:

1. Multiplication of whole numbers may be viewed as a special case of addition in which all the addends are of equal size. In set terms, 3×4 can be defined to be the cardinal number associated with a set formed by the union of three disjoint sets of four elements each, since $3 \times 4 = 4 + 4 + 4$.
2. The cross-product or Cartesian product of two sets can be used to interpret multiplication. The cross-product is formed by pairing every member of one set with every member of the other. For example, if we asked four boys and three girls to form as many mixed dancing couples as possible, we would arrive at a product set of 12 elements (couples). Thus, the product of two numbers can be associated with a rectangular pattern or array.

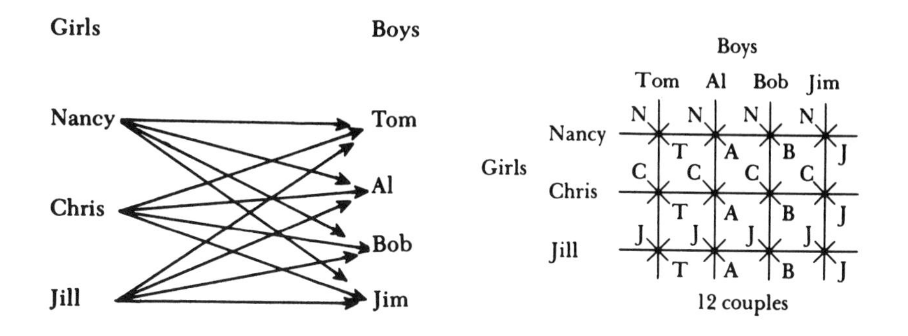

3. Multiplication may be thought of as a mapping or a function. Using a function machine, we could show multiplication in the following ways:

$$3 \times 8 = \square$$

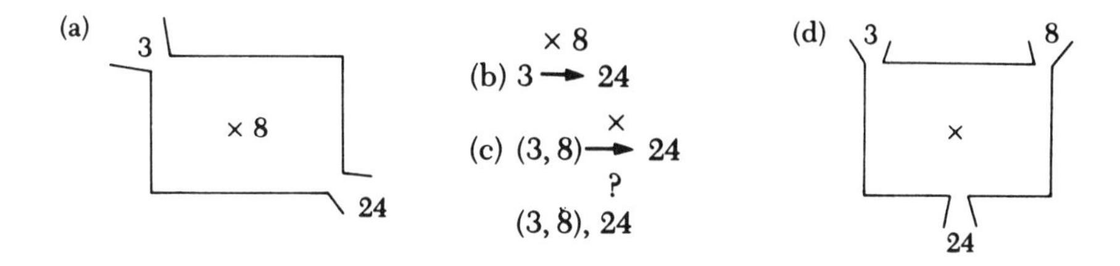

There is a great deal of difference in the three approaches described above, and in order for elementary school pupils to develop a mature understanding of the multiplication operation, they should be familiarized with the different multiplication ideas. However if maximum understanding is to be developed, a logical sequence of these ideas is important. The sequence that follows makes use of the previous mathematical experiences of the pupils to develop a mature viewpoint toward the operation of multiplication.

FOUNDATION EXPERIENCES

As in the case of addition and subtraction, the use of laboratory exploration and orally presented problems provides developmental foundation work in multiplication. The two laboratory activities and the two problems that follow are representative of the type used by successful first- and second-grade teachers to develop background for the formal study of multiplication.

Lab 1. Groups of children were given plastic tiles (about 50). They were asked to make as many rectangles as they could with 20 tiles, with 12 tiles, and so on. They made drawings on squared paper of the various arrangements.

Lab 2. Groups of three children were given a picture of a house and were asked to find the number of panes of glass in the front of the house. On completion of the lab, the groups discussed the approaches they had used.

Problem 1. Joe bought three 9¢ stamps. How much did they cost?

Problem 2. How much will five pieces of candy cost if each piece costs 4¢?

To make full use of these laboratory activities and problems, the following approaches are suggested: (1) Provide children with actual objects so that they can use counting to find an answer; (2) use tens and ones blocks and Cuisenaire rods; (3) make use of the number line for solutions; (4) encourage the children to think of as many ways as possible to solve the word problems; and (5) solve the more difficult problems by means of total-class consideration.

When children have had experience with a variety of these problem situations, they usually have little difficulty in developing the "equal-additions" concept of multiplication. Recently, groups of children with background laboratory and problem-solving experiences but no formal multiplication study scored a median of 18 correct on a 21-item test that consisted of questions such as the following ones:

1. Alice is serving ice cream at her birthday party. If one package of ice cream will serve four children, how many children will three packages serve?
2. Ann gets 2¢ each day for helping with the dishes. How much will she get for helping with dishes for four days?
3. How many are 4 fives?
4. How many are 3 sevens?
5. How many are 5 fours?

Pupils who have had an opportunity to explore number relationships will often demonstrate a grasp of numbers far beyond their level of study. One child who had not studied any multiplication heard her teacher read in *Little Women* that Amy could not think of the answer to 9×12. In about a minute, the girl asked the teacher, "Is 9 times 12 equal to 108?" The teacher answered, "Yes, how did you figure out that answer?" "I left out the twos and added all the tens. That was 90. Then I added all the twos. That was 18. I added 90 plus 18 for the answer of 108."

Such reactions are typical in programs in which an emphasis is placed on a strong foundation of orally presented problems and laboratory activities.

TEACHING THE CONCEPTS OF MULTIPLICATION

Introduction of Multiplication

One teacher introduced multiplication by giving each child a photocopied sheet containing four problems and setting up a materials table containing squared paper, number lines, rods, balances, lima beans, and paper. The teacher said, "Look at the first problem that I've given you. It says, 'Nancy pasted three pictures on each of four pages of her scrapbook. How many pictures did she paste?' Solve the problem and then show that your answer is correct in as many ways as you can. If you finish early, try to solve the other problems."

As the children worked on the problem, the teacher moved about the room offering encouragement and suggestions, often aiding pupils by asking a question that helped them see the situation involved. After noting the different means the pupils used to solve the problem, the teacher asked them to record their thinking on the board. Some pupil responses follow.

KEN: I drew a picture of the situation and then counted the number of pictures. My answer is 12.

JOE: I marked off 4 threes on the number line. This made a total of 12.

PHIL: I made use of rods. I took four rods that were equal to three each and matched them against the tens rod. I still needed two of the ones rods. My answer is 12.

MARY: There were three pictures on each page, so I added 4 threes. My answer is 12.

$$3 + 3 + 3 + 3 = 12$$

NANCY: I marked four groups of three marks and counted them. My answer is 12.

PHYLLIS: I used the balance. I used a three rod for each page and put four on one side of the balance. Then I put 1 ten rod on the other. I guessed that I would need 4 ones rods. I found that I only needed two. So, the answer is 12.

LUCY: I cut out a 3-by-4 section of square paper like the one shown in the picture. Then I made a squared-paper number line. Next, I cut the 3-by-4 section into strips and set them on the number line. My answer is 12.[1]

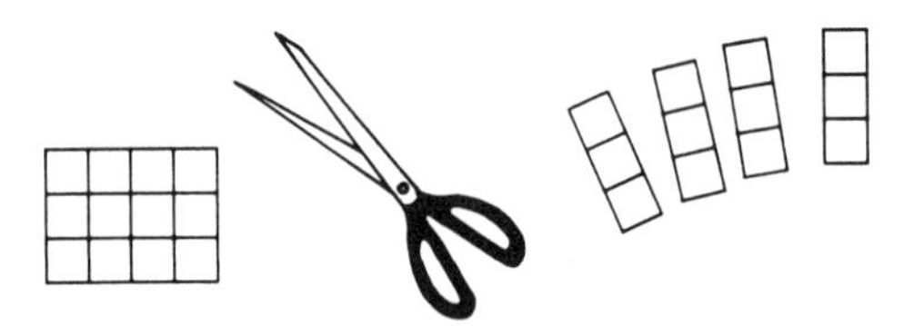

Next, the pupils identified $3 + 3 + 3 + 3 = \square$ as the mathematical sentence used to answer the first addition number question. Then they learned that the multiplication question, "How many are 4 threes?" could be written as $4 \times 3 = \square$.

The next day, a number of multiplication situations were studied. Terminology such as *factors* and *product* was developed during this time. Care was taken to ascertain that the characteristics of factors and of the product were understood before they were given names.

Developing the Cross-Products of Sets Idea

In introducing ways to develop the cross-products of sets idea, a teacher presented the laboratory sheet shown on page 181. Four other exercises were included on the sheet, involving 5×3, 4×4, 6×7, and 3×8.

Upon completion of the laboratory, the children discussed their findings. Then, using questioning, the teacher developed the table that follows for the exercises above:

DOORS IN ONE	DOORS IN TWO			
	d	e	f	g
a	a, d	a, e	a, f	a, g
b	b, d	b, e	b, f	b, g
c	c, d	c, e	c, f	c, g

[1]Note that although Lucy's approach seems to be just like the typical number-line solution, there is a very basic difference. By cutting the four strips, Lucy has moved into the multiplication idea rather than the addition idea. She is finding how many are 4 threes rather than $3 + 3 + 3 + 3$. The emphasis is thus on associating 12 with four groups of three.

Experiment

Find the number of sidewalks needed to go from each door in building ONE to each door in building TWO.

Then make a list of the sidewalks.

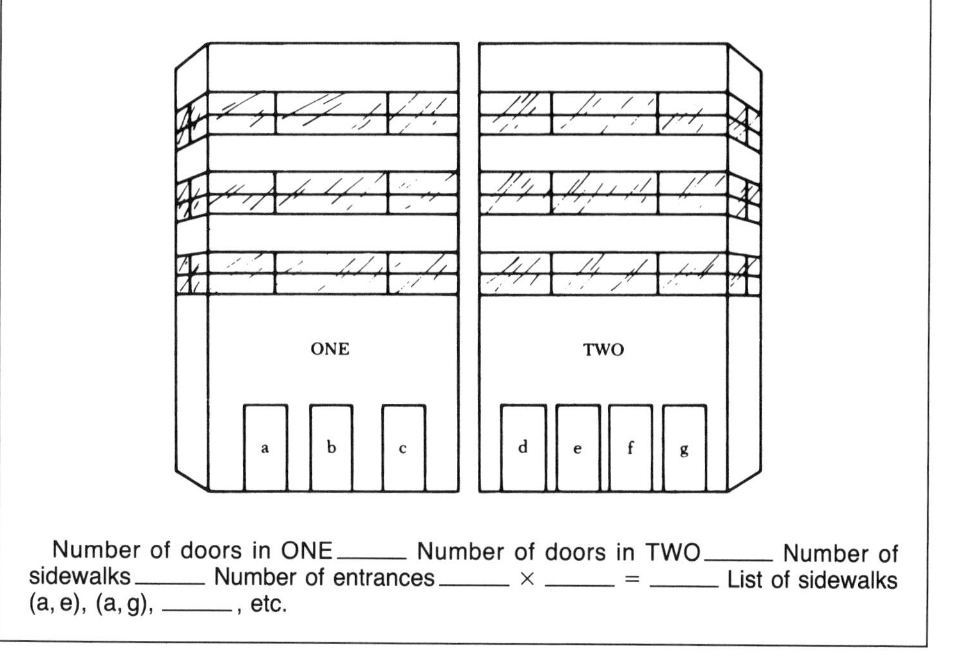

Number of doors in ONE_____ Number of doors in TWO_____ Number of sidewalks_____ Number of entrances_____ × _____ = _____ List of sidewalks (a, e), (a, g), _____, etc.

After this, the teacher gave the class the following exercises:

1. An art class was arranged with desks in pairs. There were three boys and three girls. The teacher told them to arrange themselves with a boy sitting with a girl. How many combinations were possible? What were they?
2. A woman had two books (a play and a novel) and had four bookcovers (red, green, blue, and white). She wanted to cover each of the books. How many combinations did she have to choose from? What were they?
3. A man wanted to cover three pieces of furniture (a rocker, a sofa, and an armchair). He had five different types of material (striped, print, plaid, solid, and checked). How many different combinations did he have to choose from? What were they?

Developing the Commutative Property

After children have plenty of experience in experimenting to find products and relationships between products, a lesson such as the following can be used to develop the idea of the commutative property of multiplication ($6 \times 4 = 4 \times 6$, or, in the general case, $a \times b = b \times a$).

The teacher presented the following problem: "Jane plans to plant four rows of tulips with six tulips in each row. How many tulips does she need?"

The children were asked to write a mathematical sentence for the question and then to make a diagram of the situation. The mathematical sentence was identified as $4 \times 6 = \square$, and a diagram of the type shown below was drawn by most pupils.

The diagram was drawn on the board, and then the teacher said, "How many rows do we have? How many tulips in each row?"

One pupil answered, "Four rows with six tulips in each row." Another said, "That's right, but we could say that we have six rows of tulips with four tulips in each row. Just turn the paper sideways and you can see this."

The teacher made use of this suggestion to question the children concerning the relationship of 6×4 and 4×6. The class tentatively decided that the order of the factors does not change the value of the product (the commutative property of multiplication). The teacher explained, "Patterns such as the one we have made are often called array patterns. Can you think of any other way we could show that the order of the factors does not affect the product?"

The number line, squared paper, balances, rods, and series of additions were used by class members to demonstrate this relationship. The teacher suggested that several other multiplication situations be diagramed to check the "rule" further. When the class had found the property to be true in several situations, the teacher asked them to write a sentence that would describe the "rule." After this, they were told to check their statements against those in their textbook.

The teacher asked, "How can the property we have discovered today help us with our work in multiplication?" Students suggested that it would save them time in learning the multiplication combinations, for if they learned a combination such as 7×9, they would automatically know the combination 9×7.

At this time, the "rule" was identified as the commutative law for multiplication, although no stress was placed on the learning of this terminology.

During the early study of multiplication, a wide variety of problem situations was used to give the pupils experience with multiplication combinations. No attempt was made to use only simple combinations, and no pressure was exerted to master the combinations.

Development of the Role of 0 and 1

Elementary school children do not normally need to know multiplication sentences that involve 0 and 1 until they deal with multidigit multiplication, such as $10 \times 26 = \square$. However, many children's materials develop these ideas during the study of basic multiplication facts. They can be developed in conjunction with a laboratory lesson involving the use of squared paper. A laboratory card such as the one shown below can be used to stimulate the children to consider the unique roles of 0 and 1.

Experiment

Materials: Squared paper, scissors, pencil, and paper
Procedure: Cut rectangles to the size given below.
Find the number of small squares in each.

5 by 2	0 by 5	8 by 0	9 by 9
3 by 6	1 by 6	1 by 8	0 by 7
5 by 5	0 by 6	1 by 9	

Questions:

1. What happened when you tried to cut off the 0 by 5? the 0 by 6? What could you say about the number of squares when 0 is one of the dimensions?
2. What happened when you cut off the 1 by 6? the 1 by 8?
3. How could you quickly find the number of small squares when the dimensions were 1 by another number?

From this laboratory experience, children arrive at the generalization that when 0 is one of the dimensions, there will be 0 number of squares, and that when 1 is one of the dimensions, there will be the same number of squares as the other number. The teacher can now move to multiplication sentences involving 0 and 1.

Stress on these two ideas is of great importance in later study of multiplication. Confusion over the role of 0 causes difficulty in multiplications such as 20×308. The role of 1 as the identity element for multiplication is one of the major unifying themes in the study of fractions. Different names for 1, such as $\frac{2}{2}$ and $\frac{3}{3}$, are used extensively in renaming and in fractional computation.

Developing the Distributive Property

The child's first experience with situations involving the distributive property can stem from a need to find an answer to a forgotten multiplication combination, such as 7×8, or from a multidigit situation, such as 8×23. In either case, a laboratory strategy can be used for introduction.

Note: Multiplication distributes over addition in this way:

$8 \times 23 = \square$
$8 \times (20 + 3) = \square$
$(8 \times 20) + (8 \times 3) = \square$
or, in the general form:
$a \times (b \times c) = (a \times b) + (a \times c)$
or
$(a \times b) + (a \times c) = a \times (b + c)$

In developing the distributive property, a teacher gave each child a 4-by-8 section of squared paper and said, "Try in as many ways as you can to find the number of squares on the sheet without counting." After the children had ample time to experiment, they discussed their solutions. Several children had simply multiplied 4 × 8. Others had folded the squared paper and "multiplied twice." Jill said, "I folded the paper down the middle. This gives me two 4-by-4 sections. Four times 4 equals 16. Sixteen plus 16 equals 32."

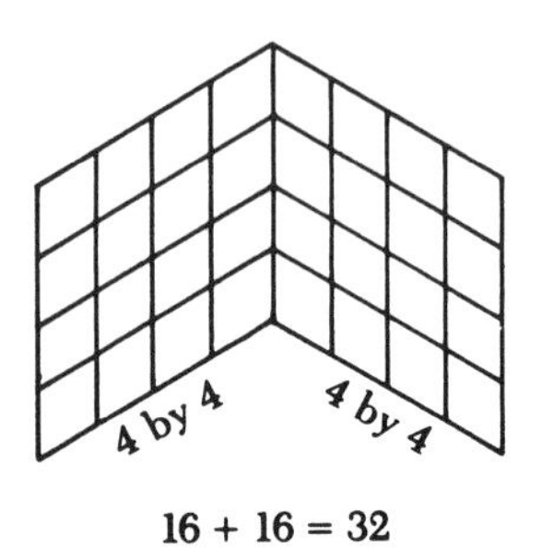

16 + 16 = 32

Class members suggested other means of folding the squared paper as shown in the following diagrams.

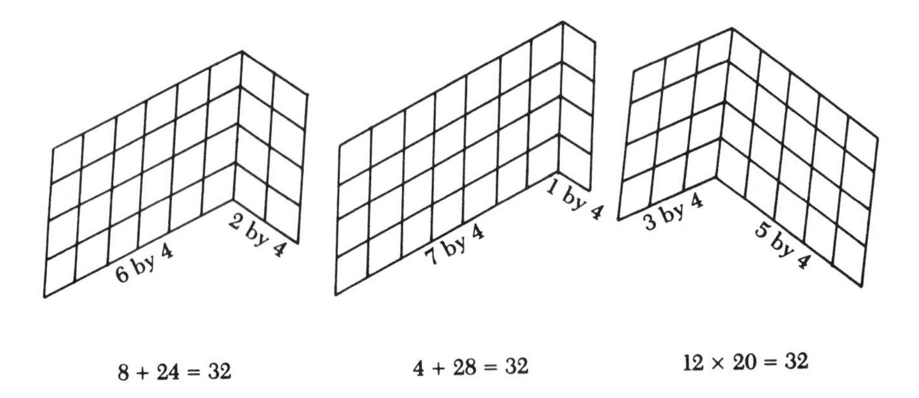

8 + 24 = 32 4 + 28 = 32 12 × 20 = 32

The teacher suggested that the experiments with the squared paper be recorded with numerals. A class member, Jill, said that parentheses could be used to clarify the record.

$$4 \times 8 = (4 \times 4) + (4 \times 4) = 16 + 16 = 32$$

$$4 \times 8 = (4 \times 3) + (4 \times 5) = 12 + 20 = 32$$

$$4 \times 8 = (4 \times 2) + (4 \times 6) = 8 + 24 = 32$$

$$4 \times 8 = (4 \times 1) + (4 \times 7) = 4 + 28 = 32$$

Various other combinations could also be used.

The teacher directed the students to try Jill's method with the examples in their book and pointed out that it should be helpful as a check and as a way to find the answer to a multiplication combination that they might have forgotten.

$$\begin{array}{r} 7 \\ \times 6 \\ \hline \end{array} \qquad \begin{array}{r} 6 \\ \times 8 \\ \hline \end{array} \qquad \begin{array}{r} 9 \\ \times 5 \\ \hline \end{array}$$

$$8 \times 5 = \qquad 3 \times 9 =$$

Frequent use should be made of this "multiplying twice" procedure, since it is an important foundation for and a check of multidigit multiplication.

Developing the Associative Property of Multiplication

The associative property is used most often by children when they are finding the product of three or more factors. Thus, $2 \times 9 \times 6$ becomes $(2 \times 9) \times 6$ and can be associated to be $2 \times (9 \times 6)$. This occurs often in measurement situations in which the problem involves finding the volume. A logical introduction is a problem in finding the number of unit cubes necessary to fill a box. It is helpful if both centimeter cubes and inch cubes are used.

One teacher gave the children a lab sheet and said, "This pictures the number of blocks taken from various boxes. How many blocks will each box hold? Try to find out in as many ways as you can."

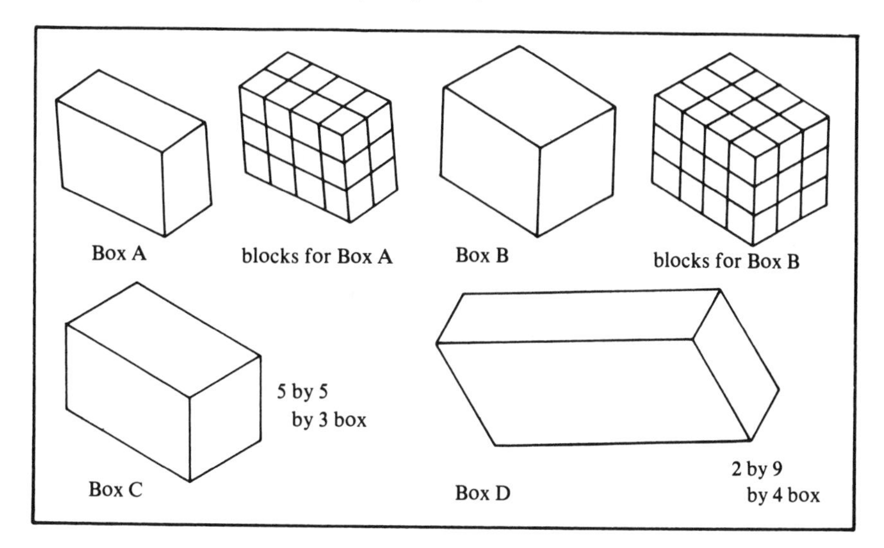

When the children had completed their search for solutions, the various approaches that had been taken were discussed. Nancy had taken cubes, produced each of the pictures with them, then counted the cubes. Bob had counted the number of cubes on the top layer and then added that number again for each row. Phyllis had counted the rows and columns on the top layer, multiplied the two numbers, and then multiplied by the number of layers.

After the children completed the cards, a group discussion was held. The children indicated that there had been several different orders of factors in each of the multiplications. Further discussion brought out the idea that the manner of grouping factors did not change the product.

Pattern Searching

After the children had wide experience with concrete settings, the following lesson was used. The teacher gave the children this sheet:

Explore: Find the products.

Which is easier, A or B?

1) A. $(6 \times 5) \times 3$	or	B. $6 \times (5 \times 3)$
2) A. $(8 \times 2) \times 5$	or	B. $8 \times (2 \times 5)$
3) A. $(4 \times 5) \times 3$	or	B. $4 \times (5 \times 3)$

Make these easier

a) $(9 \times 2) \times 5$
b) $(7 \times 5) \times 4$
c) $(2 \times 9) \times 6$
d) $(3 \times 9) \times 5$

When the children had worked the exercises, they discussed their findings. After a few arguments, there was general agreement that in the first set of exercises, the easier multiplications were (1) A, (2) B, (3) A. The children had rewritten the second set of exercises to read (a) $9 \times (2 \times 5)$, (b) $7 \times (5 \times 4)$; (c) $2 \times (9 \times 6)$, and (d) $3 \times (9 \times 5)$. Several other examples were worked to verify the idea that the order of grouping the factors does not affect the product.

Another use of the associative property can be made in conjunction with renaming. If pupils forget a multiplication combination such as $7 \times 8 = \square$, they may rename 8 as (4×2) and then use the associative property. Thus, $7 \times 8 = 7 \times (4 \times 2) = (7 \times 4) \times 2$. After pupils have a good deal of experience using the associative property, they can generalize the property by using frames $[(\triangle \times \bigcirc) \times \square = \triangle \times (\bigcirc \times \square)]$ and give the property a name.

STUDY OF MULTIPLICATION FACTS

When students have had a wide variety of experience in solving problems and exercises dealing with multiplication, the study of the facts for mastery may begin. The NCTM Standards suggests exploration of easy facts in kindergarten and the first grade, study toward efficient means of finding facts in the second grade, and study of easy facts for mastery in the third and fourth grades.

Exploration of hard facts begins in the second grade, study toward efficient solution in the third grade, and mastery in the fourth grade.

To study for mastery, the children need to have an idea of the facts they know and the facts needed to learn. The teacher may begin by handing out a photocopied sheet such as the one that follows:

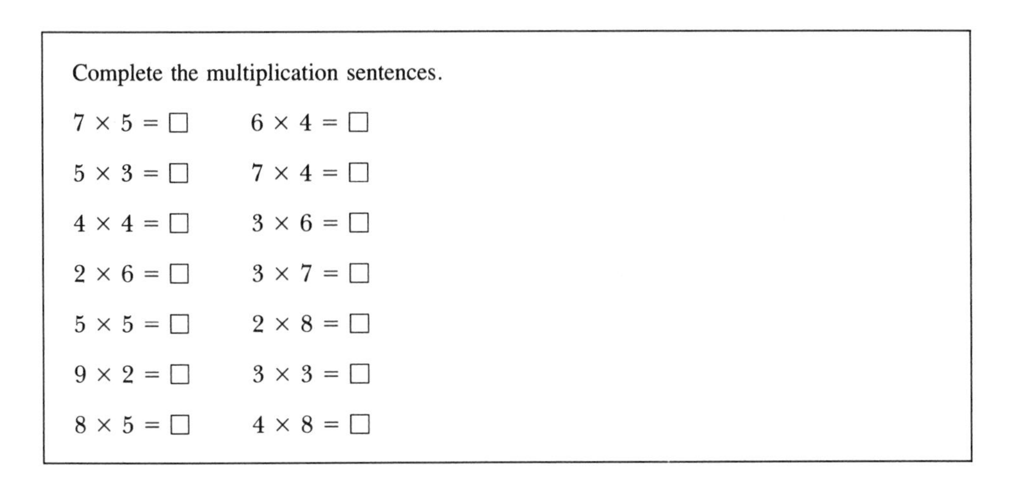

Complete the multiplication sentences.

$7 \times 5 = \square$ $6 \times 4 = \square$

$5 \times 3 = \square$ $7 \times 4 = \square$

$4 \times 4 = \square$ $3 \times 6 = \square$

$2 \times 6 = \square$ $3 \times 7 = \square$

$5 \times 5 = \square$ $2 \times 8 = \square$

$9 \times 2 = \square$ $3 \times 3 = \square$

$8 \times 5 = \square$ $4 \times 8 = \square$

The pupils may then be directed to answer as many questions as they can in 90 seconds. This procedure points out that even though all the pupils can arrive at an answer to a multiplication question if given time to use addition, an array, or counting, they may not be able to quickly answer multiplication questions. Thus, the time test is a means of showing them that they need some direct practice on the multiplication combinations.

MULTIPLICATION TABLE

One teacher developed the multiplication table in the following manner: "When I told my grandfather that we were studying the basic multiplication facts, he suggested that I use the multiplication table. He started to fill it in, but I thought you would be able to do it by yourselves." The teacher then handed out a sheet similar to the one that follows.

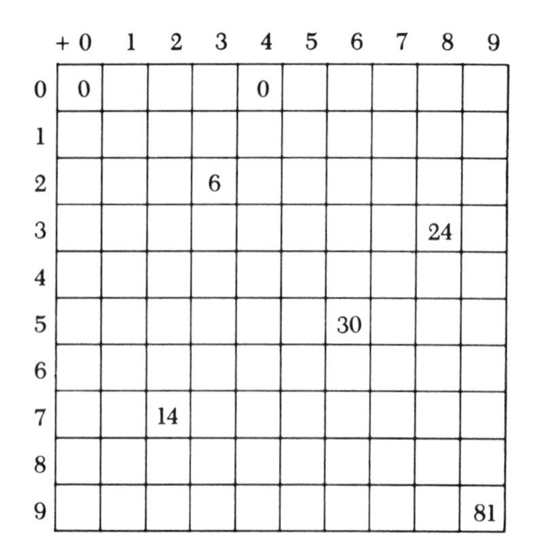

+	0	1	2	3	4	5	6	7	8	9
0	0				0					
1										
2				6						
3									24	
4										
5							30			
6										
7			14							
8										
9										81

The multiplication table may be used to study various structural patterns. The procedure of having children work independently for a time and then in groups of two or three usually proves to be effective for a laboratory lesson that simply says, "Find all the patterns you can on the completed multiplication table. Make a record of the patterns on paper."

Questions such as those that follow can be used after the children have had an opportunity to search for the patterns in the table. These questions help students to see relationships as well as to master the multiplication combinations:

1. Look at the nines column. What do you notice about the digits?
2. Draw a line from the upper left-hand corner of the table (the corner where the multiplication sign is located) to the corner containing the 81. What do you notice? (This procedure is helpful in emphasizing the commutative principle.)
3. Use the table to complete these multiplications:

   3×2 then multiply this by 8.
   8×2 then multiply this by 3.
   What do you notice?
   2×4 then multiply this by 7.
   4×7 then multiply this by 2.
   7×2 then multiply this by 4.
   What do you notice?

This procedure is helpful in emphasizing the associative and commutative properties of multiplication.

GAMES

A variety of games can be used to develop pupil facility with the basic multiplication facts. Variations on the dice-generated bingo-type game can be used. For example, three or four children are each given the card shown below. In turn, each player rolls the dice, finds the product of the two numbers, and covers this product on his or her card. A player can score only when he or she rolls the dice. (A multiplication table can be used to settle arguments as to the correctness of the product.) The first child to cover four products in a row, up or down or diagonally, wins. The object can also be to fill the card.

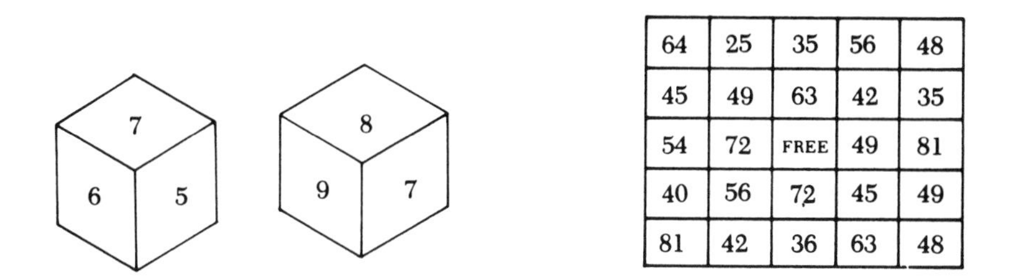

64	25	35	56	48
45	49	63	42	35
54	72	FREE	49	81
40	56	72	45	49
81	42	36	63	48

OTHER FUNCTION GRAPHS

A teacher gave the children the "times 3" graph shown on page 190, and then other "times" graphs were developed.

GIVEN: the times 3 graph.
DRAW: the times 5, 7, 8, and 9 graphs. What patterns do you find?

Multiplication used in combination with addition and subtraction can form the basis of number-line jumping activities. A series of function cards such as the two shown can be prepared. The children then roll a numbered die and use that number to replace the second box.

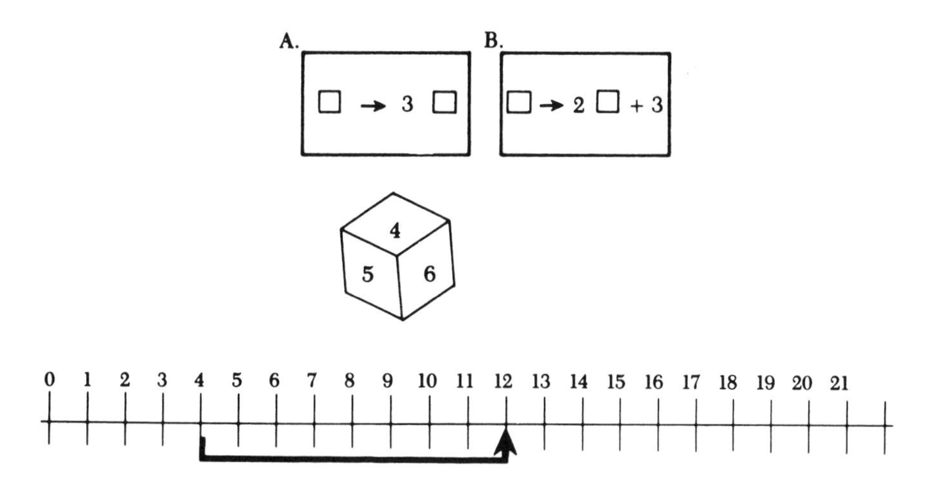

Using card A, if a 4 was rolled, then 4 would be mapped to 12.

After further experimentation, the children will be able to complete a times 3 mapping diagram as shown, which represents a magnification or stretch of 3.

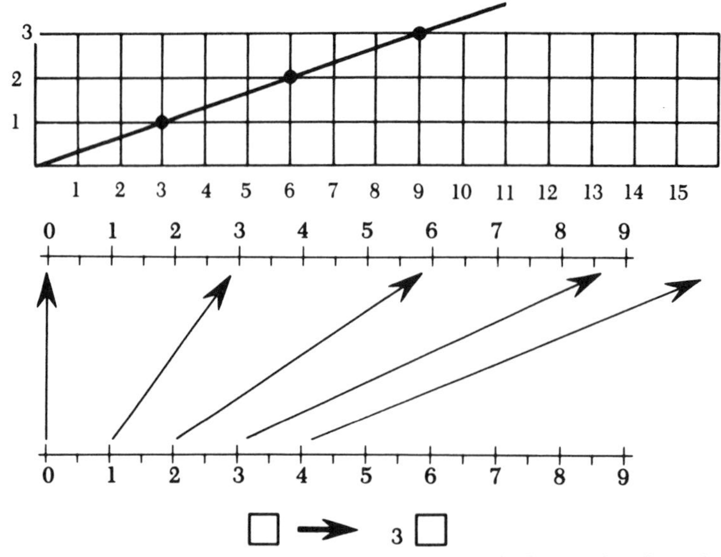

Exercises such as the one shown below reinforce the function idea as well as providing a change from the typical textbook practice page.

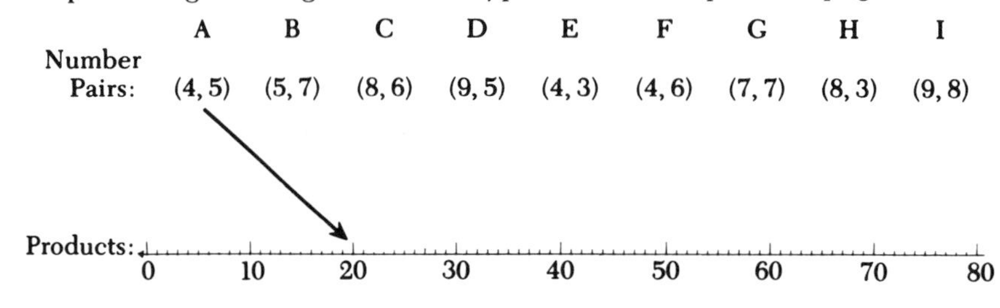

CROSS-NUMBER PUZZLES

Cross-number puzzles of the variety that follows below serve as challenging practice material. Not only do they help to fix multiplication combinations, they also further the understanding of basic multiplication principles.

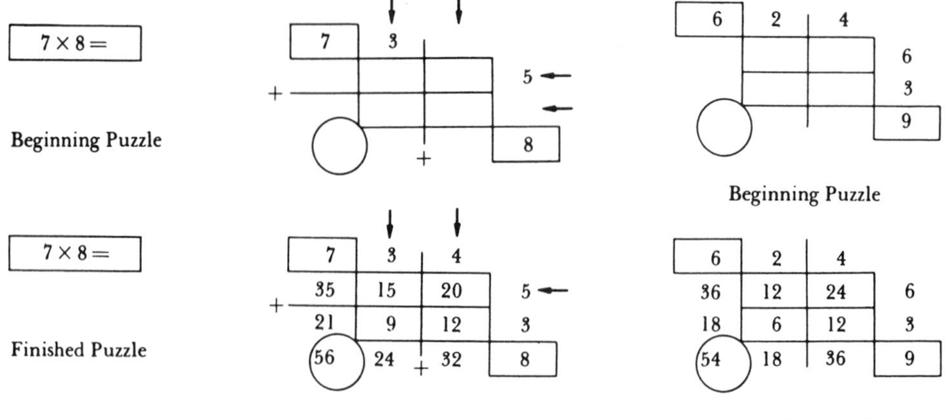

COMPUTER-ASSISTED INSTRUCTION PROGRAMS

There are a number of inexpensive drill-and-practice programs involving multiplication facts available for all of the microcomputers used in elementary schools or at home. Teachers and parents find these practice games helpful in mastering multiplication combinations.

Multidigit Multiplication

INTRODUCTION

The teacher began the study of multidigit multiplication with a problem sheet containing a number of problems similar to, "How many wildlife stamps will Jane need to fill her book if there are three empty pages and each page holds 30 stamps?" After setting up a laboratory supplies table containing squared paper, duplicated number lines, and place-value frames, the teacher directed the children to think of the mathematical sentence for each problem and then independently solve the problems in several ways. The teacher's final comment to the group was, "While there are seven problems on the sheet, I'd rather have you solve three or four of them in several different ways than work them all by using only one method."

The responses shown below were given by the children:

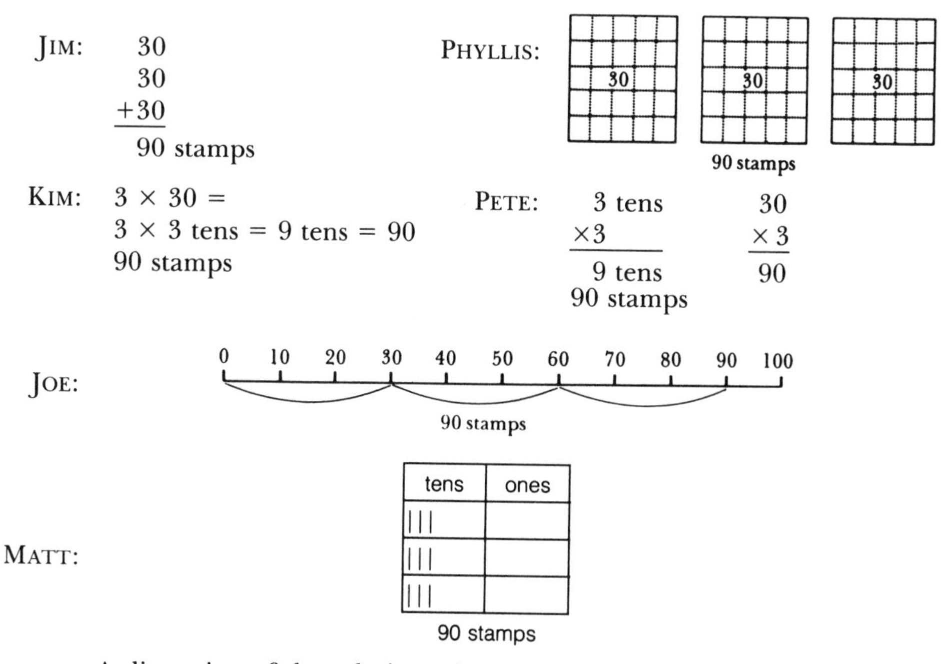

A discussion of the solutions should lead class members to formulate a generalization such as, "To multiply tens, you multiply the number of tens that you have (30 = 3 tens) by the multiplier and then you write a zero in the ones place to show that your product is tens." (3 × 4 tens = 120)

The next logical step is the multiplication of a two-digit number by a one-digit number in a situation that does not involve renaming (carrying). Some approaches that children use follow:

TED:

$$\begin{array}{r} 23 \\ 23 \\ +23 \\ \hline 69 \end{array}$$

JIM:

tens	ones
ll	lll
ll	lll
ll	lll

MAY:

$$\begin{array}{rr} & 2 \text{ tens } 3 \text{ ones} \\ \times & 3 \\ \hline & 6 \text{ tens } 9 \text{ ones} \\ & 69 \end{array}$$

Twenty-three is 2 tens and 3 ones. Three groups of 3 ones equals 9, and 3 groups of 2 tens equals 6 tens, or 60. Adding these, I got an answer of 69.

FRED: I used a squared-paper array.

JOHN: I used the number line.

HELEN: I made use of the "multiplying twice" (distributive principle) that we have used to check multiplication combinations.

$$3 \times 23 = 3 \times (20 + 3) = (3 \times 20) + (3 \times 3) = 60 + 9 = 69$$

or

$$\begin{array}{r} 23 \\ \times\ 3 \\ \hline \end{array} \qquad \begin{array}{r} 20 + 3 \\ \times \qquad 3 \\ \hline 60 + 9 \end{array} \qquad 60 + 9 = 69$$

After several days of experimental work, class discussion of the various approaches verified that the "multiplying twice" approach was the most effective, particularly when the computational shortcut was used. This shortcut was stated as

$$\begin{array}{r} 32 \\ \times\ 3 \\ \hline 6 \\ 90 \\ \hline 96 \end{array}$$

or "Think 30 + 2. Multiply 3 times 2; write the 6 in the ones place of the product. Think 3 times 30 equals 90; write 9 in the tens place of the product."

However, children were urged to experiment with various approaches.

RENAMING

Renaming is one of the more important ideas whose understanding is necessary to success with multiplication. The idea of renaming was introduced in one classroom with the following problem: "The Girl Scout troup asked Alice to bake eight dozen cookies. Remember, there are 12 cookies in one dozen. How many cookies will she need to bake? Think of the mathematical sentence and solve the problem in as many ways as you can. Then work in groups of three to figure out other solutions to the problem." The pupils later reported the following means of working the problem:

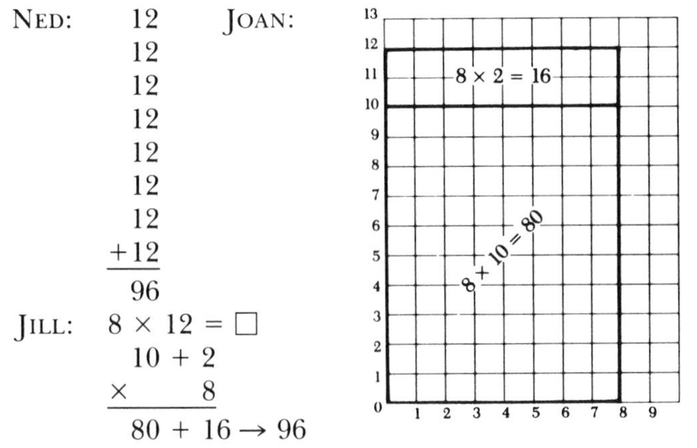

SALLY: $8 \times 12 = 8 \times (10 + 2) = (8 \times 10 + (8 \times 2) = 80 + 16 = 96$

CRAIG: You could use a number line, but it takes too long.

JEFF: Draw an 8-by-12 array. Then you can fold the array to make combinations that you know. For example:

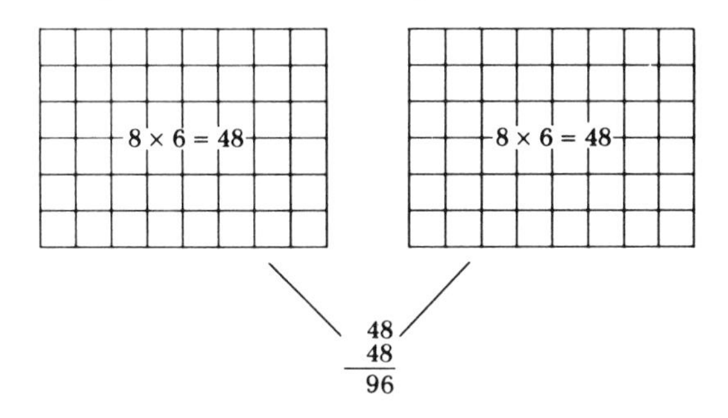

The class discussed the various approaches and decided that Sally's method and Jeff's method made use of the distributive property (studied earlier with single-digit work), as did Jill's. However, they felt that Jill's approach was easier. They also concluded that Joan and Jeff both used the distributive idea in marking off an array. Joan's method was preferred because of the ease with which they could multiply by tens.

LATTICE MULTIPLICATION

The next day, the teacher introduced the jalousie or lattice method as an alternate procedure. This is one of the older methods of multiplying and is proving to be an effective introductory procedure for most children and a usable terminal procedure for the slow learner.

The teacher used a pattern-searching strategy, beginning by drawing figure a on the board, and then writing in the product (figure b). Using a "silent strategy" the teacher drew figure c on the board and handed one of the children the chalk. The child filled in the product (figure d). The teacher continued this procedure for several other examples. When a child made a mistake, another child took the chalk and corrected it.

The teacher drew figure e and gave the chalk to Tom. The multiplication was completed (figure f). The teacher then wrote $6 \times 23 = \square$ and gave the chalk to Mary. Mary wrote 138 in the box. From the expressions on the faces of the children, the teacher gathered that most of them had discovered the relationship between "lattice" multiplication and other methods.

Then the children were given a worksheet with several examples to be completed by using lattices. The teacher moved about the room, using leading questions to guide those who were having difficulty with the lesson.

The next day, the teacher gave a worksheet containing several lattice multiplication examples to be completed. The children worked in groups of three with the direction, "Discuss the multiplication and be ready to tell the other groups the reasons that lattice multiplication 'works.'"

After the groups had time to think through the multiplication, they presented their findings.

KEN:

First we multiplied 8 times 3. That's 24. We put the 4 in the first box—that's ones, and the 2 in the second box—that's tens. Then we multiplied 8 times 2. That's, recall, 8 times 20. But we can just put the 16 (the product of 8×2) in the correct rows. The second row is tens, so we wrote in the 6 for 6 tens. Then we wrote the 1 in the last row, which is the hundreds row.

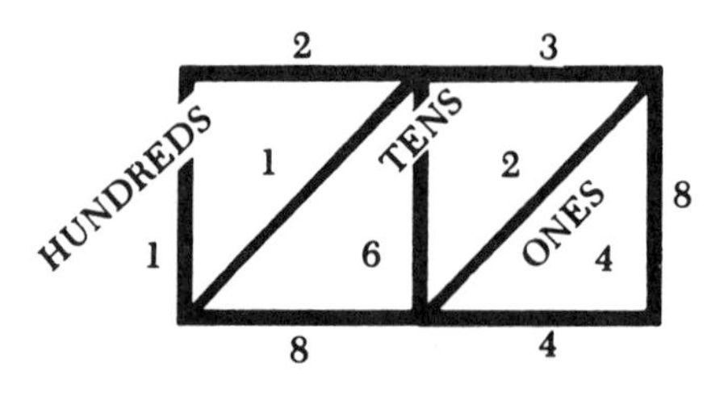

DAVE: We went through the same thinking as Ken's group. Then we also did some labeling (see the drawing I've made on the board). We also came up with these ideas that we think you can use in multiplication:

$$\text{ones} \times \text{ones} = \text{ones or ones and tens}$$

$$\text{ones} \times \text{tens} = \text{tens or tens and hundreds}$$

At this point, the teacher decided that a number of the children would benefit from continued use of some of the longer forms of multiplication or the lattice form. Those who had mastered the longer forms and lattice multiplication were directed to find shorter methods. Following some exploration time, the following forms were suggested:

$$\text{(a)}\quad \begin{array}{r} 10 + 2 \\ \times \quad 8 \\ \hline 16 \\ 80 \\ \hline 96 \end{array} \qquad \text{(b)}\quad \begin{array}{r} 12 \\ \times\ 8 \\ \hline 16 \\ 80 \\ \hline 96 \end{array} \qquad \text{(c)}\quad \begin{array}{r} ① \\ 12 \\ \times\ \ 8 \\ \hline 96 \end{array}$$

Class members could readily see the logic of forms (a) and (b) but questions were raised concerning form (c). To explain the procedure, one of the pupils said, "I multiplied $8 \times 2 = 16$; that is, 1 ten and 6 ones. I wrote the 6 in the ones place and the 10 above the tens place in the multiplicand. Then I multiplied 8×1 ten $= 80$, plus the 10 I had 'carried' equals 90. Then I wrote the 9 in the tens place of the answer." Several other examples were tested by using this method. The teacher suggested that for the remainder of the multiplication exercises, the class try the short form, checking it by one of the longer methods. Pupils having difficulty were told to use a form that they found understandable. Pupils were encouraged to remember rather than to write the numeral that was "carried."

After the pupils had had experience using the shortened form of multiplication, the teacher made use of "the arithmetic theme" to test their understanding of the procedure. Pupils who appeared to be using the "carrying" procedure mechanically were instructed to continue using one of the longer forms until they had the idea of the distributive principle firmly in mind.

TWO-DIGIT MULTIPLIERS

To introduce multiplications such as $23 \times 45 = \square$, the teacher gave each child the laboratory sheet on the next page.

The teacher suggested that in addition to grids, several other methods could be used to help find the answer. Many children had cut out the grids and folded them in different ways to find the answer. Several of the methods used are depicted on the following pages.

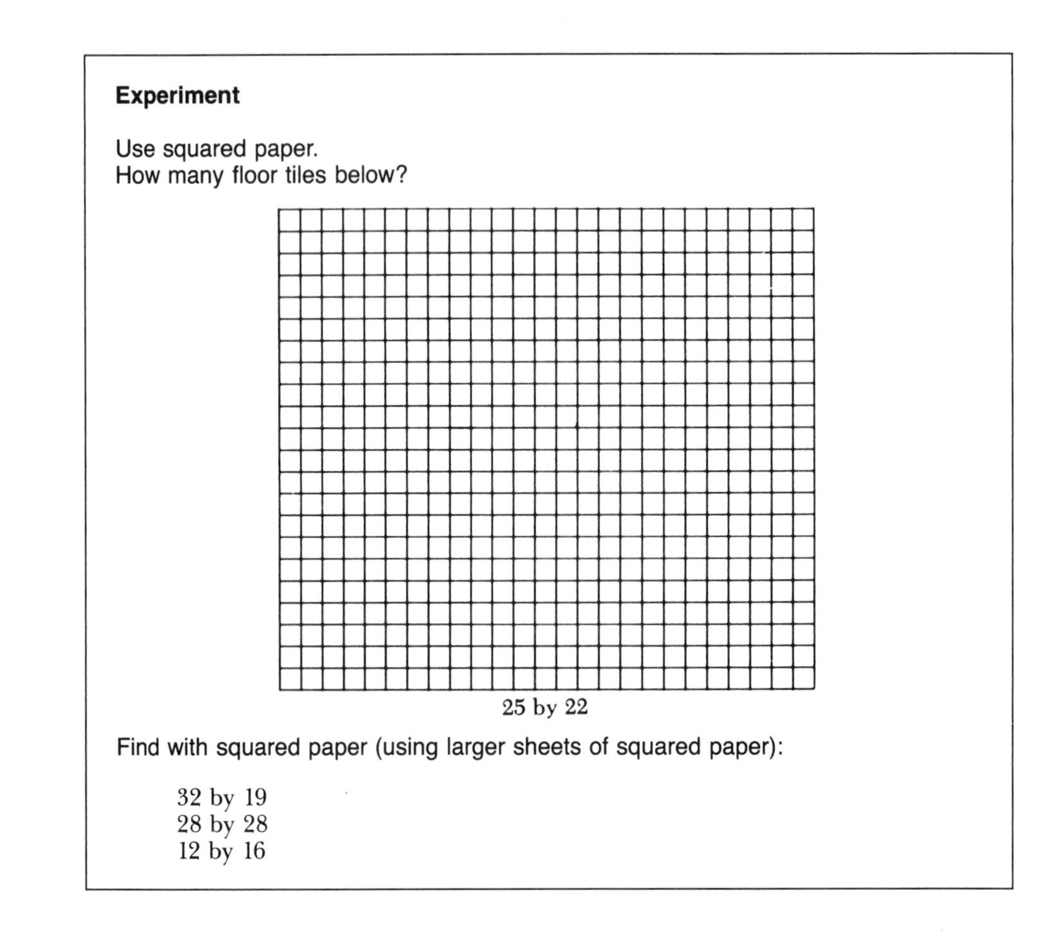

Experiment

Use squared paper.
How many floor tiles below?

25 by 22

Find with squared paper (using larger sheets of squared paper):

32 by 19
28 by 28
12 by 16

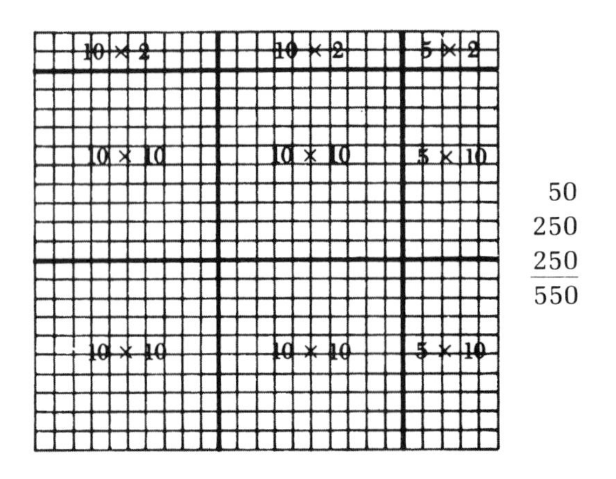

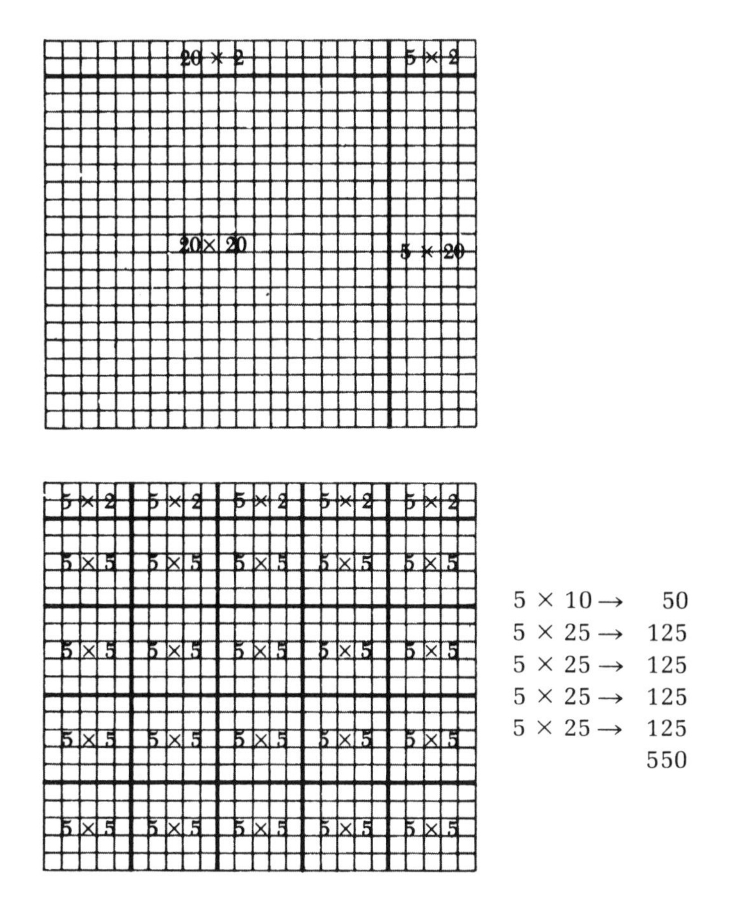

Others had remembered the use of the lattice and worked the problem in this manner:

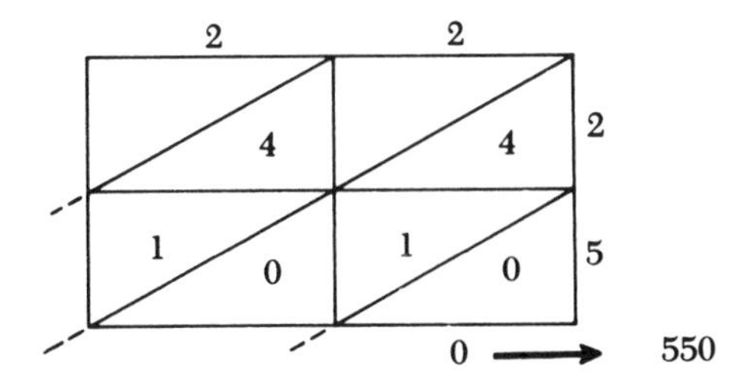

Still others had used a method close to the conventional algorithm:

$$
\begin{array}{rl}
22 & \\
\underline{25} & \\
110 & (5 \times 22) \\
\underline{440} & (2 \times 22) \\
550 &
\end{array}
$$

During the continued study of multiplication, the teacher allowed a variety of procedures. For the slow learner, lattice multiplication proved to be effective, since it requires only the use of basic facts and the lattice. The very slow learners made use of a multiplication table and the lattice. Such a procedure is quite valuable, for some children are never able to master the conventional multiplication algorithm. Later, many children moved to the standard procedure:

$$\begin{array}{r} 36 \\ \times\ 24 \\ \hline 144 \\ 72 \\ \hline 864 \end{array}$$

During the entire development, a great deal of stress was placed on place value and understanding of properties of multiplication. The questions and directions that follow are typical of those used to promote understanding.

REINTRODUCING MULTIPLICATION

As was true of addition and subtraction, there is a need to review multiplication that has been studied at an earlier grade level. Many mathematics programs present this reintroduction or review in a manner that is very similar to its original presentation. Thus, many students are not challenged, and even those who have a very poor grasp of the material do not vigorously attack the topic because they feel, "We're just doing the same old thing that we learned in an earlier grade." The suggestions that follow are designed to provide a novel setting for reintroducing topics in multiplication.

SIDE TRIP

Finger Multiplication

Even today, some people in central France do not learn the multiplication table above 5. If they wish to multiply 9×8, they bend down four fingers on the left hand (4 is the difference between 9 and 5) and three fingers on the right hand ($8 - 5 = 3$). The number of bent fingers gives the tens of the result ($4 + 3 = 7$ tens). The product of the unbent fingers gives the units ($1 \times 2 = 2$).

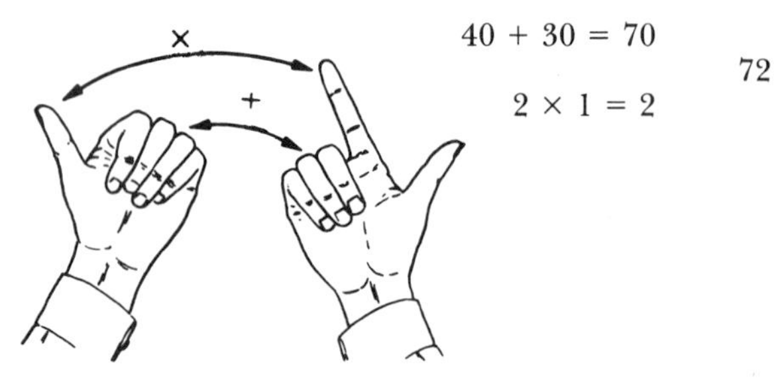

1. Try 8 × 7; 9 × 8.
2. What happens when you multiply 6 × 6?

A similar procedure for finger multiplication is to bend the fingers that equal the difference between the number multiplied and 10.

Thus, to solve 8 × 8 = □, the thinking is 10 − 8 = 2; bend down two fingers; 10 − 8 = 2; bend down two fingers; the standing fingers have a value of 10 each, and the bent fingers are multiplied.

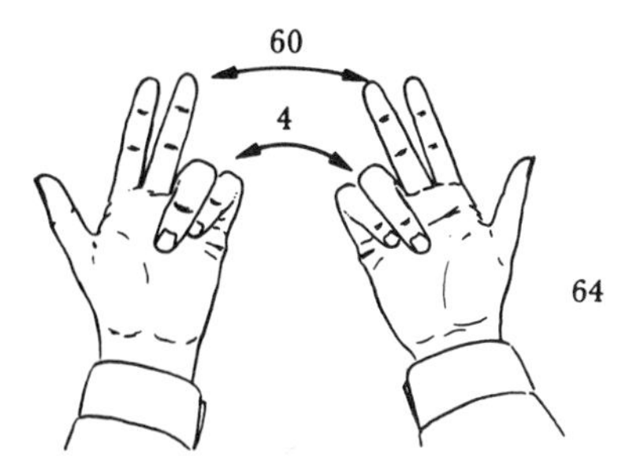

Flow-charting (grade 5 and above). As in the case of addition and subtraction, flow-charting helps children to think through the algorithms of multiplication. A laboratory lesson in which groups of two children work to develop a flow chart and then compare their chart with others provides good motivation. Also, the more able children may attempt flow charts, lattice multiplication, and other alternative algorithms.

Duplication (grade 6). The teacher stated, "Early in the Middle Ages, multiplication exercises were worked by doubling. Study the examples on the board. Why does it work?"[2]

31 × 42 =	1 × 42 =	42	25 × 36 =	°1 × 36 =	36
	2 × 42 =	84		2 × 36 =	72
	4 × 42 =	168	use a combination	4 × 36 =	144
	8 × 42 =	336	of numbers →	°8 × 36 =	288
	16 × 42 =	672	to get a	°16 × 36 =	576
	31	1302	sum of 25.	31	900
				°(Sums to 25)	

"Work the multiplication exercises in your book and check by the duplication method."

Approximation (grade 5 or 6). The teacher used an approximation game from the text.[3]

[2]D. E. Smith, *History of Mathematics* (New York: Dover 1953), pp. 114–16.
[3]Stephen S. Willoughby et al., *Real Math, Grade* Six (LaSalle, Ill.: Open Court Publishing, 1987), p. 64

APPROXIMATION GAME

Players: 3 or more
Materials: 1 calculator for the lead player
Object: To get the most points by making close approximations

Rules

1. Make a game form like this:

Round	Approximation	Point for Correct First Digit	Points for Correct Number of Digits	Score for Round
1				
2				

2. Decide how many rounds will be played. List them on the game form and add a space at the bottom for the total.
3. The lead player writes a problem on the board (for example, 73 × 59) and uses the calculator to find the answer.
4. Each player writes an approximate answer on the game form. Do not make any calculations.
5. The lead player writes the correct answer on the board, saying the first digit and the number of digits in the answer.
6. Look at your approximation and score yourself as follows: 1 point for the correct first digit and 2 points for the correct number of digits. Record your points on your game form.

If your approximation was:	And the correct answer is:	Then you would score:
3500	3652	3 points
50,000	44,370	2 points
950	9000	1 point

7. The player with the highest total score at the end of the game is the winner.

ESTIMATING AND CALCULATOR ACTIVITIES

A fourth-grade teacher asked the class to quickly guess this product, 38 × 52, and to write their guesses on the 3 × 5 cards given to them and to hold them up. Then the students were asked to check their answers with a calculator. Here are some of the guesses:

JAN: I though 40 × 50 = 200. It should have been 2,000, but I needed to remember that tens × tens can be hundreds and thousands.

JERI: I thought 40 × 50 = 2000. The exact product was 1976. I had a good estimate.

The teacher suggested using a calculator to see what two-digit numbers gave products in the thousands and hundreds and which were not as large as 1000. The children found that 31 × 31 was still in the hundreds while 32 × 32 moved over 1000. The children continued experimenting, finding 12 × 80 = hundreds; 12 × 85 = over a thousand.

There are literally hundreds of excellent calculator-estimation-multiplication exercises that you and the children can invent.

Napier's Bones (grades 5 and 6). John Napier (1550–1617) is famous for his invention of logarithms; however, he also was interested in calculating devices. Napier, building on the gelosia method of multiplication, made a set of rods to use in multiplying (sometimes called "bones" because the rods were made from horn or ivory). Each rod had the first nine multiples of a number set in the boxes used in gelosia multiplication. To multiply, select the rod of each digit of the number to be multiplied. It was very easy to multiply large numbers by a one-digit number by using his rods (figure a). To multiply a number by a multi-digit number, multiply as before, going digit by digit in the second number and then add up the partial products (figure b).

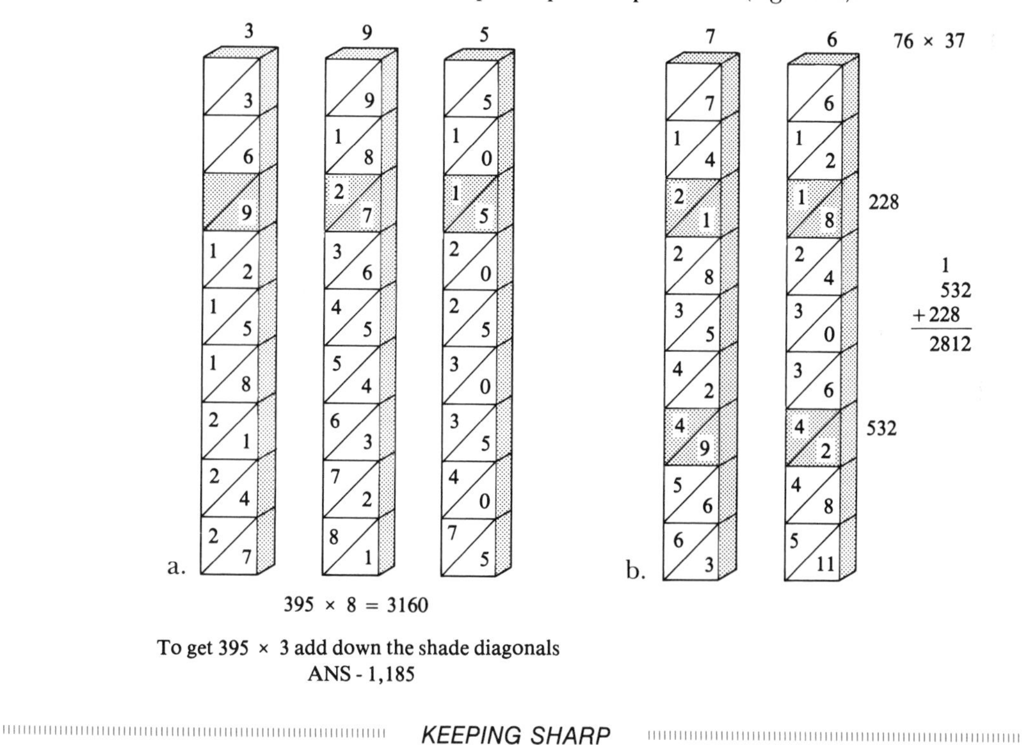

KEEPING SHARP

Self-Test: True/False

______ 1. A mature understanding of the operation of multiplication involves familiarity with different multiplication ideas.

_______ 2. Children who have developed the concepts of addition will not need further laboratory activities to understand multiplication.

_______ 3. Presenting oral problems using familiar situations helps very young children understand the uses of multiplication.

_______ 4. Squared paper can successfully be used to develop multiplication ideas.

_______ 5. Multiplication is commutative.

_______ 6. Primary-grade children can be taught only the easier multiplication combinations.

_______ 7. Children should be encouraged to memorize the basic combinations from the beginning of their work on multiplication.

Vocabulary

product
function
Cartesian products
distributive property
lattice multiplication
associative property
commutative property

THINK ABOUT

1. Showing that answers to multiplication problems are correct is emphasized in the procedures advocated in this course. What are the purposes of showing that answers are correct?
2. What are four good ways of verifying answers to multiplication questions that a pupil just beginning the study of the process might use?
3. How do basic product cards differ from basic sums cards?
4. If "multiply twice and add partial products" is used with $8 \times 3 = 24$, how many possible solutions would there be?
5. When a multiplier ends in zero, what are the objections to and advantages of setting the zero to the right of the ones position of the multiplicand?
6. What is the best transition method or premultiplication experience to be offered?
7. How could you explain two-digit multiplication and borrowing? Could you explain permutation and not give the real name for it?
8. How could you make it clear to a child that 3×5 is the same as 5×3?
9. Would there be any recommended steps in effective problem solving specifically designed for the multiplication process?

SUGGESTED REFERENCES

ALGER, LOUSIA R., "Finger Multiplication," *The Arithmetic Teacher* 15 (April 1968): 341–43.

ANGHLLERI, JULIE, and DAVID C. JOHNSON, "Arithmetic Operations with Whole Numbers: Multiplication and Division." In *Teaching Mathematics in Grades K–8: Research-Based Methods*, Thomas R. Post (ed.), Boston, Mass.: Allyn and Bacon, 1988.

BITTER, GARY G., MARY M. HATFIELD, and NANCY T. EDWARDS, *Mathematics Methods for the Elementary and Middle School: A Comprehensive Approach,* chap. 7 and 8. Boston, Mass.: Allyn and Bacon, 1989.

BROWN, STEPHEN I., "A New Multiplication Algorithm: On the Complexity of Simplicity," *The Arithmetic Teacher* 22 (November 1975): 546–54.

CLEMENTS, DOUGLAS H., *Computers in Elementary Mathematics Education,* chap. 7. Englewood Cliffs, N.J.: Prentice Hall, 1989.

COBURN, TORRENCE, *How to Teach Mathematics Using a Calculator,* Reston, Va.: National Council of Teachers of Mathematics, 1987. (contains exercises for use with children)

HAZEKAMP, DONALD W., "Components of Mental Multiplying." In *Estimation and Mental Computation, 1986 Yearbook,* Reston, Va.: National Council of Teachers of Mathematics, 1986.

KILIAN, LAWRENCE, EDNA CAHILL, CAROLANN RYAN, DEBORAH RAYAN, and DIANE TACETTO, "Errors that Are Common in Multiplication," *The Arithmetic Teacher* 27 (January 1980): 22–25.

LITWILLER, BONNIE H., and DAVID R. DUNCAN, *Activities for Maintenance of Computational Skills and Discovery of Patterns,* Reston, Va.: National Council of Teachers of Mathematics, 1980. (contains exercises for use with children)

KATTERNS, R., and KEN CARR, "Talking with Young Children about Multiplication," *The Arithmetic Teacher* 33 (April 1986): 18–21.

MORRIS, JANET, *How to Develop Problem Solving Using a Calculator,* Reston, Va.: National Council of Teachers of Mathematics, 1981. (contains exercises for use with children)

THIESSEN, DIANE, MARGRET WILD, DONALD PAIGE, and DIANE L. BAUM, *Elementary Mathematical Methods,* 3rd ed., chap. 9. New York: Macmillian, 1989.

chapter **9**

DIVISION OF WHOLE NUMBERS

OVERVIEW

Division is usually the most difficult whole-number operation for children to understand and use. It is also the most difficult for teachers to teach. This is not surprising, because division can be viewed in several ways. Also, besides being familiar with the various interpretations of division, children must be able to work effectively with addition, subtraction, and multiplication. To further complicate matters, for situations such as $122 \div 3 = n$, children must be able to guess (estimate) the quotient. For the other three whole-number operations (addition, subtraction, and multiplication), if children know the basic facts and understand the use of place value in relation to the operation, they can find an answer. This is not true for division. Division is also used much less in the daily life of the child than are addition, subtraction, and multiplication. Because of the difficulties inherent in teaching division, it is suggested that formal division be delayed until the child has a working grasp of addition, subtraction, and multiplication.

As in the other operations with whole numbers, the calculator should be a basic part of instruction. Because of the need to estimate the quotient in multidigit division, a greater use should be made of a calculator. Also, because calculators and computers de-emphasize computation, there should be an increased emphasis upon the student's discovery of mathematical principals with an emphasis on estimation.

Materials

Stirrers or Popsicle® sticks; squared paper.

1. Use sticks in bundles of 16 and 4, and singly, to help you solve these divisions:

$33_{four} \div 3_{four} = \square \qquad 123 \div 2 = \square$

$211 \div 3 = \square$

2. Use squared paper to solve these divisions (no fair dividing):

$105 \div 5 = \square \qquad 280 \div 12 = \square$

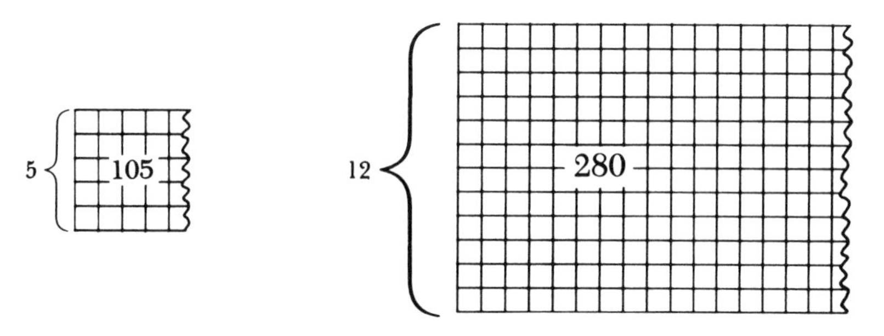

3. With two or three others, make a survey of uses of division by children and adults. What situations that require division occur most often?
4. Solve this riddle poem by Lewis Carroll. (Note the inserted clarifications.)

   Take three as the number to reason about,
   A convenient number to state.
   We add seven and ten [17], and then multiply out
   By one thousand diminished by eight [1,000 − 8].
   The result we proceed to divide, as you see,
   By nine hundred and ninety and two [992]!
   Then subtract seventeen, and the answer must be
   Exactly and perfectly true.

5. Take a length of ribbon (36 inches). Without using any mathematical operation, find out how long a piece each of six children would receive if they shared equally.
6. Take a two quart measuring cup and fill it with water. Pour the water in each of five glasses of the same size and shape. Guess how many ounces are in each glass. Check by pouring one back.

TAKE INVENTORY

1. How are measurement and partition division situations different?
2. Develop a division lesson by using a beam balance and a group-thinking teaching strategy.

3. Explain at least three ways of finding an answer to $289 \div 34 = \square$.
4. Teach a laboratory lesson on multidigit division.
5. Why is the "round-up" method of estimating the quotient better when using the "subtractive approach"?
6. Illustrate three ways of interpreting the "remainder" or "left-over part" in division.

WHAT IS DIVISION?

1. The division of whole numbers is the inverse of the multiplication of whole numbers. If a and c are identified in the expression $a \times b = c$ and if they uniquely determine b, the operation of "finding" b is called division. Thus, if $3 \times b = 12$, then,

$$\begin{array}{ccccc} \text{dividend} & & \text{divisor} & & \text{quotient} \\ 12 & \div & 3 & = & b \end{array}$$

However, if $12 \times b = 3$, then b must equal one-quarter, and thus the division operation cannot be performed by using whole numbers alone. The student must remember that any time two whole numbers are combined by the operation of multiplication, the result (the product) is a whole number. This is not true for division. Thus, for the set of whole numbers, division is not closed; that is, division of one whole number by another whole number does not always result in a whole-number quotient; it may produce a fraction.
2. The addition of whole numbers and the multiplication of whole numbers may be related, because the multiplication of whole numbers may be viewed as a special case of addition, with the addends being of equal size. Division may be related to subtraction, because the division of whole numbers may be considered as a series of subtractions in which the subtrahends are the same size. For example, $12 \div 4 = \square$ can be considered $12 - 4 = 8$, $8 - 4 = 4$, $4 - 4 = 0$; three subtractions have been made; therefore, $12 \div 4 = 3$. This thought process lends itself well to the question, "How many 4s equal 12?"
3. As in the cases of the other whole-number operations, division may be viewed as a function. If the first component is a multiple of the nonzero second component, division is a function that maps an ordered pair of numbers on its quotient. (*Note:* The first number of the ordered pair is the dividend and the second number the divisor.) To develop this mapping for all pairs of whole numbers, fractions are needed.

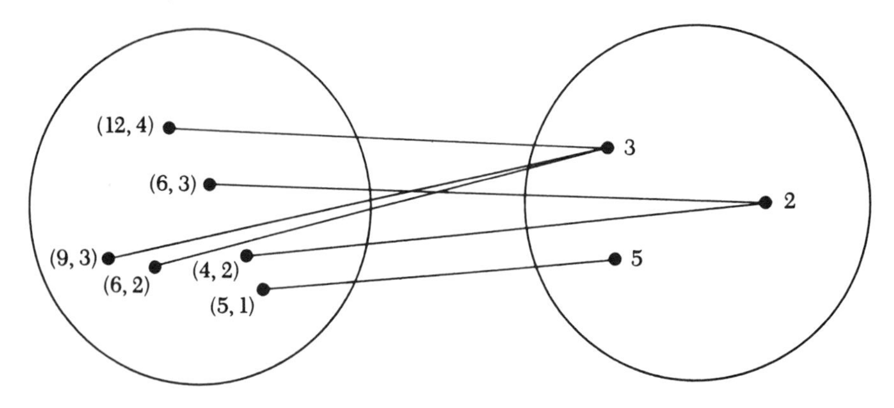

These interpretations are interrelated and should be developed from problems and laboratory activities that relate to the real world of the student.

There are basically two types of problem situations that lead to a solution by division: measurement and partition. The difference between the two can most easily be seen by analyzing two problem situations that make use of these concepts.

Measurement

For a clean-up campaign, a teacher said, "I'll give six pieces of candy to each child who brings in a basket of bottles until I have given away 24 pieces of candy." How many children will be able to win candy?

WAYS OF SOLVING

OBJECTS

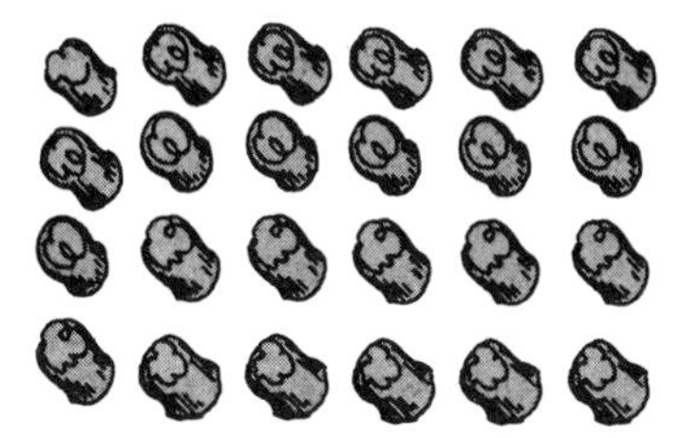

DRAWINGS

////// ////// ////// ////// 4

NUMBER LINE

1 2 3 4

0 2 4 6 8 10 12 14 16 18 20 22 24

SERIES OF SUBTRACTIONS

division
$24 \div 6 = 4$

$$\begin{array}{rl} 24 & \\ -\ 6 & 1 \\ \hline 18 & \\ -\ 6 & 2 \\ \hline 12 & \\ -\ 6 & 3 \\ \hline 6 & \\ -\ 6 & 4 \\ \hline 0 & \end{array}$$

TYPE I (MEASUREMENT)

1. We try to find "how many sets" when the number in each set is known.
2. The divisor and dividend are the same "material" (for example, pencils and pencils).
3. In terms of multiplication, we know the multiplicand. We also know the product. We are looking for the multiplier.

Partition

The teacher said, "I will give the first four of you that bring in a basket of bottles equal amounts of candy." If she has 24 pieces of candy, how many pieces will each child receive?

WAYS OF SOLVING

PASS OUT (LIKE DEALING CARDS)

Alice	*Brad*	*Chris*	*Dave*
//////	//////	//////	//////

Or associate a number with each piece of candy:

Alice	*Brad*	*Chris*	*Dave*
1	2	3	4
5	6	7	8
9	10	11	12
13	14	15	16
17	18	19	20
21	22	23	24

BY MULTIPLICATION

$$4 \times \square = 24$$

BY DIVISION

$$24 \div 6 = 4$$

TYPE II (PARTITION)

1. We try to find "how many in each set" when the number of sets is known.
2. The quotient and dividend are the same "material" (for example, pencils and pencils).
3. In terms of multiplication, we know the multiplier. We know the product. We are looking for the multiplicand.

The suggestions for teaching that follow include both measurement and partition ideas. When concrete materials are being used, both ideas are stressed. When development of basic division facts for mastery is pursued, partition situations are used, since they are more closely related to the multiplication solution of division, whereas measurement situations are more closely related to the series-of-subtractions idea.

FOUNDATION EXPERIENCES

Many elementary school mathematics programs do little to develop any understanding of division until the formal introduction, which usually occurs in third grade. This practice is to be questioned. An important yearbook of the National Council of Teachers of Mathematics stresses the idea that learning fundamental mathematical ideas is a continuous process and is facilitated by a continual development of a topic from grade to grade.[1] To implement this procedure, foundation work in division should begin in kindergarten and grade 1. However, although it is of great importance to develop readiness for division early in the primary grades, it is usually desirable to delay the formal study of it until the child has a sound grasp of multiplication.

Foundation work dealing with division can make use of laboratory experiences and orally presented work-problem situations that can be solved by the use of objects, number lines, sets, counting, subtraction, and addition. Many situations at the readiness stage can be developed through class consideration of problems; in other cases, small-group laboratories are appropriate. The situations that follow are illustrative of the types of materials that can be used in the foundation program.

Laboratory sheets containing exercises such as those below can be used. A single child or several children can work to find the solution. *Note:* The teacher must be sure that reading the material is not a problem. Thus, it is probably wise for the teacher to read the directions and the problems to the children.

1. Claudia is going to paste nine pictures on a sheet of construction paper. If she pastes three on each sheet, how many sheets of paper will she need? (The problem can be solved with pictures or drawings.)
2. Mother made 12 cookies, for three children to share. How many will each get? (The problem can be solved by taking 12 objects and "counting them out" to three pupils.)
3. The teacher can use problems such as, "How many children do we have in the class? Right, 30. How many cars will we need for our science trip to the weather station if we put six of you in each car?" (The problem can be solved by the class's counting off by sixes, by the class's forming in groups of six each, or by class considerations of the number line.)
4. Cups-and-beans activities can be used in situations such as that one cup plus five beans, three share (shown below); two cups plus four beans, each gets eight, how many share?[2]

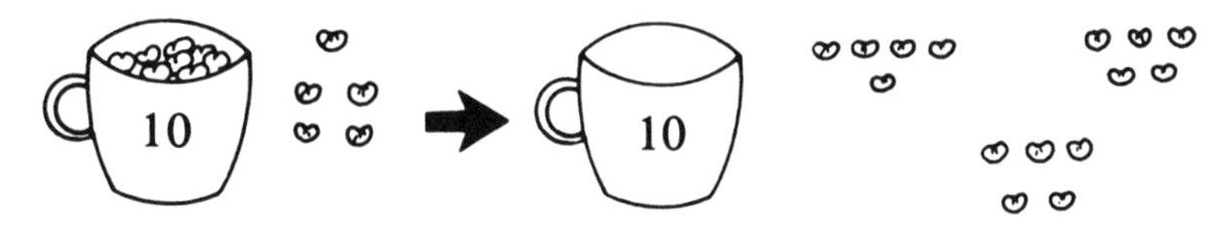

[1]*The Growth of Mathematical Ideas, Grades K–12.* Washington, D.C.: National Council of Teachers of Mathematics, 1959.

[2]See Mary Baratta-Lorton, *Mathematics Their Way* (Menlo Park, Calif.: Addison-Wesley, 1976) for readiness activities and early development activities for division.

DISCOVERING THE BASIC DIVISION FACTS

The teacher began the formal study of division[3] by giving each pupil a set of several measurement division problems with the following directions: "Study the problem. Try to determine the mathematical questions involved. Then find an answer to each problem. If you can, try to use different methods in solving each of the problems. I'll read the first problem and you can follow along. 'Jim raised Indian corn to sell for fall decorations. If he tied four ears together, how many bunches could he make from 12 ears of corn?'"

The teacher directed the attention of the children to the laboratory kits that could be used to help in solving the problems. These "kit" envelopes contained counting disks, squared paper, and number lines. Each child also had available a beam balance. When all class members had finished at least one problem, some of the pupils were asked to put their solutions on the board and to explain how they had worked the problems.

VICKI: I used Popsicle® sticks and counted off by fours.

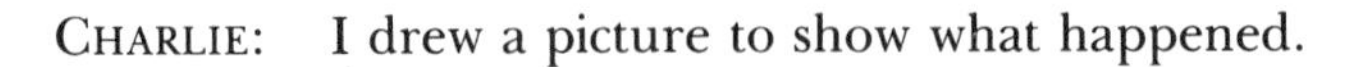

3 bunches

CHARLIE: I drew a picture to show what happened.

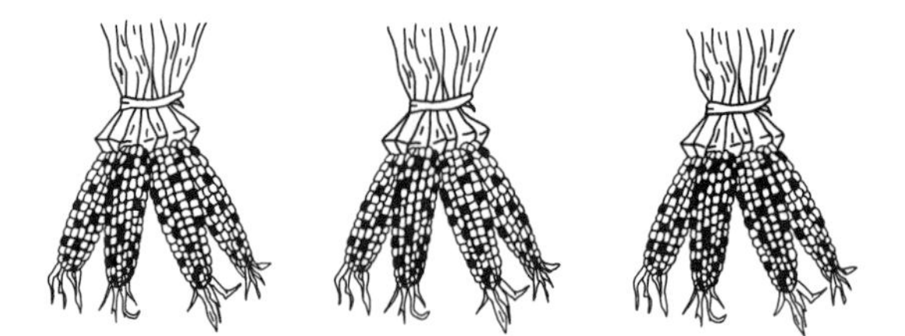

CANDY: I used squared paper. I cut off a 12 strip, then I cut it into strips of four.

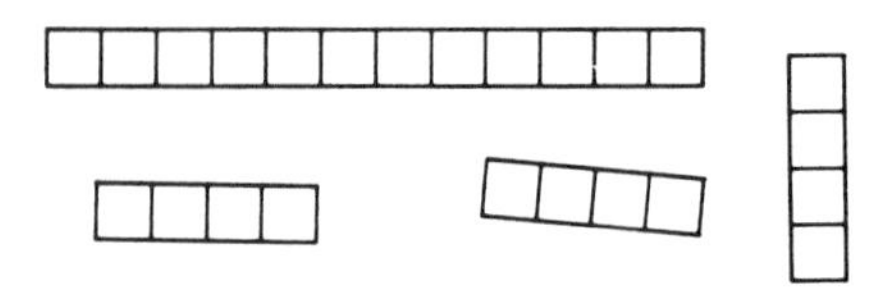

LUCY: I drew an array like we use in multiplication. I drew four dots. Then four more, then four more—12 dots in all.

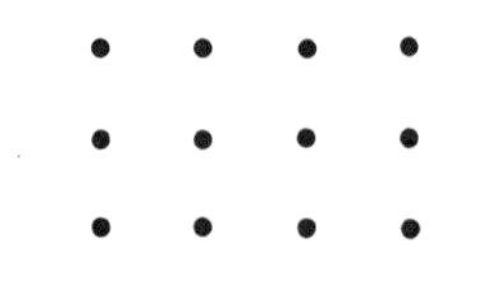

[3]As has been suggested, the formal study of division should be delayed until there has been extensive work on the multiplication facts.

LINUS: I used the number line. Also, you could start at 12 and move back by fours.

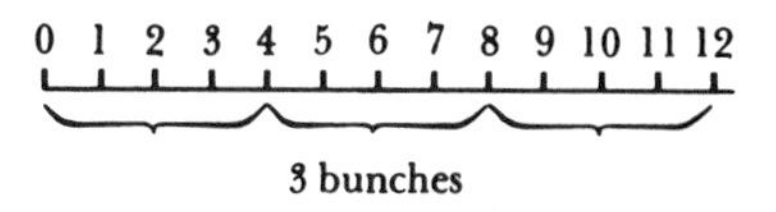

JANE: I used a balance. I put a ten weight and 2 one weights on one side. Then I experimented to find the number of four weights needed to balance. It took three.

PETE: I started with the 12 ears and subtracted four for each bunch. I had to make three subtractions to get all of the ears in bunches.

$$\begin{array}{rl} 12 & \\ -\ 4 & (1) \\ \hline 8 & \\ -\ 4 & (2) \\ \hline 4 & \\ -\ 4 & (3)\ 3\ \text{bunches} \\ \hline 0 & \end{array}$$

JOE: I started with one bunch of four and then kept adding fours until I reached 12. I had three bunches.

$$\begin{array}{r} 4 \\ +\ 4 \\ \hline 8 \\ +\ 4 \\ \hline 12 \end{array}$$

After a discussion of the methods, the teacher said, "I asked you to think of the mathematical question asked by the problem. What is it?" The pupils responded that the question was, "How many fours equal 12?" "How many fours are contained in 12?" "How many fours are in 12?" (note that these closely tie division to multiplication), and "Twelve can be broken into how many sets of four?"

The teacher noted that all the mathematical questions were representative of the situation, then said, "Let's see if we can write a mathematical sentence that represents the problem." The sentence $\square \times 4 = 12$ was suggested to replace, "How many fours equal 12?" The teacher said, "We have written addition and subtraction sentences, and you just wrote a multiplication sentence for the problem. Is this the same type of situation we have usually found in multiplication?" The class agreed that it was different in that they were actually separating a set of 12 into three equal sets of four each.

Then the teacher said, "The problem situations we have been dealing with are called *division* problems. We can write the mathematical sentence as '12 divided by 4 equals 3.' Just as we have symbols to represent the operations of addition, subtraction, and multiplication, we have a division symbol. We can write" and she wrote on the board, "$12 \div 4 = 3$." Then the teacher wrote the division sentence $15 \div 5 = \square$ on the board, and asked the questions, "How would you read this mathematical sentence? What does it mean?"

The pupils replied that it would be read, "Fifteen divided by 5 equals what number?" and that it asked, "How many groups of 5 equal 15?"

During the rest of the period, the class solved verbal problems that were concerned with division situations. In each case, they wrote the mathematical sentence that represented the problem, solved the problem, and showed the answer to be correct by solving the problem by a different method.

Because the study of division was delayed until the children were achieving well in multiplication, and because most adults never use division facts—for example, in $48 \div 6$, the large majority of adults think, "What times 6 equals 48?" and recall the multiplication fact $6 \times 8 = 48$ rather than a division fact—the teacher next used a laboratory lesson that emphasized multiplication thinking. A portion of the sheet is shown below.

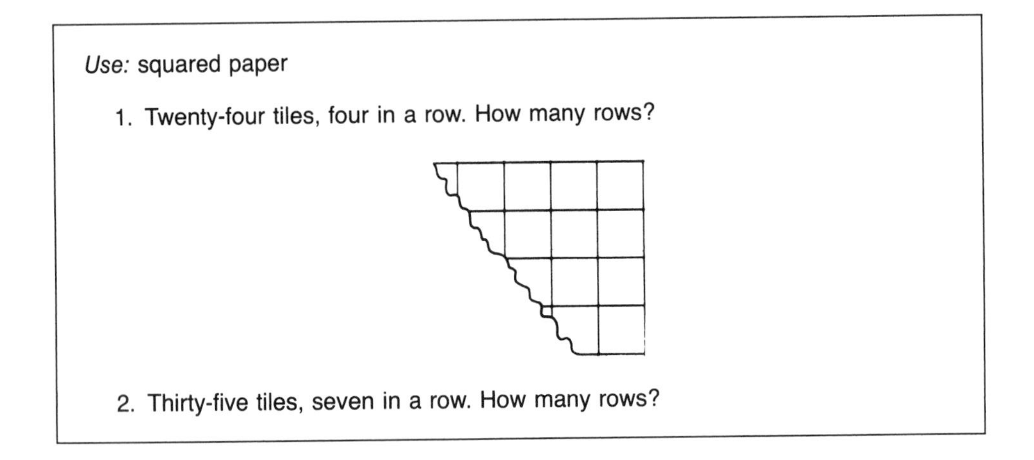
Use: squared paper

1. Twenty-four tiles, four in a row. How many rows?

2. Thirty-five tiles, seven in a row. How many rows?

The next day, the class discussed various procedures for solving mathematical sentences that involved division for the purpose of arriving at efficient procedures. The two most frequently suggested were to use a series of subtractions and to try to think of a related multiplication combination. The use of the related multiplication combination was explained in the following manner: "If I have a division question such as $15 \div 5 = N$, I can think '5 times what number equals 15, or $5 \times N = 15$.'"

In the days that followed, the class solved mathematical sentences that related multiplication and division and studied the relationship by using a multiplication table.

$$18 \div 3 = N \qquad 9 \times N = 27 \qquad N \times 5 = 15$$

$$9 \div 3 = N \qquad 10 \div 2 = N \qquad 4 \times N = 12$$

The computational forms of the division algorithm $4\overline{)12}$ were also introduced. The 12/4 form, although it is not always used in children's materials, merits consideration because it is useful in later fraction work and can be used effectively for more complex divisions.

$$\frac{12}{4} = \square$$

Such uses are described later.

The children were encouraged to think of the multiplication that would help them to solve problems and exercises. However, children who were still having trouble recalling multiplication facts made use of the subtraction forms shown below and the number line.

Example A records the subtraction above and makes some use of place-value aspects, which are helpful in later division work.

EXAMPLE A

```
   3
   -
   1
   1
   1
4)12
   4
   -
   8
   4
   -
   4
   4
   -
```

Example B handily records the count of the subtractions at the right of each.

EXAMPLE B

```
4)12
   4   1
   -
   8
   4   1
   -
   4
   4   1
   -   -
       3
```

On their own response to questions asked by the teacher, those pupils who had mastered multiplication facts soon discovered that it is not necessary to complete the entire series of subtractions. They noted that in situations such as $24 \div 4 = N$, procedures A_1 and A_2 are more efficient than B_1 and B_2.

A_1

```
4)24
  16  4
   8
   8  2
      6
```

A_2

```
   6
   2
   4
4)24
  16
   8
   8
```

B_1

```
4)24
   4  1
  20
   4  1
  16
   4  1
  12
   4  1
   8
   4  1
   4
   4  1
      6
```

B_2

```
     6
     1
     1
     1
     1
     1
     1
  4)24
     4
    20
     4
    16  8
     4  4
    12  4
     4  4
```

When the majority of the pupils were progressing well in division, the teacher introduced the partitioning concept by using problems such as, "Mark, Ken, Bill, and Gene plan to share equally 12 pieces of candy. How many pieces will each boy receive?"

The pupils thought of several procedures for solving the problems. Refer back to the measurement–partition discussion earlier in the chapter for representative solutions. The study continued with the use of measurement and partition problems along with computational situations.

When the teacher is confident that the pupils understand division, its terminology may be introduced. A logical way may be to develop these terms in relation to multiplication by means of related multiplication and division combinations. For example, how many sevens equal 14?

□	×	7	=	14
factor		factor		product
14	÷	7	=	□
dividend		divisor		quotient
(product)		(known factor)		(unknown factor)

Analysis

As in the case of the other basic whole-number operations, guided-discovery procedures were used to introduce the material. Little stress was placed on mastery of the combinations in the early stages. Since division involves many features that add difficulty, pupils should encounter a wide range of laboratory and problem situations. Also, since the division facts are seldom if ever used in isolation, the study of them should parallel the study of related multiplication facts. In fact, one current program makes little or no use of division facts, although it does stress a strong program.[4]

INEXACT DIVISION

Many physical-world situations arise in which the division is not exact. For example, "If four children are asked to share equally 10 cookies, how many cookies will each receive?" In that case, the children could each take two cookies and leave two cookies on the plate. This probably would not occur. The children probably would break the two remaining cookies in half and each eat two and a half cookies.

Inexactness in division is another stumbling block to understanding. Too often in the past, teachers have followed the procedure of having all leftover parts expressed as remainders in the early grades. In the later grades, the leftover part was always expressed as a fraction. An analysis of problem situations requiring division reveals that this procedure can be misleading. The leftover part should be expressed in keeping with the problem situation. The problems that follow require a specific decision concerning the leftover part.

1. "The pencils in a store were labeled '3 for 10¢.' How much would one pencil cost?" The answer is not 3¢ with a remainder of 1. Nor is it $3\frac{1}{3}$¢. Any time a purchase involves division, the "remainder" is rounded up. One pencil would cost 4¢.
2. "Mrs. Ronald buys 10 yards of curtain material on sale. How many curtains can she make from the piece of cloth if each curtain requires 3 yards of cloth?" In this setting, Mrs. Ronald will be able to make only three curtains, not three and a third. The answer could be expressed as three curtains with 1 yard of cloth left over.
3. "How long will each board be if I cut a 14-foot board into three parts? (Do not consider the saw cuts.)" The answer of 4 feet with a remainder of 2 feet is meaningless and therefore incorrect. In this case, the only logical answer is $4\frac{2}{3}$ feet, or 4 feet 8 inches. (*Note:* If the form $\frac{14}{3} \rightarrow \frac{12+2}{3} \rightarrow 4\frac{2}{3} \rightarrow 4\frac{2}{3}$ has been used, the leftover part naturally becomes a fraction, which can be read as an incomplete division (2 divided by 3) or as the fraction $\frac{2}{3}$.

These problems point up the difficulty of deciding what to do with the leftover part. The alert teacher will take advantage of problem situations such as these to discuss with the class the need for interpreting the leftover part in terms of the problem.

In many classes, the teacher may be able to use fractions to express leftover parts at a much earlier grade level than is common. Most third graders will give an answer of "two and a half" to the orally stated question, "If I divide five pieces of candy equally between two persons, how many pieces will each person receive?"

[4]Henry Van Engen, Project Director, *Patterns in Arithmetic* (Madison: Wisconsin Research and Development Center, 1971).

MULTIDIGIT DIVISION

Multidigit division occupies a good deal of time in the intermediate grades. It is difficult to ascertain the amount of stress that should be placed on this topic. A distinguished group of scientists and mathematicians has suggested a "downplaying" of the study of multidigit division:

> Most adults do not seem to use the long division algorithm very much. Moreover, the vast majority of interesting school units in science and mathematics do not require more than about two-figure accuracy in multiplication and division. The needed accuracy can be obtained by estimation, by rounding, by cruder and simpler algorithms, by use of arrays and grouping, or with the help of tables.... If the algorithms are to be learned, they should emerge from an investigation on the part of the students. Various useful forms they arrive at will suffice whether or not they are the most standard or the most compact.[5]

The teaching suggestions that follow are designed to provide children with basic competence in division through procedures designed to foster mathematical thinking.

Before undertaking an intensive study of multidigit division, the child should have

1. An understanding and mastery of multiplication and subtraction facts and the multiplication and subtraction algorithms.
2. An ability to supply missing factors for examples such as $3 \times \square = 27$.
3. Facility with estimations such as $6 \times \square = 8{,}248$ (either 10, 100, or 1000).
4. An understanding of measurement and partitive meanings of division.
5. Mastery of the basic ideas of place value through the thousands.[6]

Readiness with the Calculator

A variety of orally presented "guess and test" calculator activities can provide understanding and motivation when beginning multidigit division. One teacher made use of the following type of exercises in this introductory work:

1. "Guess how many twenty-fours in 350? Check with your calculator.

KIM: That's about 25; 4 twenty-fives per hundred; about 14.

2. "Guess 8500 divided by 385. Check with your calculator.

KENT: About 400 in 8500; 5 four hundreds equal 2000; 20 four hundreds in 8000; about 21.

3. "Guess what number should I multiply 35 by to get to 630? Check with your calculator."

[5]*Goals for the Correlation of Elementary Science and Mathematics,* the report of the Cambridge Conference on the Correlation of Science and Mathematics in the Schools; published for Education Development Center, Inc., by Houghton Mifflin, 1969.

[6]James W. Heddens and Beth Lazerick, "So 3 'Guzinta' 5 Once! So What!!" *The Arithmetic Teacher* 22 (November 1975): 576–78.

JO: Thirty-fives in 630; that's about 40 in 600; $10 \times 40 = 400$; $5 \times 40 = 200$; that's about 15. I checked with calculator and got 18 for an answer. Fifteen is a good estimate.

When using the calculator, an important consideration is the remainder. Any time you divide using a calculator and there is a remainder, it will be expressed as a decimal. One teacher introduced the concept by suggesting, "Divide 5 by 2 on your calculators. What do you get for an answer? Jill would you write it on the board?"

JILL: 2.5.
TEACHER: What does the 2.5 mean? What do you get for an answer when you divide with paper and pencil?
PETE: 2 with a remainder of 1. That's $2\frac{1}{2}$. .5 must be the same as $\frac{1}{2}$.

At this point the teacher introduced a decimal place-value chart, and, through group discussion, the children compared the tens and hundredth to tens of a dollar (dimes) and hundredths of a dollar (pennies). Use of experimentation helped the children to see the approximate value of each of the tenths numbers.

After the teacher felt the children had a grasp on the decimal leftover part (remainder), the teacher gave the children a number of multidigit numbers to divide. The children found that many of the answers had repeating decimals such as 5 divided by 3 = 1.666666. Experimentation and discussion helped the children discover that .6666... is $\frac{2}{3}$; .3333.. is $\frac{1}{3}$; .125 is $\frac{1}{8}$, and so on. Then the children were given the instructions, "Use a calculator to divide the number 12 by each of the numbers from 1 to 12 in turn. Record all your calculations and answers, and then discuss what these results mean." The children worked in groups of three and later shared with the entire class. Try this experiment yourself.

Teaching Multidigit Division

Multidigit division can be introduced through a situation involving a division such as $48 \div 4 = N$. A group-thinking lesson is described below.

The teacher began in the following manner: "A class made 48 artificial flowers to sell at the carnival. If they tie them in bunches of four, how many bunches will they have to sell? Write the mathematical sentence you can use to solve the problem and then solve the problem in as many ways as you can. You will see an array diagram of the flowers at the top of the duplicated sheet I've given each of you. This may help with the problem." After a few moments, the teacher said, "I see that most of you have identified the mathematical sentence. What is the mathematical sentence?—Yes, $48 \div 4 = N$ ($\square \times 4 = 48$). Some of you seem to be having trouble starting out on the problem. How can you always solve a division problem?—Right, use subtraction. Go ahead and try to find an answer." While the pupils worked, the teacher moved about the room asking questions and giving aid. As the students completed various solutions to the problem, the teacher asked class members to illustrate their solutions on large sheets of lined paper (24 inches × 30 inches) with felt-tip pens or crayons. When discussion began, the teacher had the pupils explain their methods of solving the division problem, beginning with students who used the less mature solutions and ending with those who used the more mature solutions. The children's solutions and explanations follow.[7]

SID: I used a number line made from squared paper. I cut out a 4 length and moved it along until I had counted the number of fours in 48.

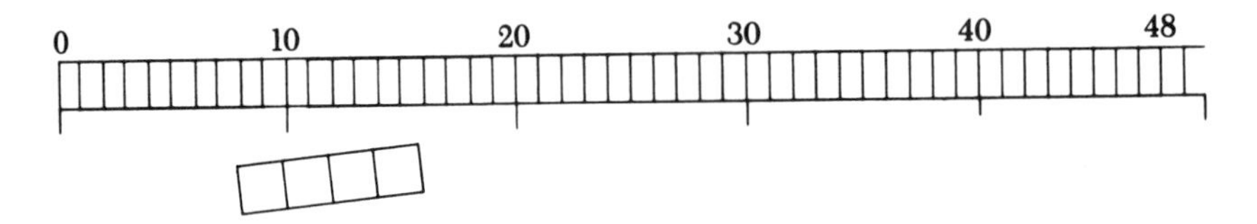

LYLE: I know that we can solve any division situation by a series of subtractions. I started to find "how many fours equal 48" by subtracting fours. Then I shortened it some by subtracting two groups of 4.

$$
\begin{array}{rr}
4\overline{)48} & \\
\underline{-\ 4} & 1 \\
44 & \\
\underline{-\ 4} & 1 \\
40 & \\
\underline{-\ 8} & 2 \\
32 & \\
\underline{-\ 8} & 2 \\
24 & \\
\underline{-\ 8} & 2 \\
16 & \\
\underline{-\ 8} & 2 \\
8 & \\
\underline{-\ 8} & \underline{2} \\
0 & 12
\end{array}
$$

JOAN: I studied the array patterns and then circled groups of 4 in the pattern.

[7]If in a lesson similar to this the children do not think of the variety of approaches suggested, it will probably be wise to use Socratic questioning to bring them out. As was previously mentioned, the knowledge that there are many ways of finding an answer usually provides children with greater confidence when they undertake a new task.

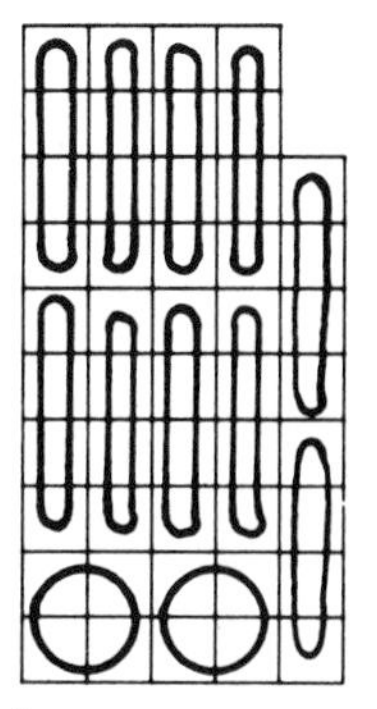

JEFF: I used the same method as Joan, but I cut a 4-strip pattern and fitted it on the squared-paper array.

MIKE: I also used the array pattern, but I shortened it some. I could see that by counting the row I could find the fours in 40. Then I partitioned the 8 into two groups of 4.

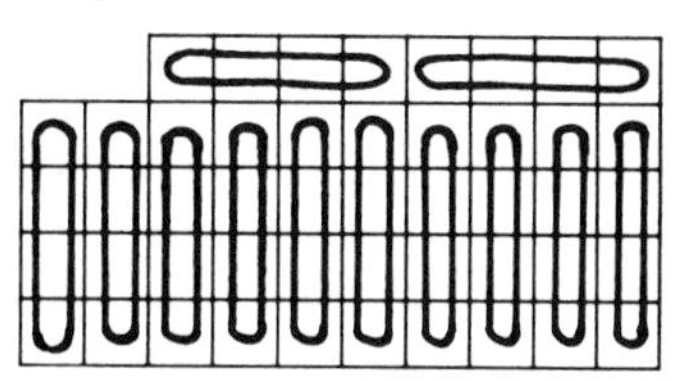

KEVIN: I thought, "My question is what times 4 equals 48." I did some thinking and decided that a good first guess was 10. I knew that $10 \times 4 = 40$. Then I just knew that there were two more fours in the 8. So my answer was 12.

$$\begin{array}{r@{}l} \square & \times 4 = 48 \\ 10 & \times 4 = 40 \\ \underline{2} & \times 4 = 8 \\ 12 & \end{array}$$

ANN: I used subtraction. I know five sets of four equal 20, so I recorded 5 and subtracted 20. I still had another group of 5 fours, so I subtracted another 20. Two fours equal 8, so I recorded the 2.

$$\begin{array}{r} \underline{12} \\ 5 \\ 5 \\ 4\overline{)48} \\ \underline{20} \\ 28 \\ \underline{20} \\ 8 \\ \underline{8} \end{array}$$

CHARLIE: I used the same procedure but a different form.

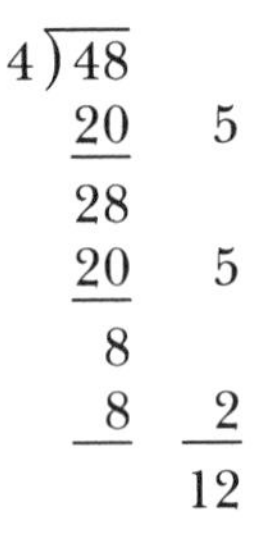

NANCY: I shortened the procedure they used. I know that 10 fours equal 40, so I subtracted 10 groups of 4. I also know that 2 fours equal 8.

```
   12
    2
   10
4)48  40
       8
       8
```

CLAUDIA: I wrote the division in a different way. Then I thought of another name for 48—40 + 8—and went ahead and divided. I think that is a good way of solving the problem.

$$\frac{48}{4} \rightarrow \frac{40 + 8}{4} \rightarrow \frac{40}{4} + \frac{8}{4} \rightarrow 10 + 2 = 12$$

The representative procedures used by the children illustrate the wide range of approaches that may be used in early division study. The approaches taken by Nancy and Claudia are quite mature in their development and might well have occurred later in the sequence of division work. The procedure suggested by Kevin merits consideration. The use of multiples of 10 in finding a quotient is usually quite productive, since children can usually multiply easily by 10. Also, his suggestion of thinking what number can be multiplied by 4 to arrive at 48 is illustrative of thinking that is necessary to mature computation with division.

During the next several days, the teacher continued the study of similar division problem situations. Often, children worked in groups of three. Normally, the task was to find as many ways as they possibly could to solve the division rather than working many division exercises in a single way.

Other procedures for solving division problems can be developed. Popsicle® sticks can be effectively used to solve partition division problems. For example, the teacher gave each child a number of bundles of 10 sticks and some single ones. They were given a laboratory-oriented sheet directing them to solve division computations by using the sticks. If they seemed to be having difficulty, the teacher suggested working in small groups or made use of a Socratic-questioning technique.

A portion of the laboratory sheet and two strategies for developing the division follow:

Experiment

Use: bundles of 10 Popsicle® sticks and single ones. Find the answers to these problems using sticks.

1. Their uncle gave Tom, Nancy, and Jill 36 stamps to share. How many stamps will each child get?
2. To make up laboratory kits, Joe has 44 pencils that he is to put in four boxes. How many pencils will he put in each box?

PROBLEM 1

KATE: I used 36 sticks to stand for the stamps. I thought, each child can have 10 sticks; I still have six left, I can give two to each one. My answer is 12 stamps for each child.

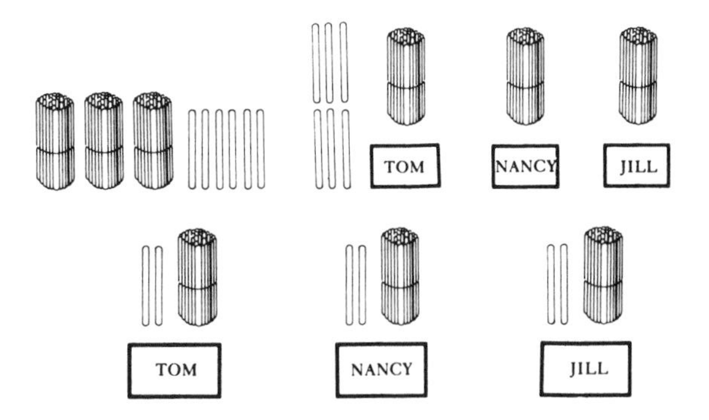

PROBLEM 2

TEACHER: How could we use the tens bundles and single sticks to solve the problem?

KEN: First we'd set out 4 tens and 4 ones for the 44 pencils.

LUCY: We'd then need to divide the sticks into four piles. We could unbundle the sticks and deal them out one at a time.

PETE: Wait. We could give each of the four boxes a bundle of 10.

TEACHER: Fine; then what?

MARY: We'd have four sticks left. We could give one stick to each box. Then there would be 11 sticks in each box.

TEACHER: Very good. Now let's see if we could write down what we did. [She wrote $\frac{44}{4}$ and $4\overline{)44}$ on the board.] Now, what did we do?

JILL: We looked at the tens first, so we really did this:

$$\frac{40 + 4}{4} \quad \text{or} \quad 4\overline{)40 + 4}$$

Then we divided.

The children went on to discuss and solve the other problems on the sheet. The teacher used discussion to stress thinking, "What number do I multiply by?" and the idea that the dividend could be renamed to help in the division.

Further discussion and exploratory work helped the children to develop generalizations concerning the division of a multidigit dividend (product). Emphasis was placed on finding the multiples of 100 and 10 that could be subtracted from the dividend (product). The teacher was able to analyze each pupil's thinking by use of a mathematics theme in which each child was asked to explain the "how" and "why" of working a division situation. The selection that follows is representative of an acceptable development.

$$5\overline{)582}$$
$$(500 + 82) \div 5 = N$$

$$\begin{array}{r} 116 \\ \hline 6 \\ 10 \\ 100 \\ 5\overline{)582} \\ \underline{500} \\ 82 \\ \underline{50} \\ 32 \\ \underline{30} \\ 2 \end{array}$$

What is the first step in finding "how many fives equal 582?" I have to think of renaming 582 so I can divide each addend by the known factor (5). I also want to use as large a multiple of 5 as I can. Looking at 582, I think that 500 can be used because it is the smallest multiple of 5 × 100 that can be used.

Now I need to find the number of fives equal to 82. Using a multiple of 10, I know that 5 × 10 = 50, so I use this.

The next question is, "How many fives equal 32?" or, "5 times what number equals 32"? I know from basic multiplication and division facts that 5 × 6 = 30. I continue to work. I subtract 30 from 32. My remainder of 2 cannot be divided by 5. My answer is 116 r2.

Multidigit Divisors

The teacher introduced the multidigit phase of division as follows: "Look at the division problems and exercises on the photocopied sheet. The first problem says, 'Packing material for electronic parts comes off the assembly line in sections that contain 168 small parts. [See the array pictured.] How many sections that hold 12 parts each can be made from the section that contains 168 parts?' Use the array and other means to solve the problem. First identify the mathematical sentence that is needed to solve the problem."

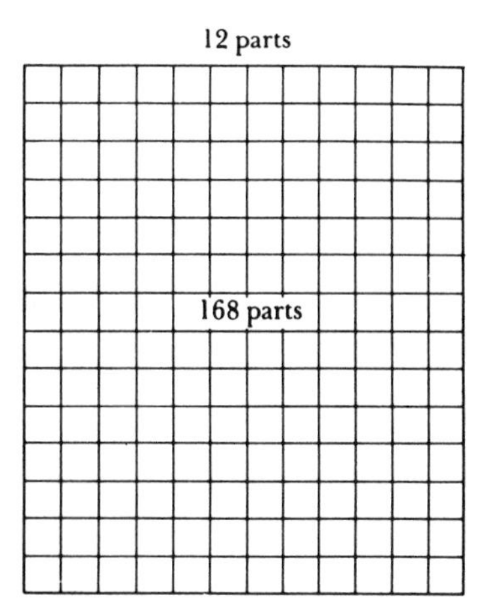

The work of the pupils on the problem revealed several solutions.

NAN: $168 \div 12 = N$. I knew there were 12 elements in each row, so I counted the number of rows: 14. This is the same as subtracting 12 from 168 fourteen times.

PHIL: $168 \div 12 = N$. I used subtraction. I knew there were at least 5 twelves. Then I took 5 more, 2 more, and 2 more.

$$\begin{array}{rr} 12\overline{)168} & \\ \underline{60} & 5 \\ 108 & \\ \underline{60} & 5 \\ 48 & \\ \underline{24} & 2 \\ 24 & \\ \underline{24} & \underline{2} \\ & 14 \end{array}$$

JEAN: I thought, "How many sets of 12 equal 168? Are there 10? Yes, $10 \times 12 = 120$. How many more? Two? Yes, more than two. Four? Yes, exactly $4 \times 12 = 48$."

$$\begin{array}{r} \underline{14} \\ 4 \\ 10 \\ 12\overline{)168} \\ \underline{120} \\ 48 \\ \underline{48} \end{array}$$

JIM: I tried to think of a way to rename 168. Since I am trying to find how many twelves in 168, I thought, $10 \times 12 = 120$. So I renamed 168 to be $120 + 48$. Then I went ahead and divided.

$$\frac{168}{12} \to \frac{120 + 48}{12} \to 10 + 4 \to 14$$

KIM: I used the same idea as Jim, but I wrote it differently.

$$12\overline{)168} \to 12\overline{)120} + 48 \to \begin{array}{r} 10 + 4 = 14 \\ 12\overline{)120 + 48} \end{array}$$

Then to direct pupil thinking toward ways to solve such division situations efficiently, the teacher used the following guided questions:

$$\begin{array}{r} \overline{25}\ r18 \\ 5 \\ \underline{20} \\ 22\overline{)568} \\ \underline{440} \\ 128 \\ \underline{110} \\ 18 \end{array}$$

TEACHER	PUPIL
Let's look at this division situation: 568 ÷ 22	
Will the quotient be in the hundreds?	No, 100 × 22 = 2,200.
Will it be more than 10?	Yes, 10 × 22 = 220.
How about 20?	Yes, 20 × 22 would be 220 + 220, or 440.
We'll try 20 as a starter. What next?	Let's try 5. 5 × 22 = 110. Good.
How does 5 turn out? Now what?	There will be no more sets of 22. Our answer is 25 r18.

OTHER DIVISION RELATIONSHIPS

As the inverse of multiplication, division has several other properties that lead to exploration by pupils. With the use of directed questions by the teacher, children can discover the following relationships:

1. There are 28 pairs of division facts in which each pair has the same dividend. For these pairs, the quotient of one equals the divisor of the other. Examples:

$$\begin{array}{r} 5 \\ 7\overline{)35} \end{array} \qquad \begin{array}{r} 7 \\ 5\overline{)35} \end{array}$$

2. For division facts having dividends that are squares (such as 36, 49, etc.), the quotient and the divisor are the same if the divisor is the square root of the dividend. Thus, $36 \div 6 = 6$.
3. Multiplying or dividing the divisor and the dividend by the same number produces no change in the quotient. Thus, $9 \div 3$, $27 \div 9$, $81 \div 27$ all have the same quotient.

   This is a form of the compensation idea and can be used quite effectively in dealing with situations involving even tens. A patterning lesson can be used to bring out this idea. Children were asked to solve the following divisions and then to look over their findings:

A. $30\overline{)90}$ and $3\overline{)9}$ and $300\overline{)900}$

B. $16\overline{)48}$ and $160\overline{)480}$ and $1{,}600\overline{)4{,}800}$

C. $50\overline{)250} \rightarrow ?$

Obviously, a greater number of examples will need to be tested. However, after a reasonable amount of exploratory work, the children will develop the idea that the quotient remains the same when both the divisor and the dividend are multiplied or divided by the same number. Note that the format that follows is helpful in developing these ideas and develops excellent readiness for using the identity element to rename fractions.

$$\frac{24 \div 2}{6 \div 2} \rightarrow \frac{12 \div 3}{3 \div 3} \rightarrow \frac{4}{1} \rightarrow 4$$

4. Multiplying the dividend by a number changes the quotient in the same proportion.

$$\begin{array}{r}4\\4\overline{)16}\end{array} \qquad \begin{array}{r}8\\4\overline{)32}\end{array} \qquad \begin{array}{l}2 \times 16 = 32\\2 \times 4 = 8\end{array}$$

5. The distributive property allows the dividend to be expressed as a sum. The quotient is the sum of the "partial quotients." Examples follow:

$$8\overline{)96} = 8\overline{)80 + 16} \rightarrow 10 + 2 = 12 \quad \text{or} \quad \frac{80 + 16}{8} \rightarrow 10 + 2 \rightarrow 12$$

$$5\overline{)125} = 5\overline{)100 + 25} \rightarrow 20 + 5 = 25 \quad \text{or} \quad \frac{100 + 25}{5} \rightarrow 20 + 5 \rightarrow 25$$

$$6\overline{)638} = 6\overline{)600 + 38} \rightarrow 100 + 6 + r2 = 106\ r2 \quad \text{or} \quad \frac{600 + 38}{6} \rightarrow 100 + 6\ 2/6 \rightarrow 106\ 2/6$$

These generalizations are important principally because they exhibit the relationship between multiplication and division, allow for pupil insight into the structure of division, provide opportunity for pupil thinking and discovery, and allow simplification in learning some of the basic division facts.

ESTIMATING QUOTIENTS

A relatively long period should be spent in allowing pupils to explore efficient methods of estimating the quotient in division situations. When the teacher feels that pupils have quite a good grasp of division the teacher may begin to develop the more standard means of estimating the quotient.

The three procedures for estimating the quotient have several names. The most common method used in textbooks of the 1960s is called the "round-down" or "apparent" method. In addition to the "apparent" method, the "round-up" and the "two-rule" procedures find favor with many writers on the teaching of elementary school mathematics.

Each of these three rather standard means of estimating the quotient is discussed next. Then a suggested teaching procedure is developed.

Round-Down or Apparent Method

The round-down or apparent method follows the procedure of using the digit to the left in the divisor as the "guide digit." In the division $24\overline{)464}$, the round-down thinking would be, "How many twos equal 4?" Or, more meaningfully, "How many twenties equal 400?" This procedure always produces a correct estimate or an estimate that is too large and therefore needs to be reduced. Also, because the tens digit of the divisor is always used to make the estimate, the guide digit is always visible.

Round-Up or Increase-by-One Method

The round-up method increases the left digit in the divisor by 1. In the division $24\overline{)462}$, the round-up thinking would be, "How many threes equal 4?" Or, more meaningfully, "How many thirties equal 400?" This round-up plan has the advantage of fitting well into the series-of-subtractions pattern that has been established earlier, because the estimate will never be too large. It also eliminates the need to erase. This aids the teacher in analyzing pupil errors. If an estimate of the quotient is too small, it still may be used and another division used. For example,

$$
\begin{array}{r}
32 \\
1 \\
1 \\
10 \\
\underline{20} \\
21\overline{)685} \\
\underline{420} \\
265 \\
\underline{210} \\
55 \\
\underline{21} \\
34 \\
\underline{21} \\
13
\end{array}
$$

A disadvantage of this procedure is that the guide digit is not written but must be remembered.

Two-Rule Procedure

The two-rule procedure makes use of both the round-down method and the round-up method. In using the two-rule pattern, a pupil rounds the divisor to the nearest multiple of 10 or 100 or 1,000, as the case may be. Thus, 54 becomes 50, 56 becomes 60, 453 becomes 500. For divisors ending in 5, the pupil determines whether to round up or down by the quotient. The two-rule procedure results in fewer occasions on which an inexact estimate is made. However, it requires a greater amount of mathematical proficiency than either of the other two methods.

The question of which of the three procedures will be used does not really arise, because, in most cases, the average and above-average pupil will discover the two-rule procedure. Those who do not figure out the two-rule procedure for themselves are probably better off remaining with a procedure they understand.

Since the round-up procedure is in keeping with an understanding of the series-of-subtractions approach to division, it can be taught as a natural outgrowth of this procedure.

A Suggested Sequence

Readiness activities can greatly aid the pupil's ability to estimate quotients. Before specific procedures are used, a series of lessons involving a "guess-and-test" strategy can be used. One teacher began this study by orally presenting this problem: "The conservation club had 175 small pine trees to plant. They planted 25 trees in each row. How many rows will they have to plant?" A brief discussion revealed that the children needed to think what number times 25 equals 175. The teacher then said, "Make a guess. Then test your answer by multiplication. I'll give you 10 seconds to guess. Now write down your guess. Then test to see how close you were."

Some of the children had correctly guessed 7 as the missing factor. Others had guessed 5 or 6. The teacher was careful to be positive about every guess made by the children. The children were then told to work in groups of four, guessing the quotient to divisions. Each groups was given several cards, such as the one that follows. One child held up the front side of the cards to the other three children, each of whom was to write a guess before the leader had silently counted to 10 (if a clock with a second hand was available, it was used for the time limit). Then the leader turned the card over, revealing the correct answer. The leader changed after every three cards.

$24\overline{)144}$ $N \times 24 = 144$ (front)	6 (back)

The next step in readiness comprised several lessons involving rounding off numbers. The teacher made use of the function idea and used a pattern-searching lesson. The children were given a sheet similar to the one partially shown below:[8]

[8]The sheet shown is representative of several types of exercises. Quite possibly, the "round-down function," the "round-up function," and the "nearest-hundred function" would not all be introduced at the same time. Also, work would be needed with tens as well as hundreds.

Break the codes:

A.	B.	C.
365 d 300 →	633 u 700 →	755 n 800 →
410 d 400 →	597 u 600 →	431 n 400 →
393 d □ →	219 u □ →	629 n □ →
642 d □ →	387 u □ →	582 n □ →
433 d □ →	354 u □ →	731 n □ →

Note: d = round down
u = round up
n = round to nearest 100

A variety of experiences will help provide children with the "rounding tools" necessary for estimating quotients.[9]

CHECKING AND REINTRODUCING DIVISION

The close relationship between division and the other fundamental operations allows for a variety of procedures for checking. A good procedure is to make use of different checks during the elementary school program, thus allowing "something new" to occur in each grade.

Inverse-relationship check (grade 3). The most common check for division is based on its inverse relationship with multiplication. Thus, to check $44 \div 5 = N$, the inverse relationship is $5 \times N = 44$. In the case above, $N = 8$ with a remainder of 4. The check is performed by multiplying $5 \times 8 = 40$ and then adding 4, because 4 has not been divided. If the answer is given as $8\frac{4}{5}$, then $5 \times 8\frac{4}{5} = 44$. This format may also be used

$$5\overline{)44}\quad 8\ r4$$

$$5 \times 8 = 40$$

$$40 + 4 = 44$$

Partial dividend check (grade 4). The partial dividend check makes use of the principle that the sum of the numbers subtracted from the dividend (plus the remainder) should equal the dividend. It should be noted that this check actually checks only the correctness of the subtraction.

[9]See Gail Spitler, "Painless Division with Doc Spitler's Magic Division Estimator," *The Arithmetic Teacher* 28 (March 1981): 26–27, for a good teaching lesson on estimating the quotient.

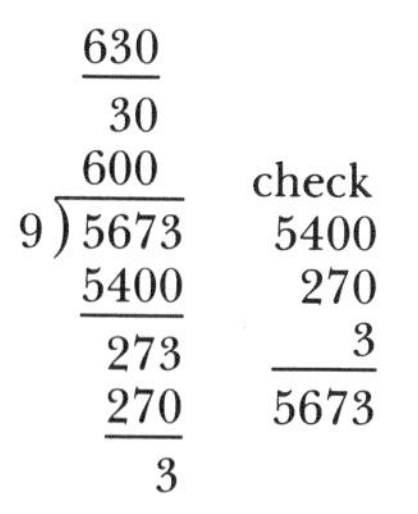

Reversing divisor and quotient (grade 5). Rather than multiplying the divisor by the quotient, one may use the quotient as the divisor in checking division. If the new quotient is equal to the original divisor and the remainder (if any) is the same, the problem has been solved correctly.

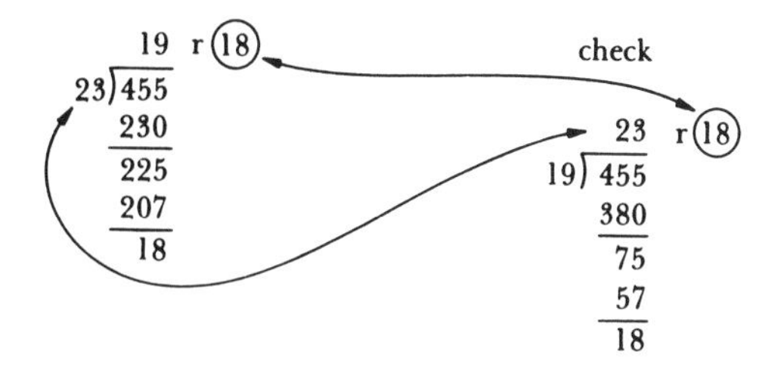

This check also has the advantage of affording further practice in division.

Casting out nines (grade 6). The excess of nines is found for each of the numbers in the same manner as for the other operations in the casting-out-nines check (by summing the digits).

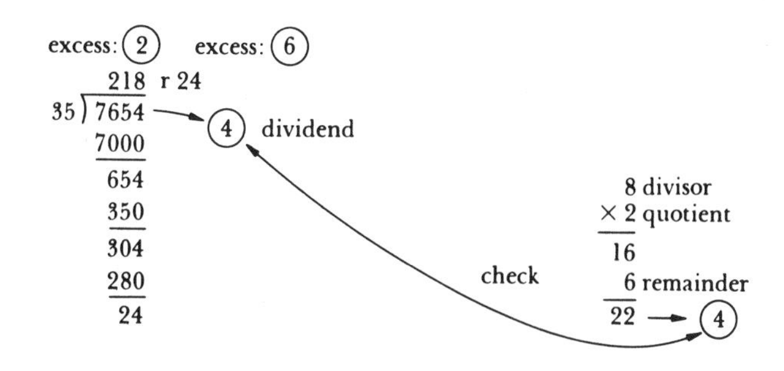

Egyptian method (grades 5 and 6). A division application of the Egyptian doubling multiplication method can be an effective reintroduction, enrichment, or even remedial procedure. If you have used the Egyptian method in multiplication, you can challenge the children to discover the "Egyptian method of division." It works as follows:

1. 105 divided by 15: double 15 until it is larger than or equals 105.

15	1
30	2
60	4
120	too large

```
 105
- 60    4
 ----
  45
- 30    2
 ----
  15
- 15    1
 ----   -
        7 answer
```

2. 235 divided by 34: double 34 until it is larger than 235.

34	1
68	2
136	4
272	too large.

```
 235
-136    4
 ----
  99
- 68    2
 ----   -
  31    6 remainder 31
```

KEEPING SHARP

Self-Test: True/False

_______ 1. Division is the inverse of multiplication.

_______ 2. Division of one whole number by another whole number always results in a whole-number quotient.

_______ 3. One way of thinking about division is as a series of subtractions where the subtrahend is always the same size.

_______ 4. The foundation experiences for division can and should be begun in kindergarten and first grade.

_______ 5. Division has two physical-world situations: measurement and "how many more are needed."

_______ 6. Formal work on division is best begun when children have an understanding of multiplication.

_______ 7. One of the undefined aspects of mathematics is division by zero.

_______ 8. Problems requiring the use of division that are related to life situations can lead to confusion (e.g., cost of one item when the price is given for three items).

_______ 9. Time spent estimating quotients provides useful experiences for children.

_______ 10. Orally presented division problems can be discussed and solved by primary-grade pupils.

_______ 11. Children can learn division concepts before they have mastered the multiplication table.
_______ 12. Partition and measurement division situations are conceptually different.
_______ 13. Division is neither associative nor commutative.
_______ 14. Research indicates that short division may be more difficult for elementary pupils because the computation is carried out with some unseen numerals.

Vocabulary

quotient	measurement division	estimation
dividend	partition division	

THINK ABOUT

1. Is the argument over whether division should be introduced with measurement or partition division situations of any real significance?
2. What three arithmetical methods of writing the basic division questions are there? What are the merits and demerits of each?
3. Which of the three rules for estimating the quotient with two-digit divisors seems best? Give a brief defense of your choice.
4. Some writers contend that the division of whole numbers cannot be taught well until pupils have a good grasp of fractions. Is this a valid argument?
5. To solve the indicated base 7 division exercise, which is more useful, the number line, serial subtraction, or addition? Illustrate the use of each.

   $3\overline{)133}$ (seven) $2\overline{)23}$ (seven)

6. Study the material on division in several textbook series. Do you think it is over-emphasized, is underemphasized, or uses about the right amount of time? Why?
7. How much should calculators be used in teaching division?

SUGGESTED REFERENCES

ANGHILERI, JULIE, and DAVID C. JOHNSON, " Arithmetic Operations on Whole Numbers: Multiplication and Division." In *Teaching Mathematics in Grades K–8,* Thomas R. Post (ed.), Newton, Mass.: Allyn and Bacon, 1988.

BATES, T., and L. ROUSSEAU, "Will the Real Division Algorithm Please Stand Up?" *The Arithmetic Teacher,* 33 (March 1986): 42–45.

HAZENKAMP, DONALD W., "Teaching Multiplication and Division Algorithms." In *Developing Computation Skills,* chap. 7. Reston, Va.: National Council of Teachers of Mathematics, 1978.

IRONS, CALVIN, "The Long Division Algorithm: Using an Alternative Approach," *The Arithmetic Teacher* 28 (January 1981): 46–48.

Mathematics Resource Project. *Number Sense and Arithmetic Skills: Multiplication/Division,* Eugene, Ore.: The University of Oregon. (available from Creative Publications).

TROUTMAN, ANDRIA P., and BETTY K. LICHTENBERG, *Mathematics: A Good Beginning,* 3rd ed., chap. 4. Monterey, Calif.: Brooks/Cole, 1987.

VAN DE WALLE, JOHN, and CHARLES S. THOMPSON, "Let's Do It: Partitioning Set for Number Concepts, Place Value, and Long Division," *The Arithmetic Teacher,* 32 (January 1985): 6–11.

VAN ENGEN, HENRY, and GLENADINE E. GIBB, "General Mental Functions Associated with Division," *Educational Service Bulletin,* 2 (1956).

WIEBE, JAMES H., *Teaching Elementary Mathematics in a Technological Age,* chap. 11. Scottsdale, Arizona: Gorsuch Scarisbrick, 1988.

chapter 10

FRACTIONS

OVERVIEW

Fractions

The intrepretation of fractions has long been a thorny problem. Historically, the distinction between numbers and numerals has been confused. The number-numeral distinction and the various possibilities of interpreting a fraction such as $\frac{3}{4}$ cause laymen and mathematics educators difficulty.

Fractions were introduced when people began to measure. The earliest treatment of fractions found was in an Egyptian manuscript, the Ahmes Papyrus of about 1550 B.C. The early Egyptians used only fractions with a numerator of 1 (unit fractions) and $\frac{3}{4}$ and $\frac{2}{3}$. This necessitated expressing fractions such as $\frac{1}{42} + \frac{1}{86} + \frac{1}{301}$. Several Egyptian notational forms follow:

$\frac{1}{10}$ = ⊂⊃∩, $\frac{1}{5}$ = ⊂⊃IIIII, $\frac{1}{20}$ = ⊂⊃∩∩

Fractions with numerators larger than 1 were developed by the Babylonians. They usually used denominators that were a multiple or a power of 60, which was the base of the Babylonian system.

Still other notational schemes were used by the Greeks, the Romans, and other civilizations. The present notation of fractions came into widespread use in the sixteenth century.

The teaching of fractions should involve much experimentation and concept development. The National Council of Teachers of Mathematics Standards suggest teaching fractions so that children can

1. Develop concepts of fractions and mixed numbers.
2. Develop number sense for fractions.
3. Use models to relate fractions and decimals and to find equivalent fractions.
4. Apply fractions to problem situations.
5. Extend their understanding of whole number operations to fractions.
6. Use mental computation with fractions.
7. Be able to solve challenging problems involving fractions.

TEACHER LABORATORY

Reprinted by permission of UFS, Inc.

Materials

Rubber bands, squared paper, ruler, and scissors.

1. Cut a rubber band. Use a marking pen to mark the rubber band into three segments of the same length. Now make a number line from 0 to 25, with the spaces between the whole numbers about $\frac{1}{2}$ inch each. Use your rubber band to find $\frac{1}{3}$ of 18, $\frac{2}{3}$ of 24, etc. (See pages 256–57).
2. Make a rubber band for fourths. Find $\frac{1}{4}$ of 12, $\frac{3}{4}$ of 16, etc.
3. Mark off squared paper in the manner shown below. Use the squared paper to find $\frac{2}{3} \times \frac{1}{4}$. Mark off squares yourself. Find $\frac{3}{5} \times \frac{1}{2}$.

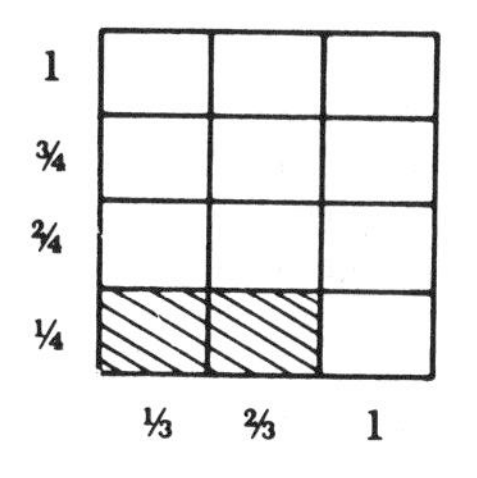

4. Find answers to these "if–then" situations:

If	xxx xxx	is 1, then	xx	is ______.	
If	xxxxxxx xxxxxxx	is 1, then	xxx xx	is ______.	
If	xxxx	is 1, then	xxxxxxxx	is ______.	
If	xxxxxx xxxxxx	is 1, then	xxx xxx	is ______ and	xx xx is ______. Find $\frac{1}{2} + \frac{1}{3}$.

TAKE INVENTORY

1. What kinds of situations can introduce fractions to the very young child?
2. What are some ways to reintroduce fractions in the upper grades?
3. Discuss the four situations that a fraction may represent, and develop an introductory activity for each.
4. Fractions are now usually presented in first-grade books, but computation with fractions is usually delayed until grade 5. How does the fraction instructional program in grades 1, 2, 3, and 4 differ from grade to grade? Use citations from one or two text series to substantiate your answer.
5. Some critics of the use of the common-denominator method of introducing fraction division criticize the method because the denominators are not used (computationally speaking). Is this a valid criticism? Why, or why not?

FRACTIONS AND RATIONAL NUMBERS

A fraction consists of an ordered pair of integers symbolized by a/b, or, more conventionally, by $\frac{a}{b}$, in which the first or top integer is called the *numerator* and the second or bottom integer is called the *denominator*. The denominator may not be zero.

Ordered pairs of numbers can have several interpretations, each of which requires a different teaching emphasis.

1. Ordered pairs of numbers may be treated as members of an equivalence class, or a *rational number*. The rational number is an equivalence class or ordered pairs of integers, $\frac{a}{b}$, where $b \neq 0$; for example $\{\frac{1}{2}, \frac{2}{4}, \frac{3}{6}, \frac{4}{8}, \ldots, \frac{a}{2a}, \ldots\}$. Normally, the rational number is named by the "simplest" member of the equivalence class, in this case $\frac{1}{2}$. The rational numbers arise from the mathematical need for an answer to the question, "If $b \cdot y = c$, what does y equal?" Thus, $b \cdot y = c$ leads to the invention of the rational number called $\frac{c}{b}$; $y = \frac{c}{b}$. For example, $5 \times y = 2$;

$y = \frac{2}{5}$. This interpretation develops a multiplicative inverse for a whole number. Thus, the multiplicative inverse for 5 is $\frac{1}{5}$, since $5 \times \frac{1}{5} = 1$ (the identity element for multiplication). In a later section of the chapter, this idea will be used to develop understanding in the division of rationals.

2. Ordered pairs of numbers may be considered the quotient of a division. Use of a whole number allows for a solution to the division $4 \div 2 = \square$. However, when a situation arises that requires an answer to $1 \div 2 = \square$ or $15 \div 6 = \square$, no whole-number answer can be obtained. In such cases, a fraction provides a solution; the answer to $1 \div 2$ is $\frac{1}{2}$, which may be read "one-half" or "one divided by two"; the answer to $15 \div 6$ is $\frac{15}{6}$ or $2\frac{3}{6}$, the latter of which may be thought of as 2 and $\frac{3}{6}$, or $2\frac{1}{2}$.

   Many historians believe that fractions were first developed in situations in which a whole-number answer to a division problem was impossible. The division interpretation of pairs of numbers is of great importance in elementary school mathematics.

3. Ordered pairs of numbers may be considered in the traditional fraction concept. Historically, the term *fraction* comes from the Latin word *frangere* or *fractio,* which means "to break."

   Most adults think in terms of pies, apples, or oranges when the word *fraction* is mentioned. For example, in the circular region in the illustration that follows, the numerator tells "how many" (3) and the demoninator tells "how large" (4ths).

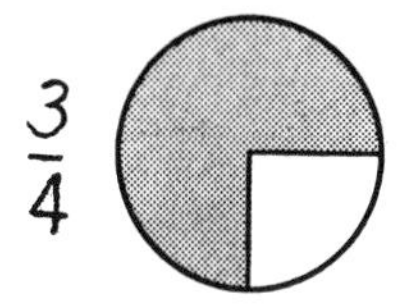

Set terminology may be effectively used to develop the "parts of a group" idea. In this context, these situations may be viewed as expressing the relationship of a subset to a set.

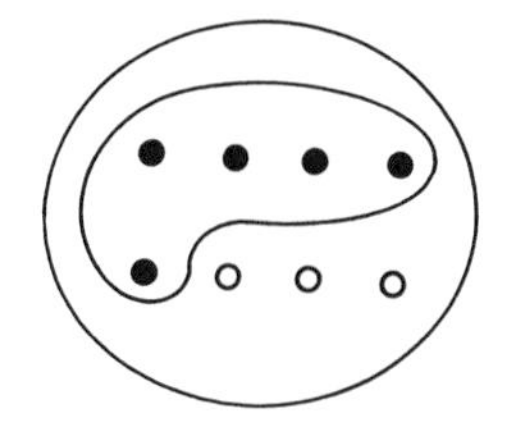

For example, there are five red and three blue marbles in a set. Bill gets the red marbles. In this case, he will get five of the set of eight, or $\frac{5}{8}$ of the marbles. Thus, the numerator (5) expresses the number size of the equivalent subsets being considered, and the denominator (8) expresses the number of equivalent subsets in the entire set. For fractions such as $\frac{5}{4}$ that indicate a number greater than 1, the numerator (5) tells how many equivalent subsets we are considering, and the denominator (4) indicates that a set contains four such subsets.

The "parts of a whole" idea can be illustrated by regions. The numerator (1) tells the number of congruent regions considered. The denominator (3) tells the number of congruent regions in a unit region.

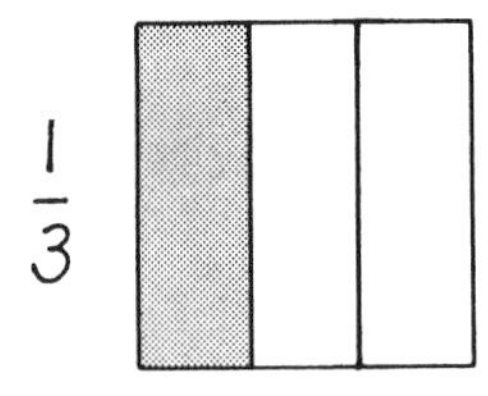

4. An ordered pair of numbers may be considered as a "ratio" or "rate pair." In such situations, an ordered pair of whole numbers indicates a rate. For example, "Claudia bought three pencils for 15 cents." The ordered pair of whole numbers is used to express a many-to-many correspondence. Such ordered pairs may be read as "3 for 15" or "3 to 15." This interpretation of ordered pairs is extremely useful in problem situations involving the comparison of prices and also leads to a valuable definition of percentage. The teaching of ratio and percentage is covered in Chapter 12.
5. Recently there have been suggestions that fractions should be thought of as operators. Using this idea, we think of $\frac{2}{3}$ as having a particular effect on a unit. The numerator (2) would be considered to be a stretcher (multiplier), and the denominator (3) would be considered to be a shrinker (divisor)

$$\frac{2}{3} \qquad \frac{\text{numerator}}{\text{denominator}} \qquad \frac{\text{multiplier}}{\text{divisor}} \qquad \frac{\text{stretcher}}{\text{shrinker}}$$

   We multiply the unit by 2 and then divide the result by 3. Or we could think of first dividing the unit by 3, getting $\frac{1}{3}$, and then multiplying by 2 and arriving at $\frac{2}{3}$.

Since these ideas are tied to mathematical operations, they are particularly useful for reintroducing fractions.

The set of rational numbers possesses the familiar properties of whole numbers. Several other ideas are also important concerning rational numbers:

1. The denominator may never be zero, because $\frac{5}{0}$ can be interpreted as 5 divided by 0, and division with zero as a divisor is undefined. If $\frac{5}{0} = \square$, then $\square \times 0 = 5$, and there is no unique number that will fulfill this necessary limitation.
2. The nonzero rational numbers are *closed* with respect to division. The system of integers does not provide an answer to $3 \times \square = 2$ or $2 \div 3 = \square$. Development of rational numbers allows an answer to these types of mathematical sentences, because $3 \times \frac{2}{3} = 2$ and $2 \div 3 = \frac{2}{3}$.
3. If the cross-products of two fractions are equal, they are equivalent fractions (they name the same rational number). Thus,

$$\frac{3}{4} \times \frac{6}{8} \qquad \frac{3}{4} = \frac{6}{8}$$

   because $3 \times 8 = 4 \times 6$. If the numerator and denominator of a fraction are multiplied by the same whole number, the new fraction and the old fraction represent the same rational number.
4. Fractions can be renamed by multiplying the numerator and the denominator by the same whole number. Thus, $\frac{5}{5} \times \frac{3}{5} = \frac{15}{25}$; $\frac{15}{25} = \frac{3}{5}$. This is true because $\frac{5}{5}$ is another name for 1, which is the identity element of multiplication.
5. The identity element for addition of rational numbers is zero. Zero can be named by any of the set of fractions $\{\frac{0}{1}, \frac{0}{2}, \frac{0}{3}, \frac{0}{4}, \ldots \frac{0}{n}\}$.

6. The rationals provide a *multiplicative inverse;* for example, the number $\frac{3}{2}$ is the multiplicative inverse for $\frac{2}{3}$, since $\frac{2}{3} \times \frac{3}{2} = 1$, and 1 is the identity element for multiplication.

Role of Fractions in the 1990s

In the first half of the 1980s there was a slow but sure movement toward the use of the metric system. There was also rapid acceleration in the use of calculators and computers in school and society. In light of these developments, suggestions have been made that the fraction form of rational numbers be stressed less and the decimal form have a much greater emphasis. This is probably wise. However, most thoughtful mathematics educators agree that children need to solve problems involving fractions, add and subtract like and unlike fractions, rename fractions and reduce them to lowest terms, and understand the concepts of multiplication and division of fractions.[1]

As you study this chapter, be sure to give particular attention to the ways that can be used to make fractions have mathematical and social meanings for the children. In today's world it would be very questionable to spend the majority of time dealing with the development of computation proficiency.

FOUNDATION PROGRAM FOR FRACTIONS

If only the mathematical criterion for inclusion of a topic in the elementary curriculum were followed, the introduction of fractions would begin following all the fundamental operations with whole numbers. In fact, this procedure was followed by most textbooks published during the nineteenth century.

However, because primary-age children have many uses of rationals and will develop concepts concerning fractions even if they are not taught them at the primary-grade level, it is important for primary teachers to begin an informal but mathematically correct development of fractional concepts early in the grades.

Children at the preschool level have heard their parents and older children make statements such as, "Nick, you can have one-half of the candy bar." "Jean, you haven't finished even a quarter of your work." "Bill, give each of the children one-third of the pieces of candy." Although the meanings that children attach to such situations are varied, the teacher may use them to develop basic ideas about fractions. The classroom vignettes that follow are designed to illustrate situations that provide opportunities for developing the meaning of fractions.

[1]See Joseph N. Payne, "Curricular Issues: Teaching Rational Numbers," *The Arithmetic Teacher* 31 (February 1984): pp. 14–17; Paul R. Trafton, "Assessing the Mathematics Curriculum Today," in *Selected Issues of Mathematics Education, 1981 Yearbook* (Berkeley, Calif.: The National Society for the Study of Education, 1980); and Zalman P. Usiskin, "The Future of Fractions," *The Arithmetic Teacher* 27 (January 1979): 18–20.

Early Instruction

One teacher began the development of rational numbers by saying, "The directions for mounting leaves given in my science book say, 'Begin with one-half of a sheet of paper.' How can we each cut a sheet of paper into halves?" Several sheets of paper were on a table at the front of the room, and several pupils demonstrated to their classmates methods of obtaining one-half of a sheet of paper. (See illustration.) Then the teacher said, "What do we mean by one-half?" Discussion brought out the idea that the sheet of paper was being divided into two parts of the same size.

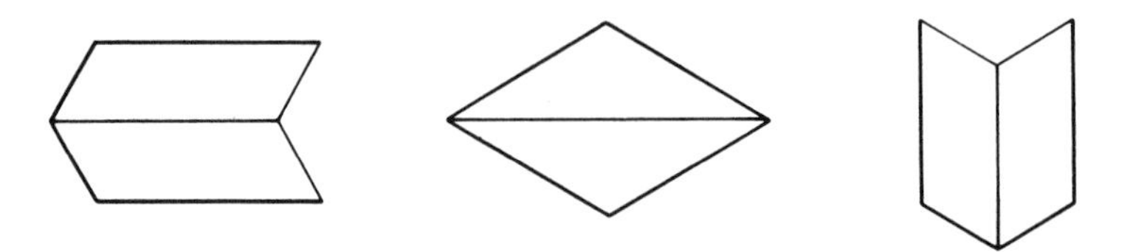

During the next few days, the teacher presented problem situations that required separating a whole into halves and identifying halves. Four such problems are these:

1. Jerry covered half his kite with white paper and the other half with gray. Which picture shows the kite painted the way Jerry colored his kite?

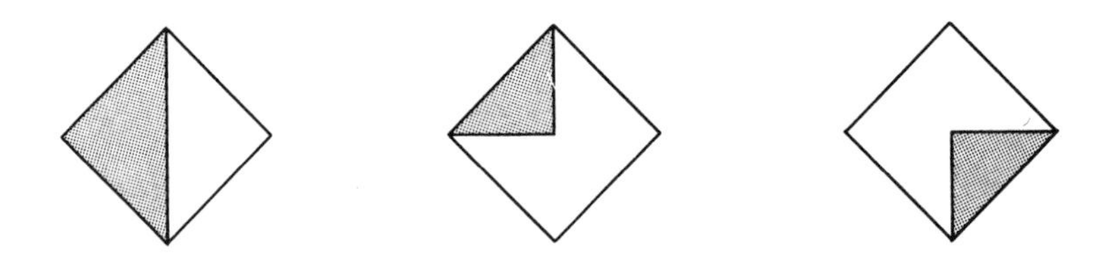

2. Jill was asked to cut this candy bar in half. Mark off the candy bar where she should cut it.

3. The directions for a science experiment indicate that a glass should be filled one-half full of water. Show how full the glass should be.

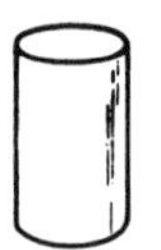

4. Jim, Jane, and Mary were arguing as to who had correctly shaded one-half of the drawing. Draw a ring around the correct drawing.

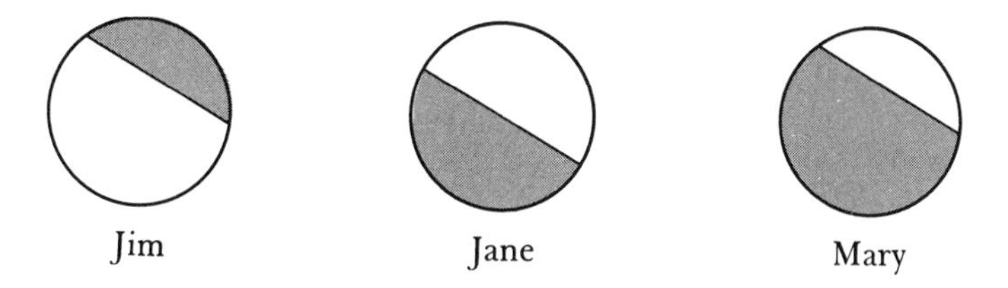

To develop another idea of fractions, one teacher used a laboratory experience involving lima beans and corn. The teacher placed five bags of corn and lima beans on each of five tables and said, "Work in pairs. Pick up a lab sheet and a crayon. You notice that the letters on the sheet are the same as the letters on the bags. I want you to take a handful of beans and corn from each bag and then count the number of beans and the total number of objects. See, I've done that. I have three beans and a total of seven objects. What fraction of the objects is beans?" The children responded that the fraction $\frac{3}{7}$ would be correct.

Then the teacher said, "See if you can use the ideas you've learned about fractions to answer this question. Alice wanted to tell her mother the fraction of pages in her workbook assignment that she had finished. She had this many pages to work [the teacher held up five pages] and had finished this many [pointed to three completed pages]. What fraction of the five pages has she worked?"

Pupils reasoned that because three of a set of four objects was stated as three-fourths, three of a set of five objects would probably be stated as three-fifths. The teacher agreed and suggested that they study the objects shown on a duplicated sheet and be ready to give a fractional name.

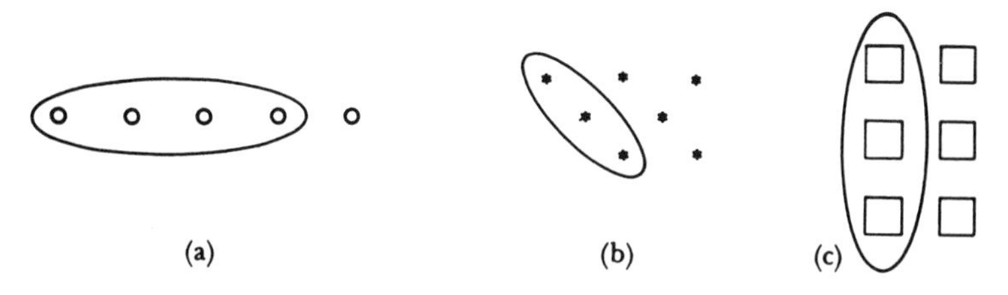

Diagrams (a) and (b) were readily identified as representing four-fifths and three-sevenths. The children generalized that to give a name to a fractional diagram, a person should first state the number of objects indicated and then state the number of objects in the set, adding *ths*. Thus, (★ ★ ★) ★ ★ ★ ★ indicated three-sevenths. Some pupils were somewhat unsure as to the name of diagram (c). Several suggested that they had previously indentified a similar diagram as representing one-half. Others agreed but said that, in keeping with the handling of four-fifths and three-sevenths, diagram (c) should be thought of as representing three-sixths. One pupil suggested that because a whole number could have more than one name, quite possibly rational numbers could also have more than one name. The teacher said, "Let's look at several other situations and see if it is possible for two fractions to represent the same idea" (the rational number).

Several geometric representations were drawn on the chalkboard, and pupils were asked to think of the possible names for the fractional ideas indicated by the drawings. A portion of the drawings and the children's remarks follow:

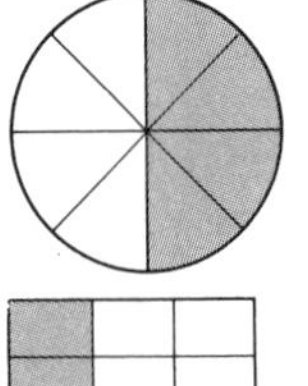

I could name the shaded part four-eighths or one-half.

We could name the shaded part two-sixths or one-third.

Pupils generalized ideas such as, "Three-sixths means the same as $\frac{1}{2}$, $\frac{5}{8}$ means the same as $\frac{10}{16}$," and so on. The teacher then referred the pupils to the number line to note the location of named sets of fractions such as $\{\frac{3}{4}, \frac{6}{8}, \frac{9}{12}, \frac{12}{16}, \ldots\}$. The pupils found that each set identified the same point on the number line.

Then the teacher said, "When fractions are names for the same point on the number line, they name the same *rational number.* The fractions are called *equivalent* fractions."

Children need a great deal of experience relating fractions to 1. One teacher used a pattern-searching lesson based on the following worksheet to bring out this idea. The children worked several minutes, moved into groups of four to discuss their ideas, and then finished the worksheet and began a second one.

Think

If □□□ is 1, then □□□□ is 1⅓

Do:

If □□□ is 1, then

a. □□ is $\frac{2}{3}$ **b. ⊞ is** 2

d. ⊞ is $1\frac{2}{3}$

This lesson was followed by a laboratory lesson in which children folded and cut to find fractions.

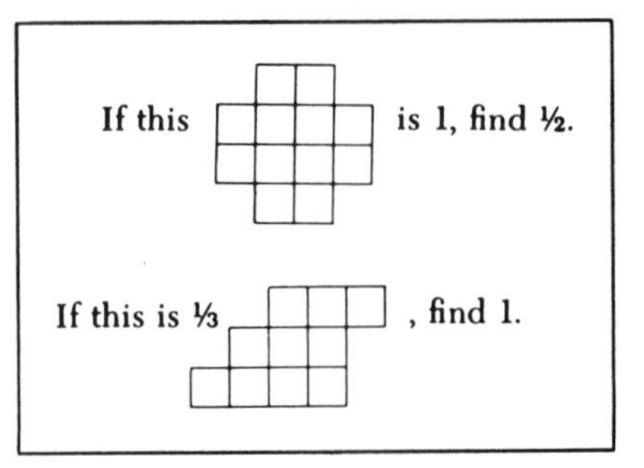

Further Work on Equivalent Fractions

The ability to convert a fraction to an equivalent form is a necessary prerequisite for all but the most simple computation involving fractions. It is a useful skill in its own right when a pupil wishes to compare two fractions with different denominators to see which represents the larger number. A use situation such as the following provides a motivation for fraction conversion: "Fred and Bill were arguing as to whose seeds had grown the most during a science experiment. Each boy measured his seedlings one evening and brought the results back to the class. Fred reported that his seedling was $\frac{5}{8}$ inch, and Bill reported his seedling to be $\frac{11}{16}$ inch. Which boy's seedling was taller?"

Pupils challenged to find an answer to this problem may determine that they must take one of two routes: Either both heights must be expressed with the same denominator, or they must be marked off on the same ruler. Use of the ruler or number line indicates that $\frac{5}{8}$ can be named $\frac{10}{16}$, and thus Bill's seedling is taller.

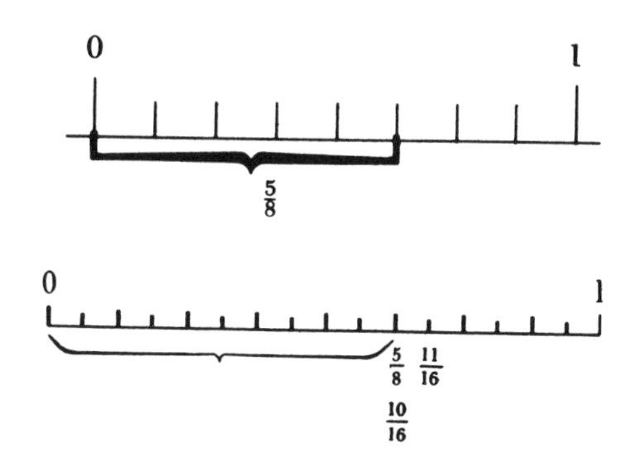

As work in fractions continued, the pupils discovered the concept that a fraction may be considered to indicate a division. Examples such as $\frac{6}{2}$ were used in this development. Pupils also noted that $\frac{3}{1} = 3$, $\frac{5}{1} = 5$, and so on.

Pupils should have many opportunities to use sets and objects divided into regions of the same size to identify names for the same rational number. Here are several questions that can be used in connection with this work:

1. There are 16 ounces in a pound. Eight ounces is $\frac{8}{16}$ of a pound. What are three fractions equivalent to $\frac{8}{16}$? [This might be a good time to compare the greater ease of using the metric system with these ideas.]
2. There are 12 inches in a foot. Four inches is ______ of a foot. What are three fractions equivalent to $\frac{4}{12}$?
3. What fraction of a yard is 1 foot? What are two other equivalent fractions that could be found by using inches in a foot and inches in a yard?

After pupils have become "comfortable" in deriving equivalent fractions by using drawings and diagrams, the class should consider other ways in which other names for a rational number can be found. The following development can be used effectively:

"Yesterday you were asked to find four equivalent fractions that are other names for $\frac{1}{2}$. You made use of drawings and the number line. Let's look at the various names for one-half [see illustration] and see if those names could be obtained more quickly. Study the names for $\frac{1}{2}$ carefully, and see if you can find any pattern. If you find a procedure, test it out by finding four different names for $\frac{1}{3}$."

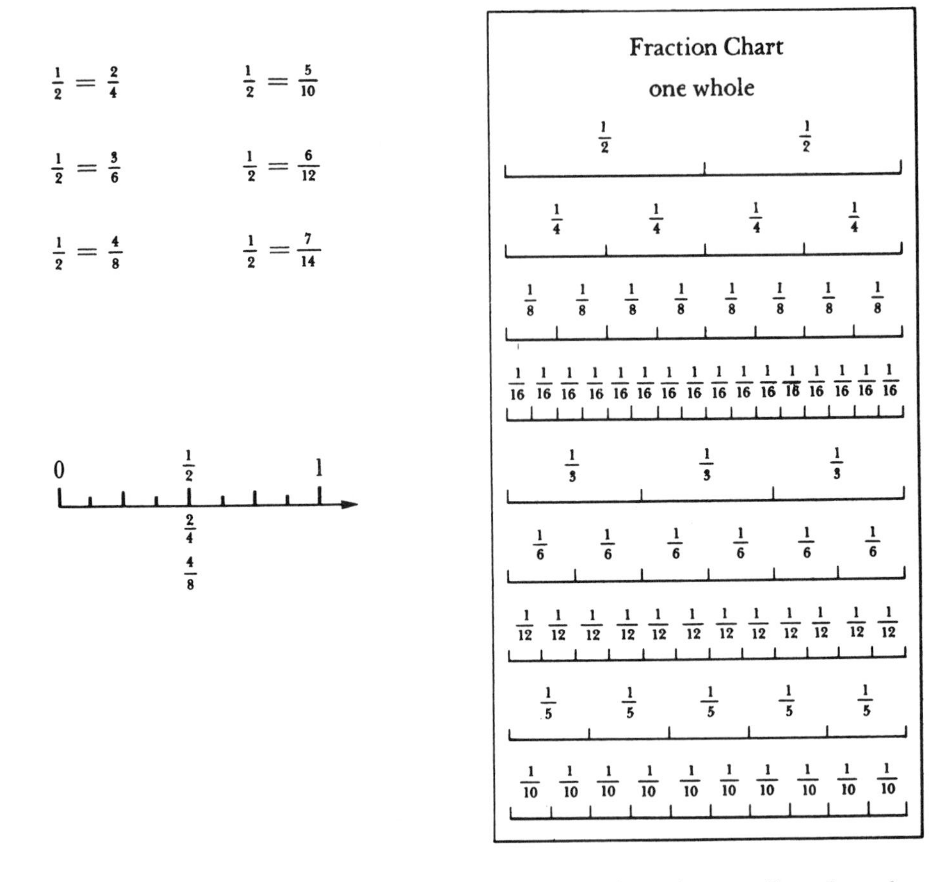

The teacher observed the students at work and mentally selected several of their approaches, then asked those pupils to suggest methods of quickly finding a number of equivalent fractions.

The first student suggested that it was possible to find other names for $\frac{1}{3}$ by continued doubling of both the numerator and the denominator. Her set of equivalent fractions was $\frac{1}{3}$, $\frac{2}{6}$, $\frac{4}{12}$, and $\frac{8}{24}$. The class members agreed that this method could be used, but several mentioned that this did not produce all the equivalent fractions; for example, $\frac{3}{9}$ is another name for $\frac{1}{3}$, but the doubling process will not produce $\frac{3}{9}$.

The next student said that in his experimentation, he noticed that all the various names for $\frac{1}{2}$ could be found by multiplying both the numerator and the denominator by the same whole number. Thus,

$$\frac{1}{2} \times \frac{1}{1} = \frac{1}{2}; \qquad \frac{1}{2} \times \frac{2}{2} = \frac{2}{4}$$

$$\frac{1}{2} \times \frac{3}{3} = \frac{3}{6}; \qquad \frac{1}{2} \times \frac{4}{4} = \frac{4}{8}$$

After testing several examples, the pupils were fairly confident that this procedure was effective.

Then the teacher asked, "Why do you think multiplying the numerator and denominator by the same whole number will give an equivalent fraction? Think about this, and then try to write a statement explaining the reason for this 'rule.' It may help to think of the ideas of whole-number multiplication."

The next day, the teacher asked several pupils to read their explanations of the "rule." One pupil said, "To find an equivalent fraction, it is possible to multiply both the numerator and denominator by the same whole number. It could look like this":

$$\frac{3 \times 2}{4 \times 2} = \frac{6}{8}$$

Drawings and diagrams were then used to help verify the findings. (In whole-number multiplication, any number can be multiplied by 1 without changing its value.)

Further work can be given in developing sets of equivalent fractions by having pupils generate sets of equivalent fractions in the form $\{\frac{2}{3}$, ________ , ________ , ________ , ...$\}$. Pupils should also be given the opportunity to discover that to change the name of a fraction to that of a given denominator—for example, $\frac{2}{7} = \frac{?}{21}$—one can think. "What number was 7 multiplied by to change it to 21?—by 3—therefore, I have to multiply 2 by 3 to produce an equivalent fraction." Later the pupils will discover that the number that both terms of the fraction are multiplied by may be found be dividing. Thus, $\frac{2}{5} = \frac{?}{20}$ may be found by thinking, "If $20 \div 5 = 4$, the 5 has been multiplied by 4. The 2 must also be multiplied by 4."

Often it is helpful to express a fraction in its lowest terms. (The lowest-terms form means that the numerator and denominator have no common factors other than 1.) Thus, $\frac{6}{8}$ is not in lowest terms, but $\frac{3}{4}$ is. The process is often called reduction, which is a misnomer because the number is not changed. Directed questions can lead pupils to discover that fractions can be simplified by dividing both the numerator and the denominator by the same whole number. As the pupils become more sophisticated in the use of simplification, they can use the idea of removing common factors. For example,

$$\frac{45}{60} = \frac{3 \times 5 \times (3)}{3 \times 5 \times (4)} = \frac{3}{4}$$

This principle is often called the "Golden Rule" of fractions.

Care must be taken to make sure that pupils understand that although

$$\frac{\not{2} \times 3}{\not{2} \times 5} = \frac{3}{5}, \frac{\not{2} + 3}{\not{2} + 5} \neq \frac{3}{5}$$

Several examples worked during the practice time given to simplification of fractions will usually suffice to settle this point.

Computer Use

There are a variety of computer assisted instruction programs that will aid children in understanding fractions as well as fractional computation. For example, "IBM Elementary Mathematics®" uses "fraction strips" to compare fractions.

Because of the graph capability of the microcomputer, a wide variety of good solid fractional instruction materials have been developed. It is well worth the time of the elementary mathematics teacher to spend several hours reviewing programs, which can aid teaching, remediation, and enrichment.

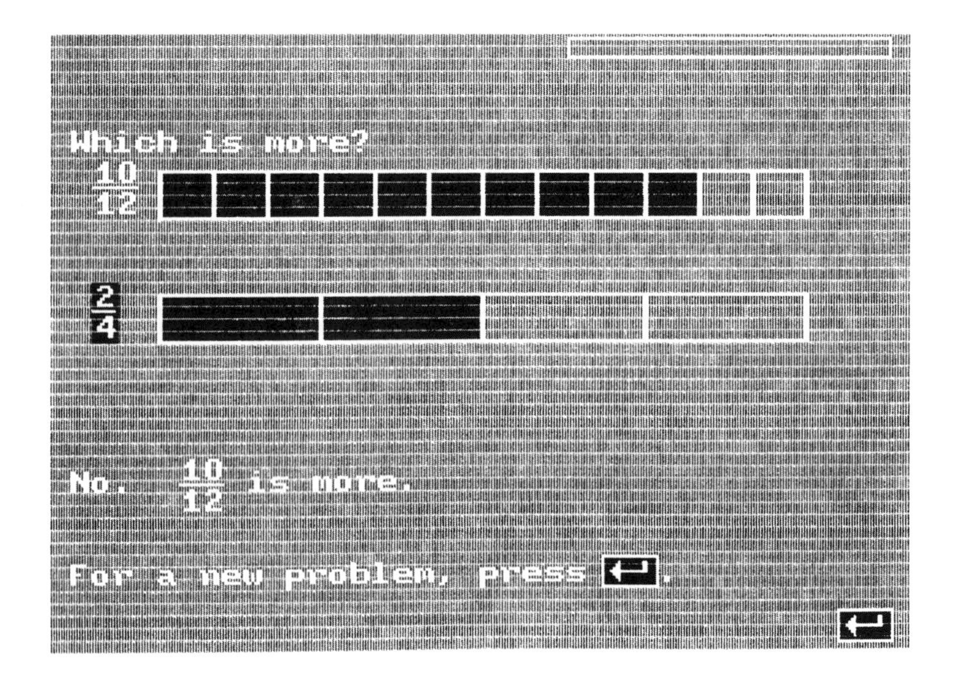

Figure 10–1 Feedback provided by the Comparing Fraction activity. (IBM/WICAT Systems Inc.)

TEACHING ADDITION OF RATIONAL NUMBERS

Foundation Work

Primary-grade pupils can readily solve problems involving the union of sets of objects that can be represented by fractions. They can also use a number line and drawings of circular or rectangular regions to solve such problems. Here is an example of each situation, with typical concrete solutions:

1. Bill and Nancy each received a set of eight pencils. After three weeks, Bill had three pencils left and Nancy had four pencils left.
   a. What part of his set did Bill have left?
   b. What part of her set did Nancy have left?

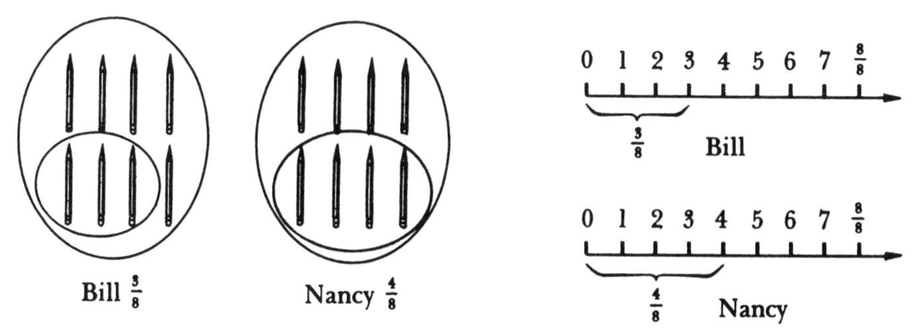

   c. If the two children combined their remaining pencils, what part of a set of pencils would they have?

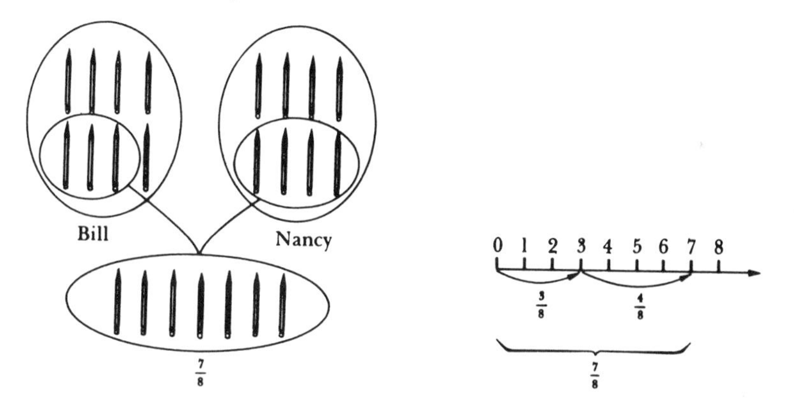

2. Joy made two pizzas and cut each into eight pieces. The shaded parts of the drawings below show the number of pieces left. What fraction of a whole pizza is left?

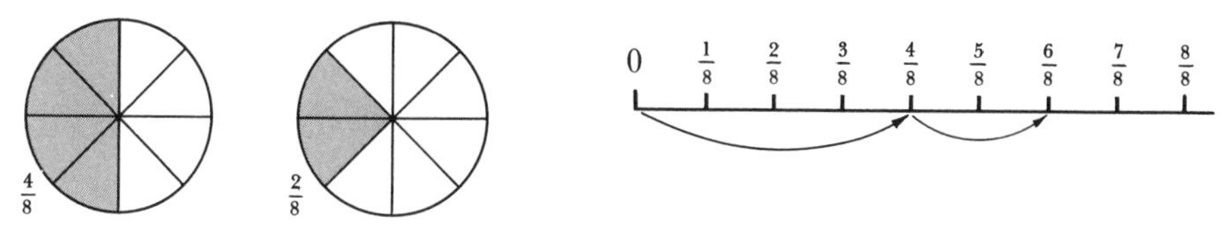

3. In baking his first cake, Ken put in $\frac{1}{8}$ of a cup of sugar and then put in $\frac{5}{8}$ of a cup more sugar. How much sugar did he put in the cake?

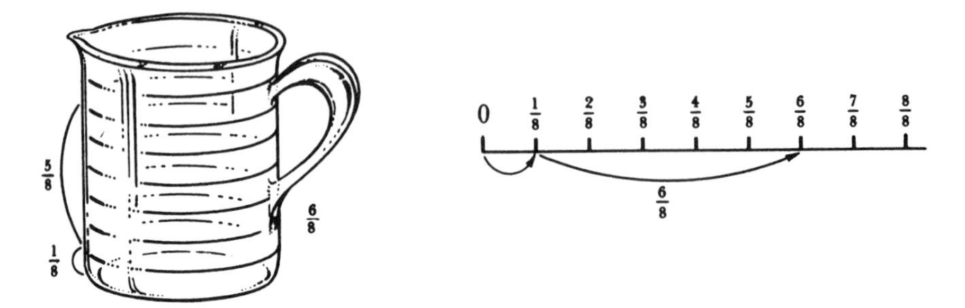

In conjunction with solving addition situations by the use of drawings and diagrams, pupils at the primary-grade level will benefit from simple problems in which they are to "think the answer" without using paper and pencil. The majority of primary-grade pupils can find the answer without pencil and paper to problems such as, "Tom worked $\frac{2}{5}$ of his problems on Monday and another $\frac{2}{5}$ of the problems today. What fraction of his problems has he worked?"

Estimation of sums. Early work in estimation is necessary if children are to develop a true understanding of fractional addition. This is not been the case. For example, Post[2] found that only 24 percent of the United States' 13 year olds are able to correctly estimate $\frac{12}{13} + \frac{7}{8}$. Given the choice of 1, 2, 19, and 21, 28 percent chose 19 and 27 percent chose 21. Obviously, they had very faulty fractional concepts.

Some activities. To develop this fraction–addition sense, a number of the following activities can be used:

1. Compare. If children are asked which is larger, $\frac{2}{5}$ or $\frac{5}{8}$, children can think that $\frac{2}{5}$ is less than $\frac{1}{2}$ and $\frac{5}{8}$ is more than $\frac{1}{2}$; thus, $\frac{5}{8}$ is greater than $\frac{2}{5}$.
2. Use an activity such as "How close can you get?"

TEACHER: Name a fraction close to 1.
JOE: $\frac{5}{6}$.
TEACHER: That's close. How close is it?
JAN: $\frac{1}{6}$ close.
TEACHER: Can you name a close fraction?
CLAUDIA: $\frac{9}{10}$; it's $\frac{1}{10}$ close.
PETE: $\frac{99}{100}$; it's $\frac{1}{100}$ close.

There are variations:

1. Get close to 1; above and below ($\frac{12}{11}$ and $\frac{11}{12}$)
2. Give a set of numerals getting close to 1.

[2]Post, Thomas P., "Fractions: Results and Implications from the National Assessment," *The Arithmetic Teacher* 28 (May 1981): 26–31.

3. Add to get close. Given ($\frac{1}{2}$, $\frac{1}{3}$, $\frac{1}{4}$, $\frac{1}{5}$, etc.), add to get close to $\frac{3}{4}$.
4. Get a fraction calculator such as the CASIO FX 350, which shows fractions with slash bar: $\frac{1}{2}$ becomes 1/2. Use the calculator to help "come close estimates" such as come close to $\frac{7}{8} + \frac{3}{4}$; come close to $22\frac{3}{5} + 6\frac{4}{9}$.

Berh, Wachmuth, and Post suggest that children need to think about fractions in their own right as well as part of a whole, ratio, or the quotient of two integers.[3]

Systematic Development

Systematic work with addition usually begins in the fourth grade. My suggestions for this introduction are at variance with many present programs for teaching the addition of rational numbers; therefore, here are several suggestions concerning addition, as a prelude to the teaching suggestions:

1. Often the introduction begins by dealing with fractions that have the same denominator. If a good foundation is provided in the primary grades, work with "like" fractions (fractions with the same denominator) offers little challenge to intermediate-grade pupils. Therefore, I suggest that "unlike fractions" be used for the first critical study of addition.
2. The understanding of congruent regions is assumed. However, the name *congruent* is not necessary to the development.
3. Because fractions represent mathematical abstractions of physical-world objects, pupils can best develop understanding if they can begin by dealing with physical-world problem situations.
4. Children should have previous experience finding "many names" for a rational number.

Teaching Development

Prior to formal study of addition, the pupils had developed number lines (see below) that emphasized the various names for rational numbers.

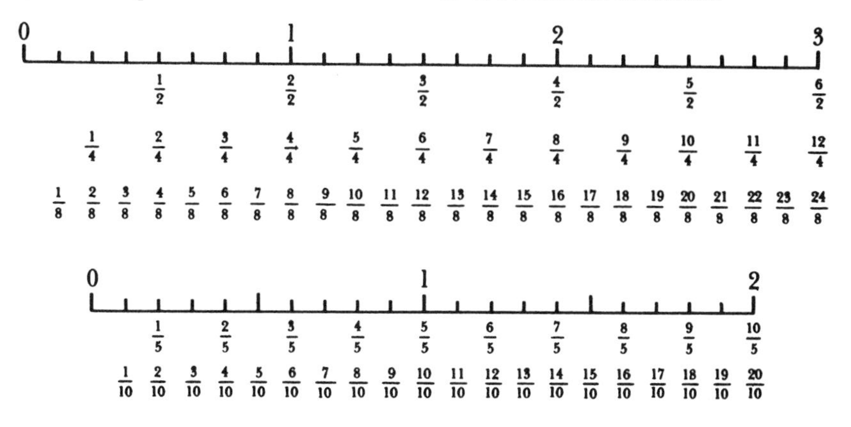

[3]Behr, Merlyn J., D. Wachsmuth, and T. R. Post, "Construct a Sum: A Measure of Children's Understanding of Fraction Size," *Journal for Research in Mathematics Education,* 15 (Nov. 1984): 323–41.

The study of addition was initiated by the pupils' consideration of the following three verbal problems:

1. Kevin lives $\frac{3}{10}$ of a mile from school. Joe lives $\frac{5}{10}$ of a mile farther from school than Kevin. How far does Joe live from school?
2. To make a depot for his model train, Harry attached a piece of wood that was $\frac{1}{4}$ inch thick to a piece of wood that was $\frac{3}{8}$ inch thick. How thick were the combined pieces?
3. On a hike, the Girl Scouts walked $1\frac{1}{2}$ miles before a refreshment break and $\frac{3}{4}$ of a mile after the break. How far did they walk?

Problem 1 caused the pupils little difficulty. The mathematical sentence representing the problem is identified as $\frac{3}{10} + \frac{5}{10} = N$. Pupils made use of the number line, drawings, or counting by tenths, or were able to just think and answer. During the discussion of the problem, the class commented that because the denominators were the same, the addition was very similar to the addition of whole numbers. "It's just like adding ones and tens: 2 tens + 5 tens = 7 tens; 3 ones + 8 ones = 11 ones."

Problem 2 required a greater amount of pupil thought. The mathematical sentence needed was identified as $\frac{1}{4} + \frac{3}{8} = N$. Pupils suggested the following solutions:

SHIRLEY: Use the number line that is marked in eighths and in fourths.

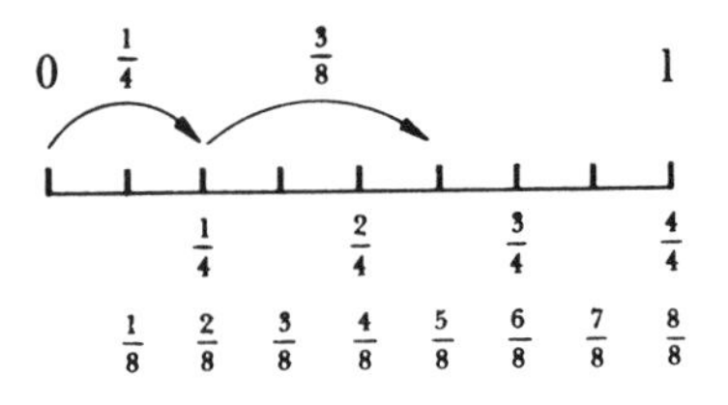

KATHLEEN: Use a diagram of a rectangular region or a circular region.

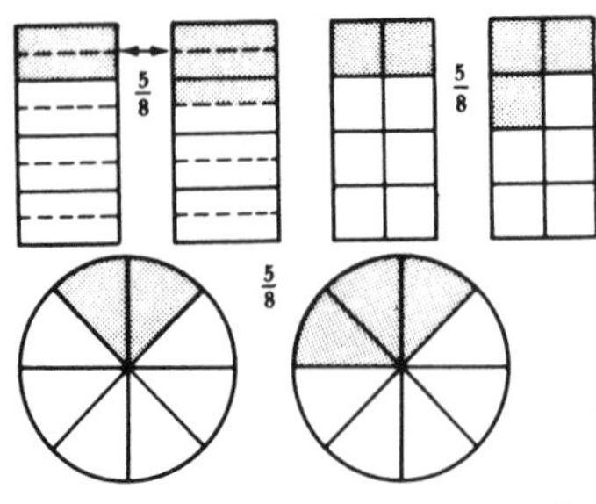

MARK: Find names for both numbers that have the same denominator.

$\frac{1}{4} + \frac{3}{8} = N$

$\frac{2}{8} + \frac{3}{8} = \frac{5}{8}$

In the discussion of the problem, the solution suggested by Mark was considered to be the most "mathematical" one. It also provided the most economical use of time, because it allowed for a solution by addition. Problem 3 was solved in a manner similar to the way problem 2 was solved.

During the lessons that follow the initial development, the teacher should give many problems and computational exercises that require the pupils to find equivalent fractions for performing the addition. The teacher should focus attention on addition involving fractions with the same denominator, addition involving fractions with "unlike" denominators, addition involving "mixed forms" such as $3\frac{1}{2} + 4\frac{2}{3} = N$, and methods involving renaming to obtain "like" fractions.

Properties of Addition of Rational Numbers

After some experience with addition of rational numbers, pupils should benefit from an inquiry concerning the properties of addition. The following questions present a challenge to make mathematical generalizations:

1. Does the commutative property of addition hold true for rational numbers?
2. Does the associative property of addition hold true for rational numbers?
3. Does the addition property of zero as the identity element hold true for the addition of rational numbers?
4. When two rationals are added, can the result always be expressed as a fraction?

In answer to these questions, the pupils should present a number of examples to verify that each of the properties works for rational numbers as it did for whole numbers. The pupils should understand that they have not mathematically proved that each property holds true. For further verification of their findings, they can compare their results with an accurate textbook.

Exercises such as these should help pupils arrive at the following generalizations concerning the addition of rational numbers:

1. If two rational numbers are added, the sum is a rational number, or may be named by a fraction.
2. If 0 is added to any rational number, that rational is the sum. Members of the following set are all names for 0: $\{\frac{0}{1}, \frac{0}{2}, \frac{0}{3}, \ldots \frac{0}{n}\}$.
3. The order of addends for rational numbers does not affect the sum. (Commutative property of addition.)
4. In finding the sum of three rationals, the first and the second may be added and the third added to their sum or the second and third may be added and the first added to their sum. (Associative property of addition.)

Mixed Forms and Renaming

Any addition situations that involve mixed forms may be performed by changing the mixed forms into fraction form ($3\frac{2}{3}$ becomes $\frac{11}{3}$). However, there are some uses for the mixed form.

In most cases, the so-called "vertical form" provides for a more efficient computation:

$$\begin{array}{lll} 3\tfrac{1}{2} & 3 + \tfrac{1}{2} & 3 + \tfrac{2}{4} \\ \underline{2\tfrac{3}{4}} \rightarrow & \underline{2 + \tfrac{3}{4}} \rightarrow & \underline{2 + \tfrac{3}{4}} \\ & & 5 + \tfrac{5}{4} = 5 + 1 + \tfrac{1}{4} = 6\tfrac{1}{4} \end{array}$$

After an understanding has been developed, the operation may be greatly shortened by handling some of the steps mentally. The pupil thinks

$$\begin{array}{lll} 4\tfrac{7}{8} & \tfrac{3}{4} = \tfrac{6}{8} & 4\tfrac{7}{8} \\ \underline{+5\tfrac{3}{4}} & \tfrac{7}{8} + \tfrac{6}{8} = \tfrac{13}{8}, \text{ which can be renamed } 1\tfrac{5}{8} & \underline{+5\tfrac{6}{8}} \\ & & 10\tfrac{5}{8} \end{array}$$

Generalizing the Addition of Rational Numbers Through Orally Presented Situations

The mathematical definition of addition for rational numbers is:

$$\frac{a}{b} + \frac{c}{d} = \frac{(a \times d) + (b \times c)}{b \times d} \quad \text{or} \quad \frac{ad + bc}{bd}$$

Numerically,

$$\frac{4}{5} + \frac{2}{3} = \frac{(4 \times 3) + (5 \times 2)}{(5 \times 3)}$$

Developing an understanding of this definition is of value for two reasons: (1) Such a procedure allows for a more effective approach to non–pencil-and-paper addition than does the method presented previously, and (2) it is the procedure that pupils will later use in algebraic work with fractions. Either a Socratic-questioning or a pattern-searching approach can be used to develop this idea. The need to add $\frac{3}{4} + \frac{2}{3}$ without pencil and paper can be used as the setting for development.

Finding the Least Common Denominator (Least Common Multiple)

With addition, many situations arise in which it is necessary to determine the least common denominator (LCD) of two or more fractions. The LCD of fractions is analogous to the least common multiple (LCM) of whole numbers. The development up to this point has not focused on efficient ways of finding common denominators for fractions. Many modern materials that are presented to elementary school pupils focus on this phase of fractions before addition. Such a procedure is to be questioned, because the first situation in which there is an extensive need for finding the LCD of fractions occurs when the addition of rationals begins.

The teacher can begin the teaching of the LCD by presenting a situation that requires the addition of several numbers, such as $\frac{3}{4} + \frac{5}{6} + \frac{3}{8} + \frac{1}{3}$. The pupils quickly realize that none of the denominators are common denominators and that multiplying all the denominators together produces an extremely large number.

Several basic approaches to this topic can be used. The traditional elementary school mathematics program suggested a "guess and check to see if you're right" approach, which at best is inefficient. Also, teachers often suggested that the pupil double the largest denominator to see if this is the LCD. More efficient procedures are suggested below and normally would be developed in the order given.

USE OF COMMON FACTORS

The teacher may remind the pupils that the LCD of several numbers will always have each of the numbers as a factor. Then he or she may suggest that the pupils factor the denominators and study their results. For the example above, the pupils find that:

$$\frac{3}{4} + \frac{5}{6} + \frac{3}{8} + \frac{1}{3} = \frac{3}{2 \times 2} + \frac{5}{2 \times 3} + \frac{3}{2 \times 2 \times 2} + \frac{1}{3}$$

Then the teacher may ask questions such as, "Will a denominator that can be divided by 2×3 also be divisible by 3? Can a denominator that can be divided by $2 \times 2 \times 2$ also be divided by 2×2?" Such questions cause the pupils to think of ways of reducing the number of factors needed for a common denominator. Usually, pupils will suggest that because the answer to the questions posed above is "yes," a common denominator for the following denominators may be found by eliminating factors that are repeated in other denominators. Thus,

$$4 = \cancel{2 \times 2}$$

$$8 = 2 \times 2 \times 2$$

$$6 = \cancel{2} \times 3$$

$$3 = \cancel{3}$$

The LCD $= 2 \times 2 \times 2 \times 3 = 24$.

Another example is

$$\frac{5}{12} - \frac{7}{18}$$

$$12 = 2 \times 2 \times 3$$

$$\rightarrow 2 \times 2 \times 3 \times 3 = 36$$

$$\text{LCD } 18 = 2 \times 3 \times 3$$

Previously it was suggested that the subtraction of whole numbers should begin several days after the addition of whole numbers. In the case of rational numbers, subtraction may well be introduced the same day as addition. This is because pupils at this level should have a good understanding of the relationship between addition and subtraction, and the techniques involved in addition and subtraction of rational numbers are very similar.

The foundation work in subtraction goes hand in hand with that of addition. Emphasis should be placed on the three types of subtraction situations: take-away, comparison, and how many more are needed.

If the additive type of renaming situation is used in whole-number work, it should be continued in work with rationals. The illustration below compares additive thinking with take-away thinking.

Take-away

$$\begin{array}{r} 17\frac{1}{2} \\ -5\frac{2}{3} \\ \hline \end{array} \qquad \begin{array}{r} 17\frac{6}{12} \\ 5\frac{8}{12} \\ \hline \end{array} \qquad \begin{array}{r} 16\frac{18}{12} \\ 5\frac{8}{12} \\ \hline 11\frac{10}{12} \end{array} = 11\frac{5}{6}$$

Additive

$$\begin{array}{r} 17\frac{1}{2} \\ -5\frac{2}{3} \\ \hline \end{array} \qquad \begin{array}{r} 17\frac{6}{12} \\ 5\frac{8}{12} \\ \hline \end{array} \qquad \begin{array}{r} 17\frac{18}{12} \\ 6\frac{8}{12} \\ \hline 11\frac{10}{12} \end{array} = 11\frac{5}{6}$$

The teacher can also develop a process for non–pencil-and-paper computation with rationals by using the same format used for addition. The algorithm for subtraction becomes the following:

$$\frac{2}{3} - \frac{3}{5} = \frac{(2 \times 5) - (3 \times 3)}{3 \times 5} = \frac{10 - 9}{15} = \frac{1}{15}$$

or

$$\frac{a}{b} - \frac{c}{d} = \frac{(a \times d) - (b \times c)}{(b \times d)} = \frac{ad - bc}{bd}$$

Subtraction such as $\frac{1}{2} - \frac{3}{4} = \square$ is not usually taught in the elementary school. However, it can be developed intuitively with the number line and verbal problem situations.

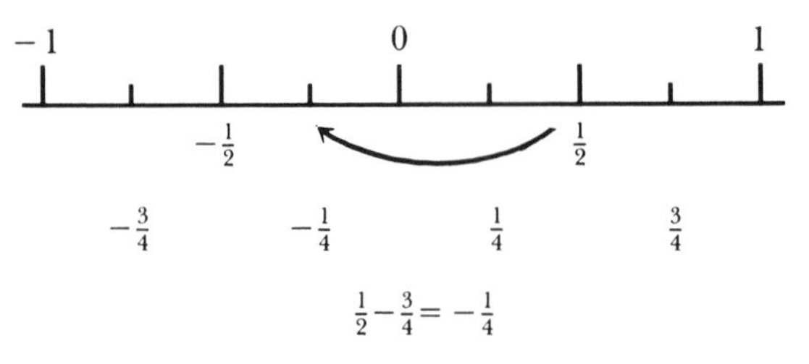

TEACHING MULTIPLICATION OF RATIONAL NUMBERS

Multiplication is the simplest of the fundamental operations with rational numbers if the major criterion is the ability to perform computation. The definition of the product of two rational numbers is

$$\frac{a}{b} \times \frac{c}{d} = \frac{ac}{bd}$$

This is also the form used for the typical algorithm. This apparent simplicity is misleading for two reasons:

1. The several fractional and mixed forms have different physical-world interpretations. They are illustrated by (a) $5 \times \frac{2}{3} = N$; (b) $\frac{2}{3} \times 5 = N$; (c) $\frac{2}{3} \times \frac{4}{5} = N$; (d) $1\frac{1}{2} \times 6 = N$; (e) $4 \times 1\frac{3}{8} = N$; (f) $4\frac{1}{2} \times \frac{3}{4} = N$; (g) $\frac{1}{2} \times 2\frac{1}{2} = N$; (h) $2\frac{3}{4} \times 5\frac{2}{3} = N$. Although (a) and (b), (d) and (e), and (f) and (g) are applications of commutativity, they still do not fit the same physical-world situation.
2. Although a series of additions interprets all physical-world situations dealing with whole-number multiplication, this is not true for work with rationals. The multiplication of $\frac{1}{2} \times 4$, or $\frac{1}{2}$ of 4, is closely related to partition division. Instead of taking the multiplicand a number of times, the problem is to find a part of the multiplicand. The form $\frac{2}{3} \times \frac{3}{4}$ is also not related to a series of additions but may be better considered in terms of arrays or areas.

Effective understanding requires that the pupil have a good grasp of three of these forms: $4 \times \frac{1}{2} = N$; $\frac{2}{3} \times \frac{1}{2} = N$; $\frac{1}{2} \times 4 = N$. I suggest that multiplication of rational numbers be taught in this order. With this understanding and further work, the other forms can be understood. An introductory lesson for each of the three types is provided next to illustrate the guided-discovery procedure for multiplication and to aid the reader in analyzing the different interpretations.

Finding Products Such as $5 \times \frac{2}{3} = N$

Multiplication involving the form $5 \times \frac{2}{3} = N$ was introduced with the following directions: "Write a mathematical sentence that illustrates each of the problems on the photocopied sheet and then solve the problems. Try to verify your answer to each problem by working it in a different way. The first problem is, 'Ann was asked to bake five batches of cookies for a club bake sale. If each batch requires $\frac{2}{3}$ cup of sugar, how much sugar will she need for the five batches of cookies?'"

Pupils were encouraged to use as many methods as possible to solve the problems. Several pupils were asked to put one of their solutions on overhead projector mounts. (The chalkboard could have been used). Their methods and explanations follow.

HOYT: The mathematical sentence was $5 \times \frac{2}{3} = N$. I decided that I could find 5 groups of two-thirds by adding.

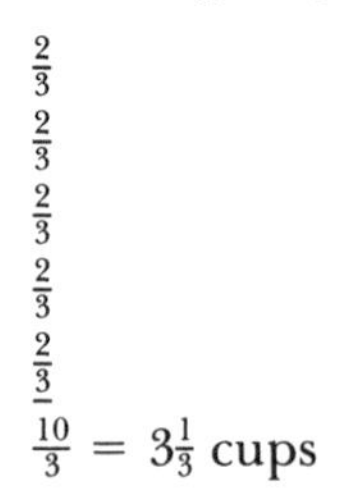

$\frac{2}{3}$
$\frac{2}{3}$
$\frac{2}{3}$
$\frac{2}{3}$
$\frac{2}{3}$

$\frac{10}{3} = 3\frac{1}{3}$ cups

JOYCE: I wrote the same mathematical sentence and then used the number line to solve the problem.

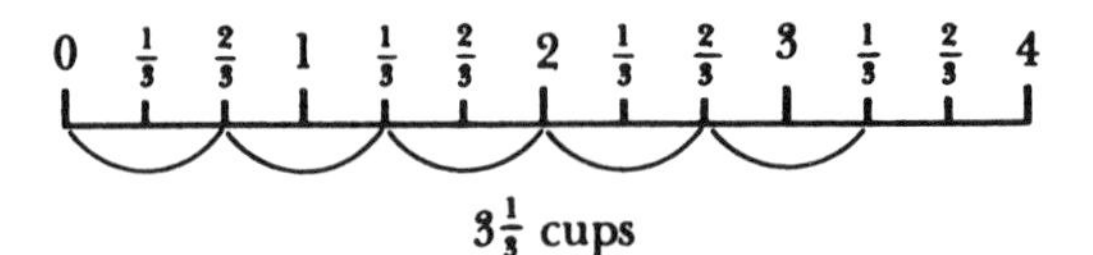

$3\frac{1}{3}$ cups

KIT: I rewrote the mathematical sentence to read 5×2 thirds $= N$. I worked the problem as we have tens and ones. So, 5 times 2 thirds = 10 thirds or $3\frac{1}{3}$.

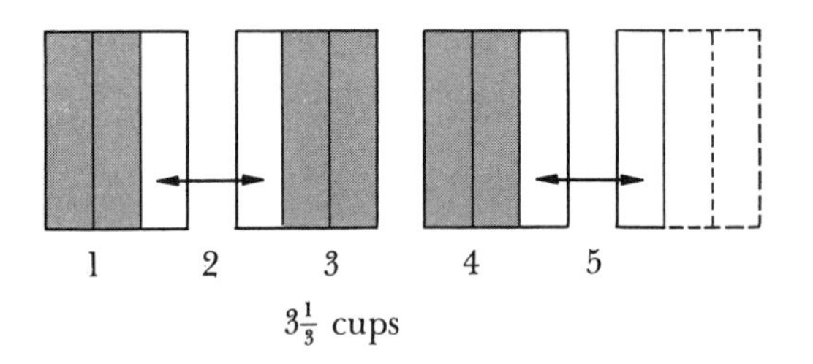

$3\frac{1}{3}$ cups

SCOTT: I used a drawing of a rectangular region. I've also shown how a circular region could be used

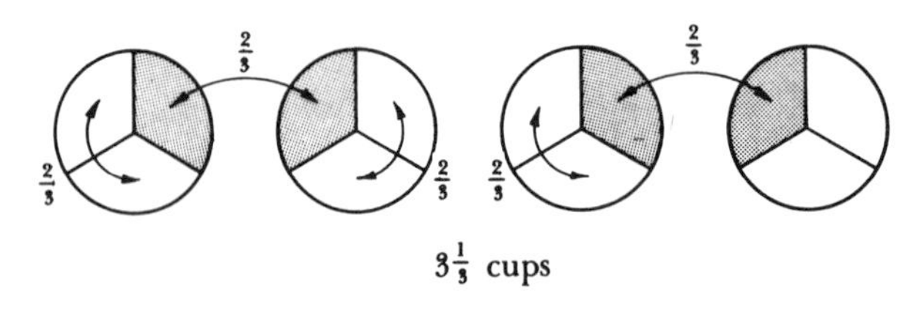

$3\frac{1}{3}$ cups

MAY: I used the number line first and then studied the mathematical sentence $5 \times \frac{2}{3} = N$. I decided that I could find the answer by multiplying the numerators and then dividing by the denominator.

The class accepted all the solutions as being logical. They preferred the use of addition and the number line to represent the situations. The multiplication method May suggested was considered to be the most efficient. The next day, the pupils used the method suggested by May to work a number of multiplication exercises and used a drawing, adding, or the number line to verify their answers.

Analysis. The process of development just described has several features: (1) The multiplication of $5 \times \frac{2}{3} = N$ can be closely identified with whole-number multiplication and is thus understandable for the large majority of pupils; (2) the use of the verbal problems provides a setting that points out the need for multiplication and also allows the pupils to abstract from a physical-world setting; and (3) the use of multiple solutions allows for individual creativity and confidence.

Finding Products Such as $\frac{1}{2} \times \frac{2}{3} = N$

The study of this multiplication type was introduced with directions to write the mathematical sentence and to solve the problem that follows in as many ways as pupils could develop.

Land is often divided into square regions that are 1 mile by 1 mile. The park is to occupy a portion of a 1-square-mile section that is $\frac{1}{2}$ mile wide and $\frac{2}{3}$ mile long. What fraction of a square mile will the park occupy?

Drawings were typically used by the pupils to find an answer to the mathematical sentence, which they identified as $\frac{1}{2} \times \frac{2}{3} = N$. Two solutions provided by pupils are given below.

RALPH: I made a drawing of the square region. The park occupies two out of six parts of the square mile. The answer is $\frac{2}{6}$ or $\frac{1}{3}$ of a square mile.

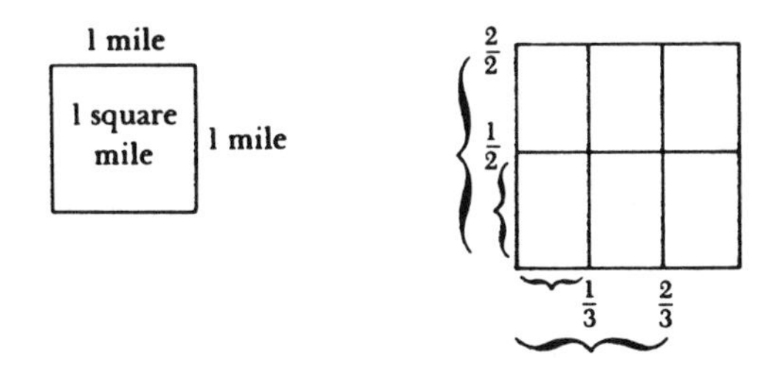

PERRY: I worked the problem the same way and then went back and looked at the mathematical sentence $\frac{1}{2} \times \frac{2}{3} = N$. I notice that by multiplying numerators by numerators and denominators by denominators, the result is the correct answer, $\frac{2}{6}$. I want to explore this further.

During the next few days, the pupils worked problems and exercises involving the multiplication of a fraction by a fraction. Their study verified that the product of a fractional multiplication could be found by multiplying the numerator by the numerator and the denominator by the denominator. The pupils generalized their "rule" by the use of frames and letters to

$$\frac{\square}{\triangle} \times \frac{\bigcirc}{\triangle} = \frac{\square\bigcirc}{\triangle\triangle}; \frac{a}{b} \times \frac{c}{d} = \frac{ac}{bd}$$

The multiplication of fractions such as $\frac{4}{3} \times \frac{3}{2}$ causes little difficulty in computation. Such multiplications are difficult to rationalize. If we consider the park problem just discussed in terms of square regions, the problem becomes that of finding $\frac{4}{3}$ of $\frac{3}{2}$ of a region. The following diagrams illustrate this situation:

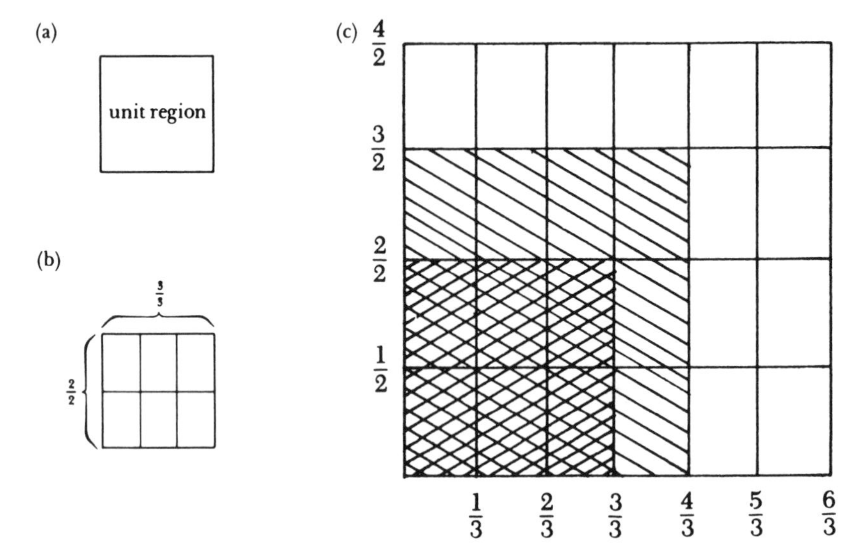

Diagram (c) illustrates that $\frac{4}{3} \times \frac{3}{2} = 12$ parts of a unit region that contains 6 parts. Thus, the denominator is named by the number of parts in a unit region.

Analysis. The approach featured the use of rectangular regions to find the products. This relates this phase of rational-number multiplication to the earlier work with whole-number multiplication in which arrays were used to find the cross-products of sets. The use of rectangular regions is superior to the use of circles, because circles cause great difficulty when fractions naming a number greater than 1 are used. Also, circles are not as representative of physical-world situations requiring multiplication in the form $\frac{a}{b} \times \frac{c}{a}$. Thus, circles do not as readily lead to pupil discovery.

This multiplication type can be experimented with using a number line and several rubber bands. The children are given an envelope with several thick rubber bands and also a duplicated sheet containing several number lines. They are directed to cut one of the rubber bands into two sections, then take one of the sections and use a pen to mark it off into two equal parts. (Most children do this by folding.) Then the teacher directs the children's attention to the number lines: "See if you can use the rubber band and the number line to solve this problem. Twelve members of the science club were visiting the observatory. The astronomer said, 'I can take only $\frac{1}{2}$ of the club at a time on the observation platform.' How many members could go on the platform at one time?"

The children experiment and quickly find that they can stretch the rubber band on the number line to partition 12 into halves.

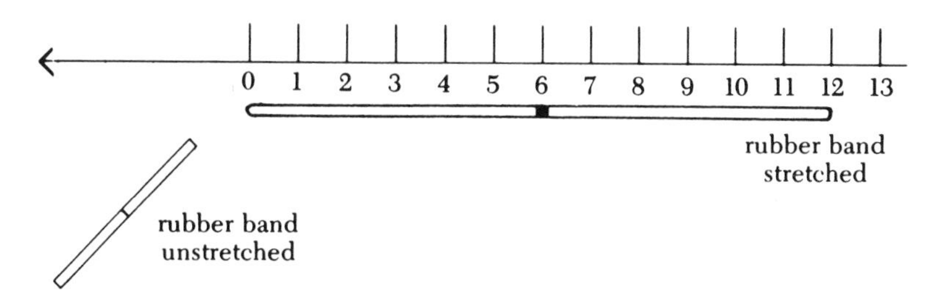

The children are then given a lab sheet with a number of similar problems and exercises requiring them to make marked rubber bands for different fractional amounts.

Finding Products Such as $\frac{2}{3} \times 6 = N$

Experimentation and discussion over a period of time help the children to recognize the following facts:

1. The multiplication $\frac{2}{3} \times 6 = N$ is the numerical equivalent of $6 \times \frac{2}{3} = N$ because of the commutative property. If $6 \times \frac{2}{3} = N$ can be thought of as $\frac{2}{3} + \frac{2}{3} + \frac{2}{3} + \frac{2}{3} + \frac{2}{3} + \frac{2}{3}$, we may think of $\frac{2}{3} \times 6$ as meaning the use of 6 as an addend less than one time. Thus, we multiply by 2 and divide by 3.
2. In the lower grades, $\frac{1}{3}$ of a number is found by dividing by 3. To find $\frac{2}{3}$ of a number, we can divide by 3 and then multiply by 2, or multiply by 2 and then divide by 3.
3. When we multiply by a number less than 1, we are looking for a part of a number. Problem situations for $\frac{1}{2} \times 6$ are usually stated: "Find $\frac{1}{2}$ of 6." When a fraction with a numerator of 1, such as $\frac{1}{5}$, is involved, the unit fraction is the multiplicative inverse of 5. Thus, multiplication by $\frac{1}{5}$ is equivalent to division by 5. The pupils can be aided in achieving an intuitive understanding of this concept. They are then taught that *of* means *times*. An effort should be made to develop an understanding of this idea. Too often, pupils multiply any time they see the word *of* in a problem.
4. Another difficulty is the need for pupils to realize that when they are multiplying rational numbers, the product is not always greater than either factor. The teacher needs to stress analysis such as $\frac{1}{2} \times \frac{1}{4} = \frac{1}{8}$. Why? Notice that the area of a region $\frac{1}{2}$ by $\frac{1}{4}$ units is less than one; remember that $\frac{1}{2}$ of 6 was 3. That was the same as dividing by 2. One-half of $\frac{1}{4}$ would be the same as dividing $\frac{1}{4}$ by 2.

Analysis. Basically, the same format was followed as in the other forms of multiplication. It is important to remember that problems involving the form $\frac{3}{4} \times 8 = \square$ are easy to compute but hard to understand. Often, many class discussions are necessary before this phase of fractional multiplication is understood.

Use of the Identity Element for Multiplication

Pupils intuitively make use of the property of the identity element for multiplication long before a formal analysis is needed. For example, pupils multiply both the numerator and the denominator of $\frac{3}{4}$ by 2 to change the fraction to $\frac{6}{8}$ before they are familiar with multiplication of rational numbers.

Soon after multiplication of rational numbers is introduced, several problem situations involving computation—such as $\frac{4}{4} \times \frac{3}{4} = N$ and $\frac{3}{3} \times \frac{2}{3} = N$—may be given. Then a guided discussion can bring out the idea that a renaming has occurred, because $\frac{3}{4} = \frac{12}{16}$ and $\frac{2}{3} = \frac{6}{9}$. Pupils should then be led to note that both $\frac{4}{4}$ and $\frac{3}{3}$ are other names for 1. Thus, what has occurred has been a change in the name of a number by multiplying, using a form of the identity element for multiplication.

Further work should provide pupils with an opportunity to discover that both numerator and denominator can be "broken" into factors and 1 "drops out":

$$\frac{9}{12} = \frac{3 \times 3}{4 \times 3} = \frac{3}{4} \times 1 = \frac{3}{4}$$

and later:

$$\frac{9}{12} = \frac{3 \times 3}{4 \times 3} = \frac{3}{4}$$

Elementary school pupils should have many opportunities to develop equivalent fractions by using various names for 1. Three representative exercises follow:

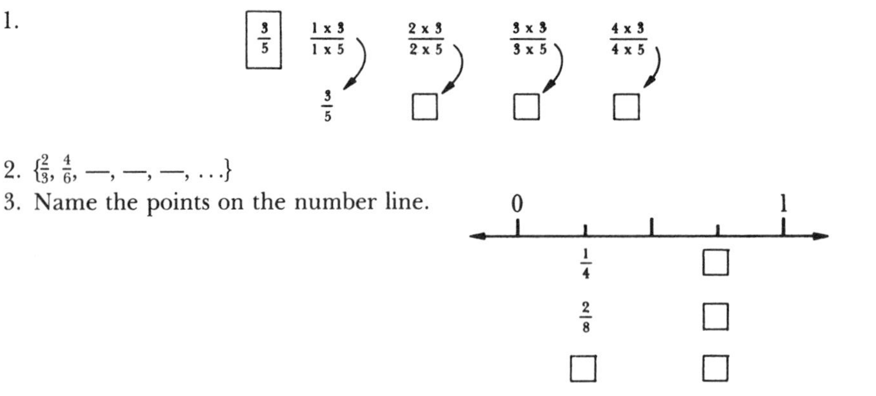

Pupils should indicate the point on a number line named by each set of equivalent fractions. Then they should find (by discussion and checking references) that each set of equivalent fractions names one rational number and one point on the number line. They should also find out that although each rational number has many names, the fraction in its lowest terms is usually used to name the number.

Estimating. One teacher gave the children several photocopied sheets containing multiplication of fractions. The students worked in pairs to find the products and discussed their findings. For example,

Find the product of

1. $\frac{1}{10}$ of $3\frac{1}{2}$
2. $\frac{1}{2}$ of $3\frac{1}{2}$
3. $\frac{3}{4}$ of $3\frac{1}{2}$
4. $\frac{9}{10}$ of $3\frac{1}{2}$

Some questions to think about were added: (1) Was the product greater or less than $3\frac{1}{2}$? (2) Which product was "larger," and which was "smaller?" (3) How can you tell whether a product is going to be a lot larger or a lot smaller?

Next, estimate the product that is closest to the exact answer.

	a	*b*	*c*	*d*
1. $\frac{9}{10}$ of $7\frac{5}{6} \approx$	$7\frac{1}{2}$	$3\frac{1}{2}$	$8\frac{1}{3}$	$15\frac{3}{4}$
2. $\frac{1}{2} \times 22\frac{3}{4} \approx$	$20\frac{1}{2}$	$32\frac{3}{4}$	$23\frac{1}{2}$	$11\frac{3}{8}$
3. $\frac{3}{20}$ of $17\frac{1}{2} \approx$	$2\frac{5}{8}$	$15\frac{1}{2}$	$20\frac{2}{3}$	$30\frac{3}{4}$
4. $\frac{3}{4} \times 97\frac{7}{8} \approx$	$96\frac{1}{3}$	$73\frac{2}{3}$	$20\frac{1}{2}$	$121\frac{1}{3}$

TEACHING DIVISION OF RATIONAL NUMBERS

Teaching the division of rationals has always caused difficulty for both teachers and students. In fact, at one time, many who were interested principally in the social utility of arithmetic topics suggested that division using fractions might well be deleted from the arithmetic curriculum. In recent years, most authorities on the teaching of elementary school mathematics have agreed that the study of division is valuable for its mathematical contribution. Knowledge of the process is essential to the study of algebra, and it is often used in the scientific fields.

As in the case of multiplication, six forms of division involve fractions, mixed forms, and whole numbers. The six forms can be illustrated by the following examples: (1) $6 \div \frac{3}{4} = \square$; (2) $\frac{3}{4} \div 6 = \square$; (3) $\frac{3}{4} \div \frac{1}{6} = \square$; (4) $6\frac{1}{2} \div \frac{3}{4} = \square$; (5) $\frac{3}{4} \div 6\frac{1}{2} = \square$; and (6) $6\frac{1}{2} \div 1\frac{1}{2} = \square$. The most universal type is that of form 3, $\frac{3}{4} \div \frac{1}{6} = \square$. If the development of computational skill in division were the only criterion, this form would be the most logical for introductory purposes, for all the other forms can be converted to it. However, if we believe that pupil discovery and understanding of the relationship between rationals and whole numbers are important, study should begin with form 1, $6 \div \frac{3}{4} = \square$.

Two major methods of teaching the inversion procedure are in use today, the common-denominator approach and the reciprocal approach. Each procedure has its advocates and valid reasons to support it.

Because the use of multiple solutions to problems and exercises is a valuable learning procedure, it is suggested that neither the common-denominator method nor the reciprocal method be discarded; both can be used in developing the process of division. The illustrative lessons in Chapter 1 describe a means of making use of both procedures. (See Chapter 1 for the major developmental work on division with fractions.)

Division Without Pencil and Paper

The multiplication of rationals without pencil and paper is relatively simple. However, because of the necessity of either finding a common denominator or inverting, the division of rationals without pencil and paper is often difficult. But there is a procedure that eases such work. Rather than invert, the pupils can take the cross-products of the number pairs. For example,

$$\frac{5}{9} \div \frac{3}{4} = N \qquad \frac{3}{5} \div \frac{8}{9} = N$$

$$\frac{5}{9} \div \frac{3}{4} = \frac{20}{27} \qquad \frac{3}{5} \div \frac{8}{9} = \frac{27}{40}$$

(a) (b)

This procedure is actually the mathematical definition of the division of rationals:

$$\left(\frac{a}{b} \div \frac{c}{d} = \frac{a \times d}{b \times c}\right)$$

Pupils can normally discover this procedure if the teacher writes several division computations on the chalkboard and then quickly writes down the answers. In trying to figure out why the teacher can solve the division this quickly, the pupils begin to note the cross-product relationship.

A Problem Situation

The problem shown here is a good example of the type of situations needed to further children's thinking about division with fractions. The teacher's manual suggests using estimation as a key element in such problems. It is further suggested that you try to develop several other division problems.

> Mr. Griffin has $32\frac{1}{2}$ yards of material from which to make suits. Each suit requires $1\frac{2}{3}$ yards of material.
>
> How many suits can Mr. Griffin make from $32\frac{1}{2}$ yards of material? Work in groups. Find at least three different ways to solve this problem.
>
> Does each way of solving the problem give you the same answer? How can you check to be sure your answer is right?[4]

[4]From Stephen A. Willoughby et al., *Real Math, Grade Six* (LaSalle, Ill.: Open Court Publishing, 1985).

SIMPLIFICATION OF MULTIPLICATION AND DIVISION

The simplification of multiplication and division (after inversion has occurred) is usually referred to as *cancelation.* Two examples of simplification or cancelation are

$$\frac{\overset{1}{\cancel{3}}}{2} \times \frac{7}{\underset{3}{\cancel{9}}} = \frac{7}{6} = 1\frac{1}{6} \qquad \frac{\overset{1}{\cancel{2}}}{\underset{1}{\cancel{3}}} \times \frac{\overset{2}{\cancel{6}}}{\underset{4}{\cancel{8}}} = \frac{2}{4} = \frac{1}{2}$$

Pupils often indiscriminately cross out numerals in inappropriate situations, such as in division before the divisor has been inverted. Difficulties with cancelation may be caused by incomplete handling of the reason for cancelation in textbooks and by poor foundation work in multiplication of rational numbers.

A careful study of cancelation reveals that applications of the basic properties of mathematics are involved. In multiplying $\frac{2}{3} \times \frac{3}{4} = N$, it should be remembered that when the multiplication involving rationals is performed, whole numbers are used; that is,

$$\frac{2 \times 3}{3 \times 4} = \frac{6}{12} = \frac{1}{2}$$

Factoring is an excellent means of canceling. It provides another look at cancelation and is also an important preparation for algebra. Factoring may be used in the following manner:

$$\text{First time: } \frac{2}{3} \times \frac{3}{4} = \frac{3 \times 2}{3 \times 4} = \frac{3 \times 2 \times 1}{3 \times 2 \times 2} = \frac{3}{3} \times \frac{2}{2} \times \frac{1}{2} = \frac{1}{2}$$

$$\text{Second time: } \frac{2}{3} \times \frac{3}{4} = \frac{\overset{1}{\cancel{2}}}{\underset{1}{\cancel{3}}} \times \frac{\overset{1}{\cancel{3}}}{\underset{1}{\cancel{2}} \times 2} = \frac{1 \times 1}{1 \times 2} = \frac{1}{2}$$

To avoid many of the problems of cancelation, I suggest that teaching it be delayed until the pupils understand operations on fractions with all the fundamental operations. The use of cancelation may well be delayed for several months after the multiplication of fractions is introduced. Also, developing cancelation in a guided-discovery manner should help pupils identify the significant aspects of the process. In conjunction with this development, the teacher can ask the pupils to make use of the basic mathematical properties to verify that cancelation is "legal."

The Multiplication–Division Relationship

The inverse relationship of division to multiplication provides a ready check for division of rational numbers. Pupils should be encouraged to discover methods of checking division problems and exercises. The previous work on whole-number multiplication and division should provide students with a lead in developing the idea that $8 \div \frac{2}{3} = 12$ may be checked by multiplying $12 \times \frac{2}{3} = \square$. If pupils do not discover this relationship, understanding of it may easily be developed by a series of guided questions.

KEEPING SHARP

Self-Test: True/False

_______ 1. Although there are several possible interpretations for ordered pairs of numbers ($\frac{a}{b}$), the teaching strategy remains the same for all cases.

_______ 2. Informal introductions to fractions should be an important part of the primary-grade mathematics program.

_______ 3. Any rational number may be represented in many ways.

_______ 4. The ability to convert a fraction to an equivalent form may be mathematically interesting for bright students, but it is not a necessary skill for all students.

_______ 5. The use of fractional number lines may help children visualize the relative sizes of various common fractions.

_______ 6. Multiplying the numerator and the denominator of a fraction by the same number makes use of the identity element for multiplication.

_______ 7. Computations requiring the use of fractions can best wait until the middle or upper grades, when pupils have some facility with the algorithms for the four operations.

_______ 8. Addition is not closed for rational numbers, because if two rational numbers are added, the sum cannot always be expressed as a fraction.

_______ 9. The need to find the least common denominator efficiently occurs in the addition of rational numbers and should be taught at that time.

_______ 10. There are no "real-world" situations requiring the multiplication of a fraction by a fraction.

_______ 11. Because division of rational numbers is the inverse of multiplication of rational numbers, multiplication is a good check for correctness of division problems.

_______ 12. Division of rational numbers can be performed by using cross-products rather than inverting the divisor.

_______ 13. Division of rational numbers is a mathematical curiosity and has no practical value.

_______ 14. Fractions are equally useful in the metric and English systems of measurement.

Vocabulary

reciprocal	prime number
LCD	multiplicative inverse

THINK ABOUT

1. How did the use of such schemes as "unit fractions only" and "sexagesimal fractions only" make for easier understanding of fractions?
2. What schemes for rationalizing the inversion procedure in fraction division are used in various textbooks? Use at least four series of books.
3. If the simplification of a complex fraction procedure is used to introduce fraction division, what prerequisite knowledges are most essential?
4. Do all children need to learn such computations as "Add $\frac{3}{4}$, $2\frac{1}{6}$, $3\frac{9}{15}$, and $1\frac{4}{25}$"?
5. What are some ways to present fractions other than illustrating pie section on a flannel board?

SUGGESTED REFERENCES

BITTER, GARY G., MARY M. HATFIELD, and NANCY T. EDWARDS, *Mathematics Methods for the Elementary and Middle School: A Comprehensive Approach*, chap. 10. Boston, Mass.: Allyn and Bacon, 1989.

BEHR, MERLYN J., and THOMAS R. POST, "Teaching Rational Number and Decimal Concepts." In *Teaching Mathematics in Grades K-8: Research-Based Methods*, Thomas R. Post (ed.). Boston, Mass.: Allyn and Bacon, 1988.

COXFORD, ARTHUR F., and LAWRENCE W. ELLERBRUCH, "Fractional Numbers." In *Mathematics Learning in Early Childhood, Thirty-Seventh Yearbook*, pp. 191–204. Reston, Va.: The National Council of Teachers of Mathematics, 1975.

DIENES, Z. P., *Fractions: An Operational Approach*. New York: Herder and Herder, 1967.

DILLEY, CLYDE A., and WALTER E. RUCKER, "Division with Common and Decimal Fractional Numbers," *The Arithmetic Teacher* 17 (May 1970): 428–32.

ELLERBRUCH, LAWRENCE W., and JOSEPH N. PAYNE, "A Teaching Sequence from Initial Fraction Concepts through the Addition of Unlike Fractions." In *Developing Computational Skills*. Reston, Va.: National Council of Teachers of Mathematics, 1978.

Experiences in Mathematical Ideas, vol. 1, chap. 5. Washington, D.C.: National Council of Teachers of Mathematics, 1970.

FEHR, HOWARD F., "Fractions as Operators," *The Arithmetic Teacher* 15 (March 1968): 228–32.

PAYNE, JOSEPH N., "Curricular Issues: Teaching Rational Numbers," *The Arithmetic Teacher* 31 (February 1984): 14–17.

POST, THOMAS R., "Fractions: Results and Implications from National Assessment," *The Arithmetic Teacher* 28 (May 1981), pp. 26–31.

POST, THOMAS R., MERLYN J. BEHR, and RICHARD LESH, "Interpretations of Rational Number Concepts." In *Mathematics for the Middle Grades (5–9)*, chap. 7. Reston, Va.: The National Council of Teachers of Mathematics, 1982.

PRIELIPP, ROBERT W., "Teaching One of the Differences between Rational Numbers and Whole Numbers," *The Arithmetic Teacher* (May 1971): 317–20.

THIESSEN, DIANE, MARGRET WILD, DONALD D. PAIGE, and DIANE L. BAUM, *Elementary Mathematical Methods*, 3rd ed., chap. 8 and 9. New York: Macmillan, 1989.

chapter 11

DECIMALS

OVERVIEW

Understanding positional notation and the concept of 10 as the base of our system are two of the most important ideas of our number system. These concepts come into play in all phases of work with decimals. Also, decimals are of increasing importance in everyday life as the calculator becomes more into play and the metric system gains wider use.

It is much more important to stress problem solving and concept understanding in decimal work than computational proficiency. However, the calculator can add greatly to both computational proficiency and concept development.

The National Council of Teachers of Mathematics suggest including decimals so that children can

1. Develop decimal concepts.
2. Develop decimal number sense.
3. Use models related to decimals and notation.
4. Use models to explore decimals.
5. Apply decimals in problem situations with and without calculators.
6. Extend whole number operations to decimals.
7. Use mental computation with decimals.
8. Be able to solve challenging problems involving decimals.

TEACHER LABORATORY

1. Break the code to complete the chart:

a. 6 → 6	d. 16 → 20	h. 15 → □
b. 9 → 11	e. 34 → 42	i. 42 → □
c. 14 → 16	f. 64 → 100	j. 68 → □
	g. 29 → □	

   What have you done?

2. *Materials:* squared paper. You are working in a candy factory. The squared paper represents pieces of candy. Think of the pieces this way:

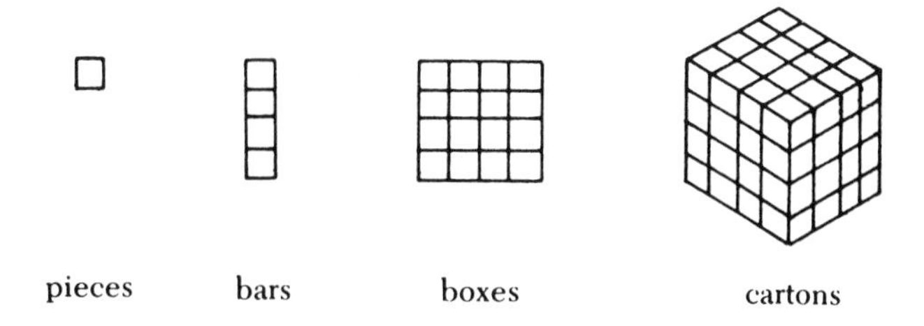

   Use a place-value frame:

cartons	boxes	bars	pieces

   Cut out the following number of pieces of paper: 23, 16, 31, 35. Categorize them by using the place-value frame. What do you find? How does this compare with dealing with another base? Which base?

3. Write the following fractions as decimals:

$$\frac{1}{11} \quad \frac{2}{11} \quad \frac{3}{11} \quad \frac{4}{11} \quad \frac{5}{11} \quad \frac{6}{11} \quad \frac{7}{11} \quad \frac{8}{11} \quad \frac{9}{11} \quad \frac{10}{11}$$

   How many digits recur in the decimals? What pattern can you find?

TAKE INVENTORY

1. Illustrate a decimal by using an abacus, squared paper, and a place-value frame.
2. Develop a lesson by using ancient systems of numeration.
3. Tell how the ideas of factors and the need to practice multiplication could be used to develop exponents.

4. Collect a variety of suitable materials, then develop a lesson by using the laboratory approach to teach scientific notation using exponents.
5. Should the decimal point be used in defining a decimal fraction? Why or why not?
6. What are the basic differences between the same-decimal-unit and the power-of-10 methods of dividing decimals?

DECIMALS AND NUMERATION

Historical Development

The Babylonians were the first to develop a system of fractions that used exclusively a multiple or power of the base as the denominator. Because the Babylonians used a base of 60, a sexagesimal (sek-sa-jes-a-mal), the denominators of their fractions were multiples of 60, such as $\frac{3}{60}$ and $\frac{45}{3,600}$. The sexagesimal fraction was still used for scientific purposes during the Middle Ages. During that period, 2 hours 20 minutes 45 seconds was written $2 + \frac{20}{60} + \frac{45}{3,600}$ hours, rather than $2 + \frac{1}{3} + \frac{1}{80}$ hours.[1]

The decimal method of writing fractions was developed during the Middle Ages. The Middle Eastern astronomer and mathematician al-Kashî (early 1400s) used the characters *sah-hah* 1415926535898732. Translating the *sah-hah* gives:[2]

Integer

3 14159. . . .

Several European mathematicians used a form of decimals in the early sixteenth century. The first published explanation of the decimal fraction appeared in 1585 in the book *La Disme* by Simon Stevin, a Flemish mathematician. Stevin would have written 7.564 as $7\,_0\,5\,_1\,6\,_2\,4\,_3$. Later writers used the following forms:

123 . . .

$7,\ 5'\ 6''\ 4'''$; $7/\overline{564}$; 7/564; $7\underline{/564}$

The writing of decimals is still not standardized. The following different forms of decimals are still in use:

United States: 7.564

Great Britain: 7 · 564

France and Scandinavia: 7,564

[1]From D. E. Smith, *History of Mathematics* (New York: Dover, 1953), vol. 2, p. 228.
[2]*Ibid.*, pp. 238–39.

Modern Interpretation

In the context of today's mathematics for the elementary school, decimals may be considered as another name for rational numbers. Thus, the numerals 3/4 and 0.75 represent the same rational number. As children progress up the mathematical ladder, they learn that decimals have a wider use; they are used to convey the idea of irrational numbers. As you probably recall from mathematics courses, a rational number can be written as a repeating decimal; for example, .50000... or .33333.... An irrational number does not produce a repeating decimal.

EXTENDING PLACE VALUE

The extension of place value to many places is an important readiness activity to decimals, because it helps to develop the type of thinking necessary for success with decimals. Today, with the vast distances involved in space travel, the enormous cost of government, movement to the metric system, and the infinitesimal measures used with instruments such as the electron microscope, decimal notation is of real interest to the elementary schoolchild. In the material that follows are suggestions for introducing the reading of large numbers and a movement into decimals.

Illustrative Situation by Using Place Value

In explaining the extension of place values, a teacher said, "I read in the paper that the cost of the development of a solar cell is the amount I've written on the board. [$210,564,542] How much is this? Let's indicate the value of each digit. What are the values of the various units?" As the pupils indicated the value of a digit, the teacher wrote the value above the numeral.

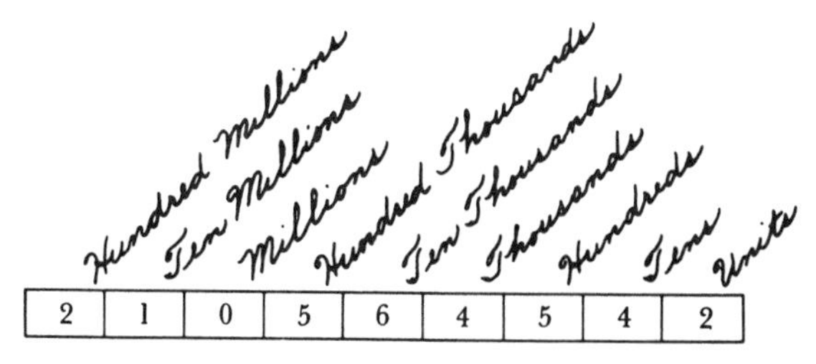

In cases where the pupils were unsure of the value, the teacher led the class to the answer with a question. If the children did not know, the value was supplied by the teacher.

Next, the teacher said, "Now let's read that amount. What would we have?" A student read, "Two hundred millions, one ten million, zero million, five hundred thousands, six ten thousands, four thousands, five hundreds, four tens, two ones."

The teacher asked, "Is the way you read the numeral the way we normally read numerals?" The pupils agreed that it was not. They described the procedure of reading the million only once (two hundred ten million), the thousand only once (five hundred sixty-four thousand), and the 542 as five hundred forty-two (not reading the tens and ones).

After further questioning by the teacher, the chart below was developed:

millions	thousands	ones
210	564	542

Then the pupils were asked to think out and to write a short statement concerning the reading of large numbers. One pupil's statement was, "To read a large number, separate it into groups of three digits, starting with the ones. Name each group of three. Read the numerals in each group as though they were ones, but give the period name after each, with the exception of ones. It's just understood that it's the ones. For example, 23,456,782 is 23 million, 456 thousand, 782."

From experience the children were familiar with *deka, hecto,* and *kilo.* With the use of resource materials, the chart that follows was developed by the children:

1,000,000	100,000	10,000	1000	100	10	1
mega	—	myria	kilo	hecto	deka	unit (meter, gram, etc.)

Billion — 1,000,000,000 / giga

Trillion

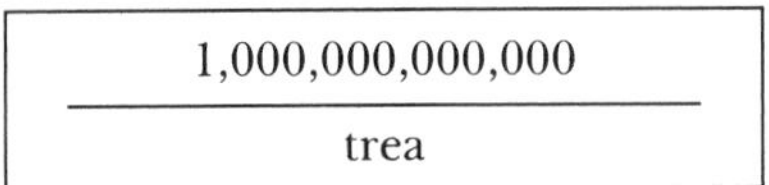

FOUNDATION WORK WITH DECIMALS

Decimals are used extensively in real-world situations. Therefore, it is not difficult to develop good use situations for introducing them. In fact, the teacher should exert caution to make sure that the decimal idea is really understood. The child's familiarity with dollars and cents may obscure his or her mathematical thinking concerning decimals. A basic concept that can be developed by a combined use of money and metric measure is the importance of 1. As in the case of the study of fractions, 1 should be emphasized in decimal study. At the lower-grade level, children report their measures in terms of centimeters and millimeters. At the upper-grade level, they can experiment to find that 2 centimeters plus 3 millimeters can be written as 2.3 centimeters.

In primary grades, pupils should have opportunities to deal with many fractions that have denominators of a power of 10. Particular emphasis should be given to the manipulative and oral phases of the study. This work naturally leads to notation in a natural manner. Below are several activities that primary teachers find useful in beginning the development of the ideas of decimals:

1. Children were given a ruler marked in centimeters and asked to draw a line a little more than .07 meters; a line a little less than .14 meters; etc.
2. "Look at your set of rods" (Cuisenaire or similar rods measuring from 1 cm to 10 cm). "The orange rod [10 cm] will be called *one*. Which rod would be one-tenth?" (white); "two-tenths?" (red); etc.
3. Children were given squared paper and directed to cut out a number of 10-by-10 squares. They were also given a lab sheet, part of which follows:

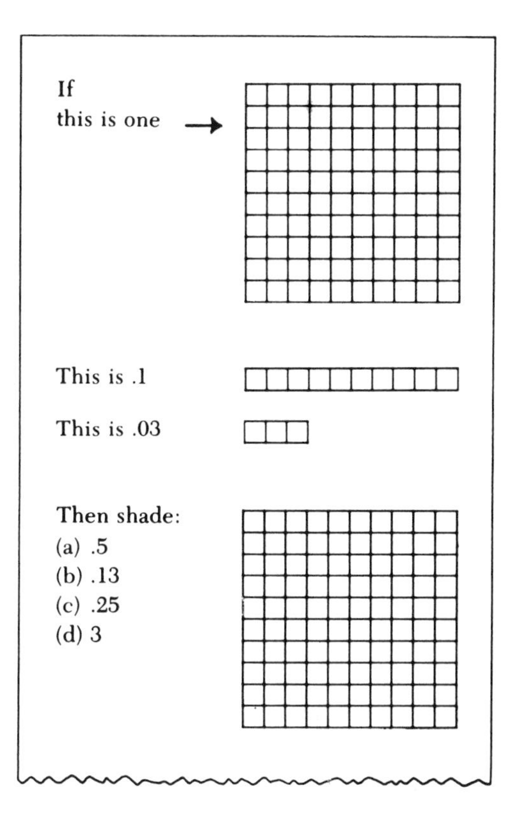

4. If a meter is 1, a decimeter is ______, and a centimeter is ______.
5. Alice used rods to show the value of three-tenths. Is she right?

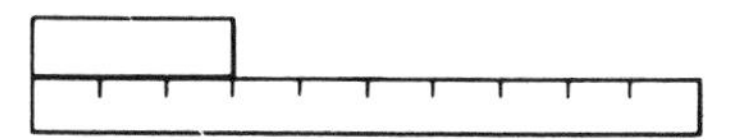

6. How many hundredths are three-tenths? Use the chart above to determine this.
7.

HECTOMETER	DEKAMETER	METER	DECIMETER	CENTIMETER
100	10	1	1/10	1/100

What do these mean?

8. Jill said that there had been 18 centimeters of snowfall. How could you say this using meters? (.18)
9. Ken reported that his father told him the right-hand column of the odometer (often incorrectly called the speedometer) measures in tenths of a mile. If the tenths-column odometer was on zero when Ken's father left home, how far has he traveled?

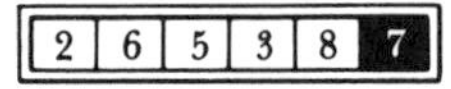

10. Larry says that it's 500 meters from school to the dairy store. Orin says that Larry's wrong and that the distance is one-half a kilometer. How could you settle their argument?
11. How is this abacus different from others we've used? How could you show $3\frac{5}{10}$ $6\frac{23}{100}$?

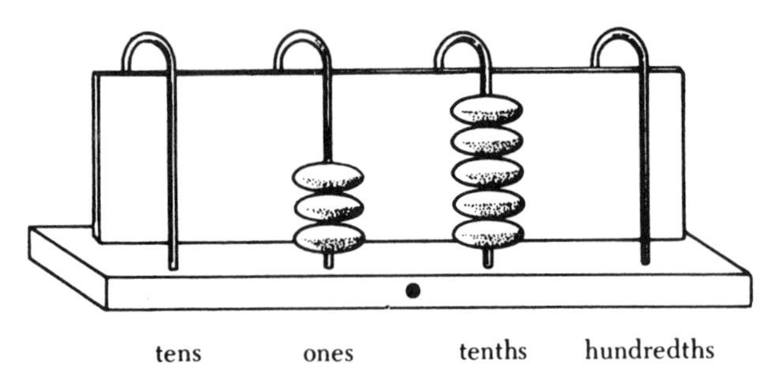

READING AND WRITING DECIMALS

Previous everyday use of and instructional background in the reading and writing of decimals should allow the teacher to begin with a situation that permits a rather thorough analysis of the meaning of decimals. A duplicated sheet containing several situations that used decimals was given to the pupils. The pupils were then instructed, "Read these statements to yourself and be ready to discuss the value of the numbers involved in the statements." The first three statements follow:

1. The population in the United States in 1900 was 76.212 million.
2. Jim walked 3.4 kilometers.
3. The length of the electronic part is 2.5 cm.

The children were then asked to read the statements aloud and to comment on the meaning of each. When the teacher felt that they were grasping some of the ideas of decimals, individual children were given an uncompleted place-value chart (see table) and a short time to try to complete it. Then in small groups they discussed their findings, with help from the teacher if they had difficulty.

1000 (10 × 10 × 10)	100 (10 × 10)	10 (10)	1 (1)	$\frac{1}{10}$ $\frac{1}{(10)}$	$\frac{1}{100}$ $\frac{1}{(10 \times 10)}$	$\frac{1}{1000}$ $\frac{1}{(10 \times 10 \times 10)}$
0 thousands	1 hundreds	7 tens	9 ones	6 3 tenths	2 hundredths	3 thousandths

When the groups had completed the charts, the teacher conducted the following discussion:

TEACHER: What is the purpose of the decimal point?
EVE: To tell us which are decimals.
TEACHER: Anything else?
ANN: It does that, but I think the main purpose is to tell us where the ones place is located. Look at the chart.
TEACHER: Good. How do we name the part that is on the right-hand side of the decimal point?
PAUL: Let me try an example: .7 is called tenths; .77, hundredths; .777, thousandths. Actually, the digit farthest to the right of the decimal point gives the decimal its name.
TEACHER: What if I want to write .534 as a fraction? How can I be sure I'm right?
RAY: Check on the chart. It would be $\frac{554}{1,000}$
TEACHER: Do you see any other pattern?
PHYLLIS: The decimal .534 has three numerals on the right-hand side of the decimal point, and there are three zeros in $\frac{554}{1,000}$. The number of decimal places tells us the number of zeros in the denominator when it's expressed as a fraction.

In addition to verifying the suggestions made in the previous material, the following points should be developed before proceeding further into work with decimals:

1. A numeral such as 4.56 may be read in several ways: "four and fifty-six hundredths," "four point fifty-six," or "four point five six." The use of the word *point* in the reading of a decimal has the distinct advantage of indicating the suggested decimal notation. For example, if someone asks that the numeral

indicating "seven and three-tenths" be written, the listener does not know whether $7\frac{3}{10}$ or 7.3 is wanted. "Seven point three" indicates that the decimal form is preferred for this case.

2. The use of *and* should be reserved for decimals. In common usage, 149 is often read as "one hundred and forty-nine." In this case, no error will be made. However, it is difficult to ascertain the meaning of "two hundred and twenty-eight thousandths." If *and* is used with whole numbers as well as fractions, the pupil does not know whether the meaning is 200.028 or .228.
3. Decimals and fractions are two types of symbolism for rational numbers. Fractions have numerators and denominators; with decimals, the denominator is not written. A number may be named either a *fraction* or a *decimal,* which terms are preferable to the older terms, *common* or *vulgar fractions* and *decimal fractions.* Thus, $\frac{3}{10}$ is called a fraction, and .3 a decimal. (It should be noted that there is not complete agreement among some in mathematics education; others do refer to decimal fractions.)
4. It is often helpful to place a zero before the decimal point for decimals with a value less than 1 (0.234). It is, in fact, required that one do so under SI, the International System of Units.
5. Pupils often read .01 as one-tenth because $\frac{1}{10}$ has two digits in the denominator. This misconception should be carefully discussed.
6. Emphasis should be placed on the *ones* place as central to decimal notation. This usage emphasizes structure, balance, and symmetry. Too often, the decimal point is taught as the center of the system. (See illustration.)

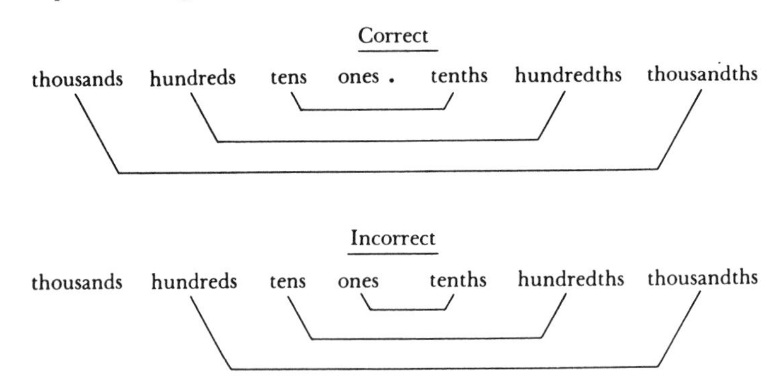

7. The steps in reading a decimal are read the part of the numeral to the left of the decimal point, read the decimal point as "and" or "point," and read the entire part of the numeral to the right of the decimal point as though it represented a whole number following that statement with the place value of the digit on the far right.

Addition and Subtraction with Decimals

Children have many experiences adding dollars and cents in the early grades, which makes addition with decimals computationally simple but may increase the difficulty of understanding it. Because of this early experience with the addition of hundredths (cents), I suggest that the introductory work in decimal addition begin with problems involving the addition of tenths or thousandths.

As a readiness for formal decimal addition, one teacher gave each child a paper ruler marked off in tenths and the lab sheet that follows:

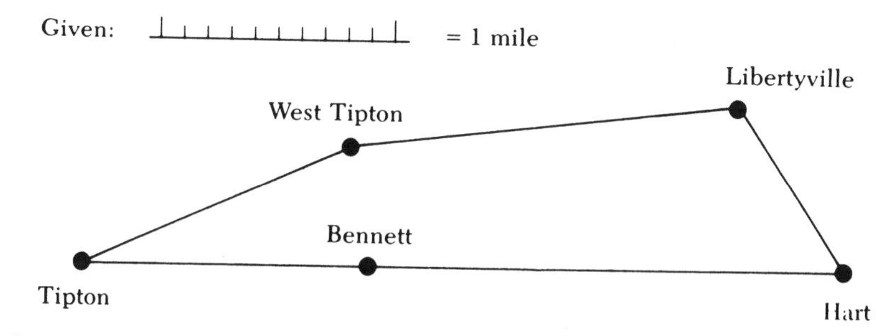

Find these distances

1. from Tipton to Libertyville through West Tipton
2. from Tipton to Hart
3. from Tipton to Libertyville through Hart

After the children had worked the problems, the teacher saw that a number of them had found the typical method of adding decimals, but she made no mention of the fact at the time. After several experiences such as the one before, the teacher began a more formal introduction with the following problem: "As part of a physical fitness program, the pupils of Kennedy School were to keep a record of the kilometers and tenths of a kilometer that they walked each week. Ralph walked 4.3 km one day and 3.4 the next. How far did he walk on the two days?" Directions were given to write a mathematical sentence that could be used to solve the problem and to verify the answer by solving the problem in another way. The pupils were also asked to think of ways in which addition varied when performed with decimals, whole numbers, and fractions. Some of the methods commonly used for solution by the pupils follow.

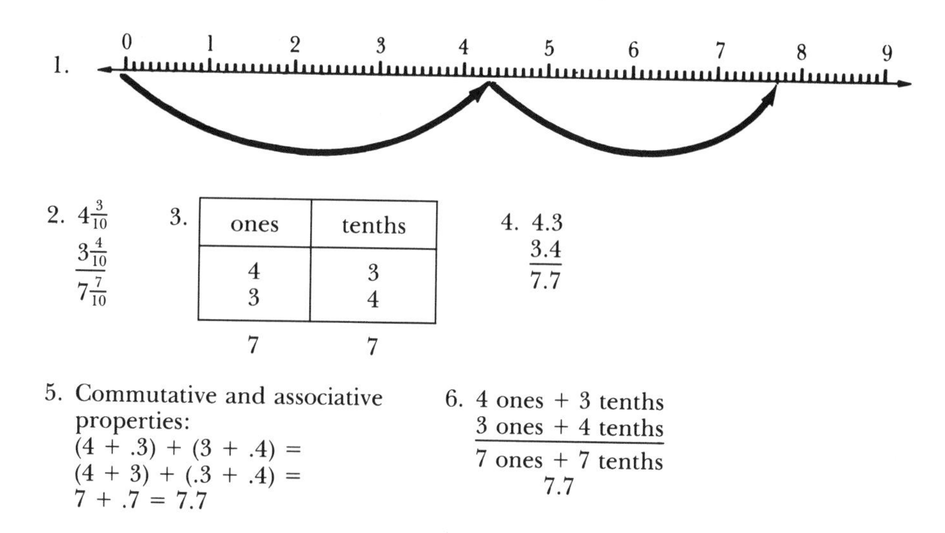

2. $4\frac{3}{10}$
$3\frac{4}{10}$
$7\frac{7}{10}$

3.

ones	tenths
4 3	3 4
7	7

4. 4.3
3.4
7.7

5. Commutative and associative properties:
(4 + .3) + (3 + .4) =
(4 + 3) + (.3 + .4) =
7 + .7 = 7.7

6. 4 ones + 3 tenths
3 ones + 4 tenths
7 ones + 7 tenths
7.7

In a following lesson, the pupils added numbers such as 34.532 and 6.724. The challenge was given: "Analyze your process of renaming in adding. Use expanded notation and other means to show your thinking." The pupils used a form similar to the following to illustrate their thinking:

$$\begin{array}{rl} 34.532 \rightarrow & 34 + \ \ .5 + .03 + .002 \\ \underline{6.724} \rightarrow & \underline{\ \ 6 + \ \ .7 + .02 + .004} \\ & 40 + 1.2 + .05 + .006 = 41.256 \end{array}$$

The addition of numbers such as $3.6 + 5.8 = N$, which requires regrouping, was solved by pupils in the following ways:

1. 3 ones + 6 tenths
   5 ones + 8 tenths
   8 ones + 14 tenths = 8 ones + 1 one + 4 tenths = 9 ones, 4 tenths = 9.4

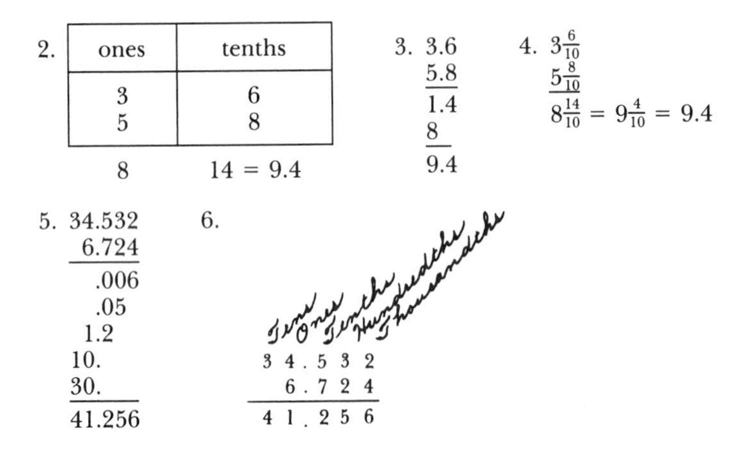

Other examples such as $.06 + .08 = N$ may be developed as 6 hundredths + 8 hundredths = 14 hundredths = 1 tenth, 4 hundredths = .14.

Place-value charts, pocket charts, and the abacus should be available at the mathematics table for those who need to work with concrete materials to understand the place-value concepts of decimals.

The teacher's use of leading questions should help pupils discover the following ideas concerning addition with decimals:

1. Although decimals and fractions name the same number, decimal addition is more closely allied to whole-number addition. The significant idea to remember is that computation is most easily handled when tenths are added to tenths, hundredths to hundredths, etc. Thus, it is extremely important to keep the place value of the addends in mind.
2. If the sum of a column is 10 or more, we have to rename. This is analogous to a fraction greater than 1.
3. The "carrying" or renaming process works from any column to the next, not just from ones to tens or from tenths to ones. This is true because any position is 10 times as great in value as the position to its right.
4. If the addends are expressed as tenths, the sum will be expressed as tenths, etc.
5. All the basic properties of whole-number addition (commutative property, associative property) and the role of the identity element (zero) hold true with decimals.

Subtraction with decimals can be generalized from addition with decimals. At this stage of their mathematical development, pupils understand the inverse relationship between addition and subtraction. Thus, they see the need for subtracting tenths from tenths and so on. They also generalize the renaming situations in the same way they did in addition. Place-value frames probably provide the best means of rationalization. Pupils who use varying procedures for multidigit subtraction (primarily additive and take-away thinking) should explore the rationale of their particular method.

Addition and Subtraction with "Ragged Decimals"

"Ragged decimals" occur in an addition or subtraction situation in which the elements are expressed in differing decimal units; for example, $2.34 + 3.4 + 1.564 = N$, or $3.45 - 2.3 = N$. There have been heated arguments among mathematics educators concerning the validity of using addition or subtraction of this type. Mathematically, such computations offer no problem because, in the addition example, the addend $3.4 = 3.400$, and thus all the addends can be easily converted to the decimal unit of the addend having the greatest number of decimal places. In this manner, the addition becomes

$$\begin{array}{r} 2.34 \\ 3.4 \\ +1.564 \\ \hline \end{array} \quad \text{or} \quad \begin{array}{r} 2.340 \\ 3.400 \\ 1.564 \\ \hline \end{array}$$

The bone of contention is the preciseness with which the decimals represent a physical-world measurement. If the addition above involved finding the total weight of several chemicals weighed in grams, questions such as the following could be asked: Were all the chemicals weighed with the same precision? We know that the 1.564 represents a chemical weighed to the nearest one-thousandth of a gram; has the chemical weighing 3.4 been weighed to the nearest thousandth of a gram, or only to the nearest tenth? If it has been weighed only to the nearest tenth of a gram, it may actually weigh anywhere between 3.35 and 3.45.

For enrichment, the study of significant digits can be undertaken.

MULTIPLICATION WITH DECIMALS

The major difficulty in teaching multiplication with decimals is determining the number of decimal places in the product. Traditional mathematics programs simply taught the following rule: "Count the number of decimal places in the factors. This is the number of decimal places that should appear in the product." At best, such a procedure produced only computational proficiency and left pupils wondering why a computation such as $3 \times .3 = N$ produced a product less than 1.

As in whole-number addition, the use of the grid is a valuable tool in introductory multiplication. Also, children will have used grids involving units less than 1 in their study of multiplication involving fractions. Problems such as the one on pp. 269–70 provide a lead into this type of solution.

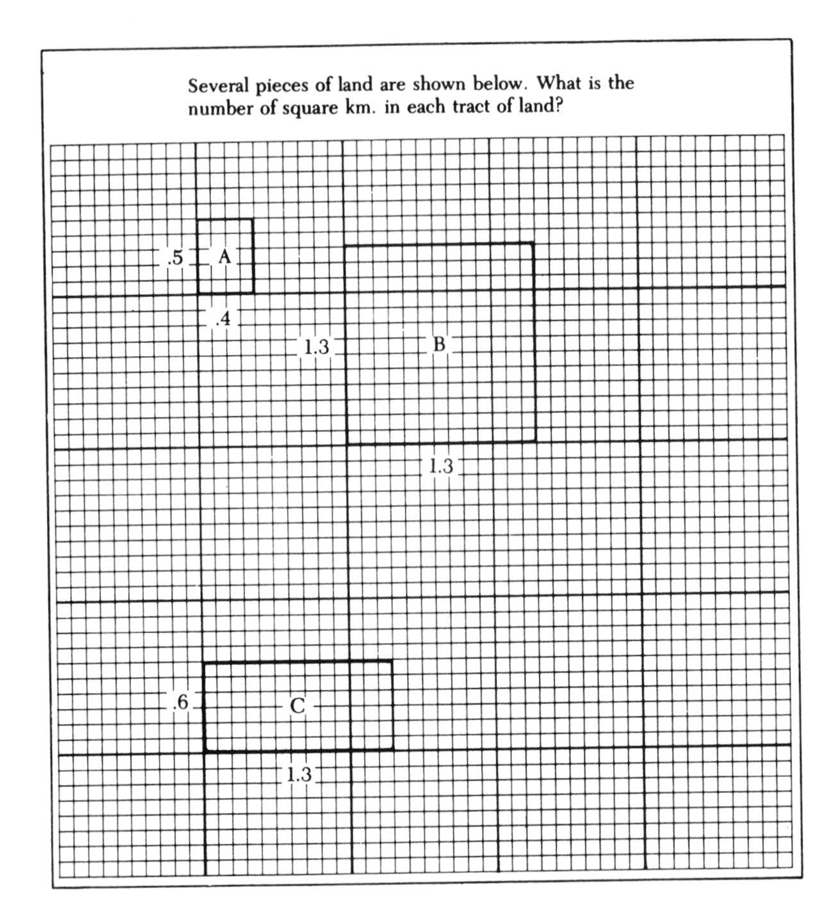

This type of problem can be followed with a problem such as this: "The pupils in the science club are going to perform the same experiment three times, to triple-check their results. If the experiment requires 3.7 grams of sulfur, how many grams of sulfur will they need for the three tests?" The pupils should write the mathematical sentence, estimate the answer, work the problem, and verify the results.

In one class the common solutions involved changing the decimal to a fraction and then multiplying, adding the decimals, and using the number line.

$$3 \times 3.7 = 3 \times 3\tfrac{7}{10} = 3 \times \tfrac{37}{10} = \tfrac{111}{10} = 11\tfrac{1}{10} \text{ or } 11.1 \text{ grams}$$

$$\begin{array}{r} 3.7 \\ 3.7 \\ +\ 3.7 \\ \hline 11.1 \end{array} \text{ grams}$$

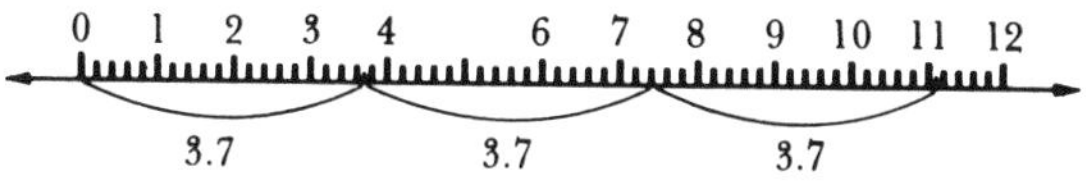

Several of the pupils had estimated the product as "a little over 10." They multiplied by using the decimals and placing the decimal point according to their estimate. One pupil explained, "I knew that the answer would be more than 10 but less than 15. I multiplied and got an answer of 111. I knew that this would mean that the answer should be 11.1."

Several other problems and computational exercises involving various types of decimal multiplication were given to the class. It was suggested that the assignment be worked by using fractions. The pupils were also to attempt to determine a means of placing the decimal point in the product of the decimal form.

During a discussion that followed the work period, the teacher asked, "If I multiply $\frac{2}{10} \times \frac{3}{100}$, my answer will be in—right, thousandths. If I multiply tenths times tenths, my answer will be in—right, hundredths. Let's look at these ideas in decimal form. If I multiply $.23 \times .2$, my answer should be in ______."

The pupils explained that $.23 \times .2$ was multiplying hundredths times tenths, and, therefore, the product should be expressed in thousandths. The teacher gave several other decimal multiplications and then asked if the pupils could develop an efficient means of placing the decimal point. The pupils worked out the generalization, "The product will have as many decimal places as there are in both factors. For example, $.237 \times .45 = N$ will have five decimal places in the product because the combined decimal places in the factors are five." The students then consulted references to check the validity of their statement and compared it with the statement given in the book.

The procedure detailed is helpful in developing the decimal multiplication algorithm. As work continues on multiplication with decimals, the teacher should make use of study procedures that review and solidify the rationale of decimal multiplication. Procedures that can be used are periodic checks of decimal multiplication by fractional multiplication, mathematics themes on "How and why I multiply by using decimals," and (after exponents have been introduced) the development of the rationale of decimal multiplication using the basic number properties and exponents.

DIVISION WITH DECIMALS

Division with decimals is one of the more difficult phases of elementary school mathematics because many pupils have difficulty in understanding the rationale of the process. Four types of structural situations that are taught at the elementary school level call for decimal division:

1. Division of a decimal by an integer.
2. Division of an integer by an integer that is not a factor of the dividend.
3. Division of an integer by a decimal.
4. Division of a decimal by a decimal.

Current children's textbooks and books on the teaching of elementary school mathematics usually introduce decimal division with situations involving the division of a decimal by an integer ($1.6 \div 4 = N$). This is probably a holdover from the time when decimal division was taught by a rather mechanical placement of the decimal point. Thus, in $4\overline{)1.6}$, the pupil placed the decimal point in the quotient directly above the decimal point in the dividend. Problem situations involving that form of decimal division are usually of the partition division variety.

It was previously emphasized that *measurement* division situations are better for introductory work than *partition* situations because the measurement type can be solved in several ways, thus providing a greater opportunity for pupil discover. In keeping with the guided-discovery principles advocated in this book, I suggest that a measurement division problem of the type $2 \div .4 = N$ be used for introductory purposes. This is analogous to a fractional division problem of the type $6 \div \frac{1}{2} = N$ and allows for a greater variety of pupil solutions than does the division of a decimal by an integer.

An illustrative problem and several possible solutions to it follow. As in previous introductory lessons, the use of multiple solutions of the problem should be emphasized. The teacher can also remind the pupils of the relationship between decimals and fractions. The teacher should not expect that all the solutions that follow will occur on the same day. The pupils should have a good deal of experience in using several of the approaches to ensure an understanding of decimal division.

Problem. A road sign said, "Monona 3 miles." How many kilometers is this? If a kilometer is about .6 of a mile, how many kilometers is 3 miles? (Note that the problem involves measurement division, because the size of each unit, .6 mile, is known.)

Solutions. The mathematical sentence needed was identified as $3 \div .6 = N$, which could be written $.6\overline{)3}$.

HERB: I used the number line to find how many .6s = 3. There are 5.

0 1 2 3

ALICE: We have solved other division situations by using subtraction, so I used a series of subtractions.

$$\begin{array}{rl} 3.0 & \\ \underline{.6} & (1) \\ 2.4 & \\ \underline{.6} & (2) \\ 1.8 & \\ \underline{.6} & (3) \\ 1.2 & \\ \underline{.6} & (4) \\ .6 & \\ \underline{.6} & (5) \end{array}$$

HERMAN: I made use of the division of fractions and the common denominator. [Inversion might also have been used, but it doesn't add as much to the decimal thinking.]

$$3 \div \frac{6}{10} = \frac{30}{10} \div \frac{6}{10} = \frac{5}{1} = 5$$

CARY: I used a different form of the common denominator, 3 = 30 tenths.

$$6 \text{ tenths} \overline{)3} = 6 \text{ tenths} \overset{5}{\overline{)30 \text{ tenths}}}$$

WALLY: I experimented with several mathematical ideas. I found that I could use the identity element to change the problem to whole-number division.

$$\frac{10}{10} \times \frac{3}{\frac{6}{10}} = \frac{30}{6} = 5$$

In the discussion that followed, all the solutions were accepted as being mathematically correct. The pupils felt that the solutions by Herb (the number line) and Alice (a series of subtractions) were good ways to demonstrate the thinking involved, but that both solutions were rather cumbersome. The students thought that division by using the common denominator was a very useful method, particularly when written in the form suggested by Cary: $6 \text{ tenths} \overline{)30 \text{ tenths}}$. Many were intrigued with the method used by Wally (changing a fractional division to whole-number division by using the identity element).

A worksheet containing several division problems and computational exercises was then assigned to the class. It was suggested that each of the methods discussed be used and that pupils attempt to identify further improvements on the solution.

During the discussion that followed, the teacher referred to the two completed division computations that follow and asked whether these two approaches could be combined.

$$.4\overline{)2} = 4 \text{ tenths} \overset{5}{\overline{)20 \text{ tenths}}} \qquad \frac{2}{4} = \frac{2}{\frac{4}{10}} \times \frac{10}{10} = \frac{20}{4} = 5$$

or

$$\frac{2}{.4} \times \frac{10}{10} = \frac{20}{4} = 5$$

Class members noted that it would not be necessary to change both the divisor and the dividend to tenths. Multiplying both by 10 and making the problem a whole-number division problem was effective and an application of mathematical principles (use of the identity element). The class agreed that an effective method for solving the division-of-decimal problems was to rename the numbers with whole numbers. As the pupils continued their work, they found that in division situations such as $2.3\overline{)5.43}$, the useful form of the identity element was $\frac{100}{100}$ rather than $\frac{10}{10}$. Thus, the division became $230\overline{)543}$.

The next division problem introduced was of the type $.4 \div 2 = N$. The problem used was, "For a science experiment, Nancy is to give each of four students an equal portion of 1.6 grams of copper sulfide. How much copper sulfide will each pupil receive?" The pupils generally wrote the division as $4\overline{)1.6}$ and then changed the problem to whole-number division: $40\overline{)16}$. This created difficulties for many of the pupils. The most common solution involved writing the division as a fraction and reducing the fraction: $\frac{16}{40} = \frac{4}{10} = .4$. Bill said he had estimated that the answer would be less than 1. He then divided without thinking of the decimal point, getting an answer of 4. He reasoned that the correct answer should be .4.

Several other division computations were worked by an estimation method. During this time, several pupils noted, "When you divide tenths, the answer will be expressed in tenths. Or you can use place value. Place the decimal point in the quotient directly above the decimal point in the divisor." For example,

$$\begin{array}{r} .82 \\ 4\overline{)3.28} \\ \underline{32} \\ 8 \end{array} \qquad \begin{array}{l} \phantom{\text{check } 4)}82 \text{ hundredths or } .82 \\ \text{check } 4\overline{)328} \text{ hundredths} \end{array}$$

After pupils have a good understanding of decimal division, they can shorten the writing required by using an arrow or a caret to indicate the multiplication they have performed. Thus, $3.4\overline{)6.25}$ becomes

$$3.4_{\curvearrowright}\overline{)6.25_{\curvearrowright}}$$

or

$$3.4_{\wedge}\overline{)6.2_{\wedge}5}$$

The use of the caret has often been criticized as being mechanical and without meaning. This is a factor of the method used in teaching division rather than of the device itself. It is strictly a computational shortcut, one that can be used instead of writing out the multiplication by the identity element.

There are methods of placing the decimal point in the quotient other than those suggested in the illustrative teaching situations. One of the more common makes use of the inverse relationship between multiplication and division. The sum of the number of decimal places in the factors of a multiplication equals the number of decimal places in the product, and since the quotient times the divisor equals the dividend, it follows that the number of decimal places in the quotient can be found by subtracting the number of places in the divisor from the number in the dividend. To illustrate,

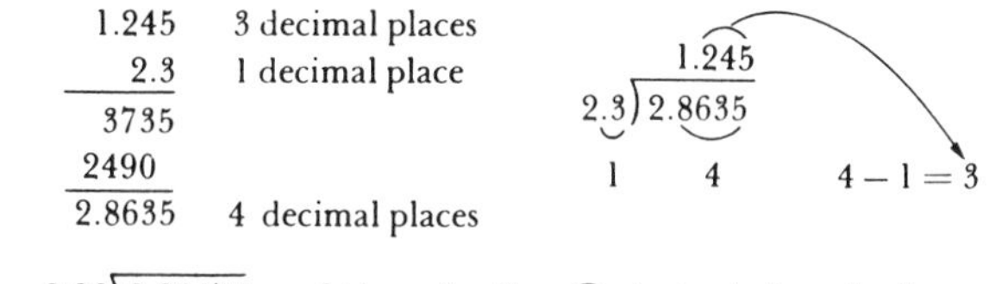

Thus $2.36\overline{)2.36472}$ would have $5 - 2 =$ (3) decimal places in the quotient.
(2 decimal places in 2.36; 5 decimal places in 2.36472)

This subtractive procedure presents difficulties when the division is not exact and the person computing wishes to add several decimal places to the dividend ($.8\overline{)10}$). Problems also arise when there are more decimal places in the divisor than in the dividend, although several decimal places can be added to the dividend: ($2.06\overline{)41.2}$ can be changed to $2.06\overline{)41.200}$).

A study of the two methods of placing the decimal point—making the divisor a whole number of multiplying by a power of 10, or subtracting the number of decimal places in the divisor from the number of decimal places in the dividend—showed that (1) in general, pupils taught to make the divisor a whole number more often placed the decimal point correctly, (2) the subtractive approach was effective with above-average learners, and (3) children could be taught to understand both procedures.

CONVERTING FRACTIONS AND DECIMALS

Usually, pupils experience little difficulty in converting decimals to fractions. Actually, if children can read decimals, they are able to convert them to fractions with a denominator of some power of 10. After the fraction is written with a denominator as a power of 10, the only task that remains is the simplification of the fraction.

Changing a fraction to a decimal is not as simple, for, to do this, a pupil often needs to perform decimal division. Because it is sometimes necessary to convert fractions to decimals before decimal division can be taught, the teacher can use two other techniques. Early in the study of decimals, the pupils may use a table of decimal equivalents. Pupils find it beneficial to memorize several of the more common decimal equivalents, such as $\frac{1}{8} = .125$, $\frac{3}{4} = .75$, and so on. Later in decimal study, but before the division of decimals is studied, pupils may handle conversion as a problem in finding equivalent fractions that have a denominator of 10, 100, or 1,000.

After the techniques of decimal division are mastered, the pupil can use division to find decimal equivalents. For example, to change $\frac{2}{5}$ to a decimal, the pupil views $\frac{2}{5}$ as meaning 2 divided by 5, or $5\overline{)2}$. He performs the division

$$\begin{array}{r} .4 \\ 5\overline{)2.0} \end{array}$$

and states, "$\frac{2}{5} = .4$."

For decimals that are nonterminating, such as $\frac{1}{3} = .3333\ldots$, the child may round the decimal form off to a finite number of decimal places for an approximation of $\frac{1}{3}$. Children should realize that this is an approximation and that they can approximate as accurately as they wish. However, they should *not* write $\frac{1}{3} = .33$.

TEACHING EXPONENTS

The study of exponential notation allows pupils to probe further the subtleties of place value. It also allows them to gain insight into the place of decimals within the number system.

The study of exponents can be introduced through the contemplation of extremely large numbers. One teacher, for instance, said, "Yesterday I was talking with a boy who said, 'You're a teacher and supposed to know a lot about mathematics. I bet I can stump you. Let's see you read this for me.'" (The teacher wrote on the board, 13,000,000,000,000,000,000,000,000.) "That's the approximate weight of the earth in pounds. Can you read this number?" Several pupils made attempts but were unable to name the number. Then the teacher said, "You can find out by checking an unabridged dictionary."

"Now," said the teacher, "can you read the numeral I wrote on the board?" The number was identified as 13 septillions.

The teacher than commented, "We seldom use the period names like *septillions* today; and yet we make greater use of very large numbers today than did people in the past. Today we talk about distances in space, the number of bacteria in a particular area, and the like. How do we avoid using such large numbers?" The pupils suggested that new terms had been developed, such as "light years," "astronomical units" (the distance from the earth to the sun, 93,000,000 miles), and "parsecs" (a parsec is 3.26 light years, or 206,283 astronomical units).

The teacher said that mathematicians, other scientists, and almost all who deal with mathematics make use of another scheme, one that involves the idea of factors. The teacher wrote 100, 1,000, and 10,000 on the chalkboard and asked, "How else could each of these be written?" The pupils suggested several methods, but the teacher focused attention on the following method:

$100 = 10 \times 10 \qquad 1{,}000 = 10 \times 10 \times 10 \qquad 10{,}000 = 10 \times 10 \times 10 \times 10$

Later, a chart such as the one on p. 283 was developed.

NUMERATION SYSTEMS WITH BASES OTHER THAN 10

The teaching of numeration systems with bases other than 10 was contained in some of the books of the 1800s.[3] They dealt with base 12 (a duodecimal system). During the strong emphasis on "modern mathematics," the majority of

[3]Benjamin Greenleaf, *The National Arithmetic on the Inductive System* (Boston: Robert S. Davis & Co., 1861), pp. 421–24.

programs started nondecimal bases in grade 4 or 5 and had one or two units per year on the topic. This was probably an overemphasis. At the upper-grade level, an introduction to other bases can help children (1) extend their knowledge and understanding of place value, (2) understand that it is not the choice of a base of 10 but rather the basic number properties and place value that are the major factors in causing our number system to operate effectively, (3) realize that using a base of 10 was an arbitrary choice made by early peoples, probably because of their 10 fingers (another base, such as 8 or 12, might be just as effective, in fact, more effective),[4] and (4) develop further insight into the fundamental operations. Operations that use another base require the pupils to think in terms of number properties.

A number of experiments concerning the effect of teaching other bases on the understanding of base 10 have been conducted. The typical finding is that such study does not contribute any more to base 10 understanding than does direct study of base 10. Thus, the teacher should take care not to overemphasize the study of other bases. It should be considered a study that is interesting to most children, but not crucial; and slow learners should not be required to master the ideas of the other bases.

Expanded Notation

10^3 $10 \times 10 \times 10$ THOUSANDS	10^2 10×10 HUNDREDS	10^1 TENS	 1 ONES
	2	4	5
3	6	7	2

$245 = (2 \times 10^2) + (4 \times 10^1) + (5 \times 1)$

$3672 = (3 \times 10^3) + (6 \times 10^2) + (7 \times 10^1) + (2 \times 1)$

Numeration in Base 4

Numeration in base 4 can be introduced by using the procedures that follow.

The teacher randomly placed 23 flannel representations of arrowheads on the flannel board and said, "A Native American of early California had this number of arrowheads. How many did he have?" The pupils tried to count the number of arrowheads, but found it difficult. The count ranged from 21 to 26.

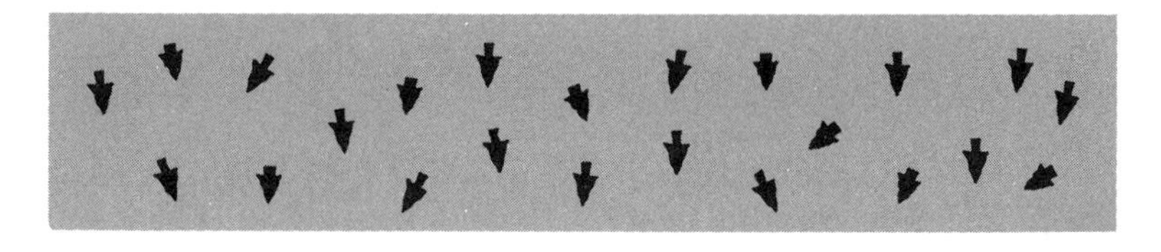

Then the teacher said, "We've had difficulty counting the number of arrowheads. How could we arrange them so that we could tell quickly how many arrowheads there are?" Several different methods of grouping were suggested:

[4] R. E. Andrews, *New Numbers: How Acceptance of a Duodecimal (12) Base Would Simplify Mathematics* (New York: Harcourt, Brace & World, 1935).

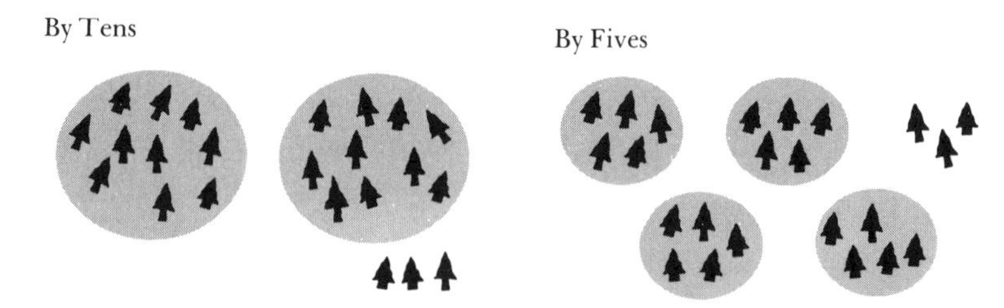

Then teacher said, "Civilizations haven't always grouped by tens. The Mayans of Central America grouped by twenties. Can you think why?" After a pause, a student answered, "Yes, it was probably because they used both fingers and toes to count. They lived in a warm climate and didn't wear shoes of the sort we do."

The teacher added, "The California tribe that I mentioned, the Yukes, grouped by fours. Can you think of a reason for this?" None of the pupils was able to. Therefore, the teacher explained that it is believed they used the space between the digits on one hand as a representative group.

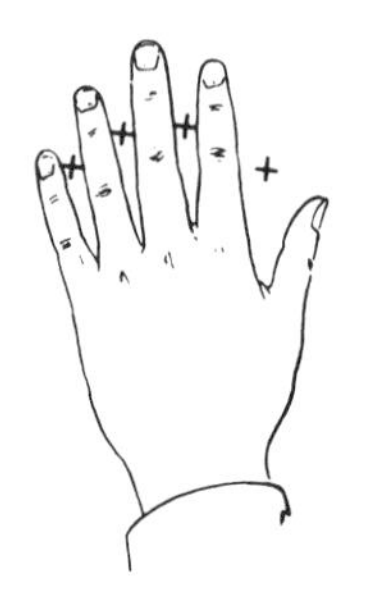

The teacher said, "How would they have grouped a set of 13 arrowheads?" The pupils grouped the arrowheads into fours, three groups of four with one unit remaining.

The teacher asked, "How would it look on a place-value frame?" and the pupils filled out this frame:

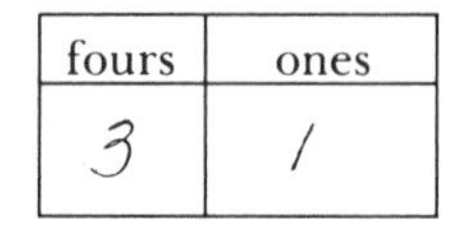

fours	ones
3	1

The teacher suggested that as a check, the students look at the meaning of the place-value frame. They noted that the digit to the left of the ones place indicated the number of fours (in this case, 3). Thus, 3 fours and 1 can be considered to be another name for 13 in base 10.

Then the pupils were given groups of sticks of varying sizes and were asked to arrange them into groups of fours and ones and to indicate the amount on the place-value frame. (All these amounts were less than 16 base 10). Then the teacher said, "We can write numerals in other bases without the place-value frame by indicating the base at the right and slightly below the numeral. How many objects are there in a set represented by 13 in base 4? Right, seven—1 four and 3 ones. Write the numeral in base 4 that represents the collections of sticks we've shown in the place-value frame." Number lines to compare base 4 and base 10 were developed by class consideration.

Base 10

0 1 2 3 4 5 6 7 8 9 10 11 12 13 14 15 16 17 18 19

Base 4

0 1 2 3 10 11 12 13 20 21 22 23 30 31 32 33 100 101 102 103

Some difficulty was experienced in moving from 33_{four} to 100_{four}. Several pupils who suggested that 40 should follow 33_{four} were corrected by others, who reminded them that the numeral 4 is not used in base 4. By inspecting 16 dots in groups of fours, the class reasoned that 16 dots represented the base times base in a scale of 4. The teacher asked, "How do you represent the base in base 4?—right, 10_{four}. How would we represent the base times base?" The class agreed that to make use of place value, the base times base should be 100_{four}.

The class was challenged to indicate the number of its original set of arrowheads (23) with a base 4 numeral. Several pupils mistakenly wrote the numeral as 53_{four}. However, the majority had found methods of deriving the correct base 4 numerals. Several of their methods follow:

(a)

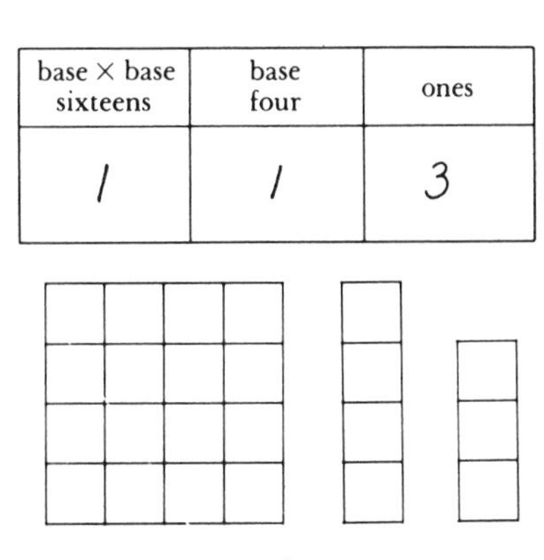

squared paper

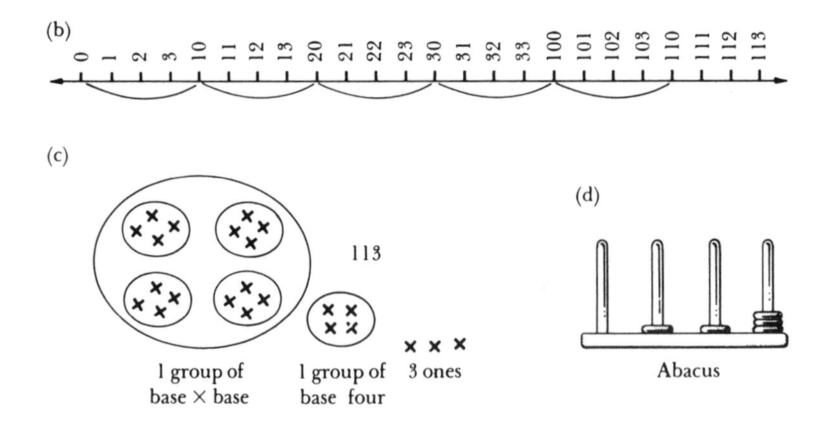

The class worked further on expressing base 4 and base 10 numerals and then considered the development of a place-value chart for base 4. Their completed chart follows:

Base 4

B^3	B^2	B^1	B^0
Sixty-four 1^3 $(4 \times 4 \times 4)$	Sixteens 4^2 (4×4)	Fours 4^1	Units 4^0

The pupils then worked with several other bases and developed place-value charts for them. They found the chart to be helpful in converting numerals from one base to another. From the use of the chart and further discussion, the pupils formulated the following generalizations and procedures:

1. When we're working in another base, we're still working with the same number, but we give the number a different name. For example, 13_{four} and 7_{ten} represent the same number, but they are different numerals for this number.
2. If we understand place value, we can rather easily convert a numeral in another base to base 10. For example, 233_{five} means (2 of the base × base) + (3 of the base) + (3 ones). It can be written as

25s	5s	1s
2	3	3

$$233_{five} = (2 \times 25_{ten}) + (3 \times 5_{ten}) + (3 \times 1_{ten})$$

$$233_{five} = 50 + 15 + 3$$

$$233_{five} = 68$$

3. To convert a numeral from base 10 to another base, we can develop a chart of powers of the base. Then we group in powers of the new base. For example, to convert 68_{ten} to base 5, we develop the chart and then think: Any groups of $5 \times 5 \times 5$, or 125s? Any groups of 5×5, or twenty-fives? Yes. How many? Two. Two twenty-fives = 50. Thus, we have 18 remaining. How many fives in 18? Three. How many ones are left? Three. $68_{ten} = 233_{five}$.

B × B 5 × 5 25	B 5	1s
2	3	3

$$\begin{array}{r} 68 \\ \underline{-50} \\ 18 \\ \underline{-15} \\ 3 \end{array} \begin{array}{l} \\ (2 \times 25) \\ \\ (3 \times 5) \\ \text{ones} \end{array}$$

Binary and Duodecimal Notation

Although they are not suggested for introductory work, the duodecimal and binary systems are used more often than the other nondecimal bases. They follow the same rationale as base 4. However, it should be noted that for base 12, two new symbols (usually T and E) are used to replace the base 10 ten and eleven.

Older Numeration Systems

A study of numeration systems used by ancient peoples can be both interesting and useful for providing insight into present-day problems. Typically, schools study the system used by the Romans because it is used on clocks, in cornerstones on buildings, in outlines, and so on. The Roman System can be of greater value to the pupils if its structure is also studied. In addition, pupils can benefit from studying the systems used by the Babylonians, the Egyptians, the Mayans, the Greeks, and possibly the Chinese. (Such a study can often be used effectively as a portion of interdisciplinary units.)

KEEPING SHARP

Self-Test: True/False

_______ 1. *Decimals* is another name for integers.

_______ 2. Both extremely large and extremely small numbers can be written in a simplified form by using decimals and rounding-up.

_______ 3. Squared paper aids children's understanding of decimal quantities by graphically illustrating their relative sizes.

_______ 4. A number line is not useful for teaching addition and subtraction of decimals.

_______ 5. Addition of decimals has *all* the properties of whole-number addition.

_______ 6. The expression 10^4 shows that 10 is multiplied by 4.

_______ 7. The product of $10^5 \times 10^4$ is 10^9.

_______ 8. One way to express the quotient for $3^5 \div 3^2$ is 27.

Vocabulary

exponent	period
power	nonterminating decimal

THINK ABOUT

1. Explain how the metric system could be used as a vehicle for teaching decimals. Give at least five suggestions.
2. Develop a lesson in which a laboratory approach is used to teach scientific notation using exponents.
3. Relate the concept of decimals to the pupils' experiences with fractions. Develop three activities by using this relationship.
4. Develop a set of activity cards for each of the four operations with decimals, using a variety of approaches.
5. What two procedures presented in pupil texts on the division of decimals seem to be the best for giving pupils an understanding of division with a decimal divisor?
6. How many procedures for teaching pupils how to read decimal fractions are advocated in modern pupil textbooks? Use at least five series to provide data for your answer, and identify what you think are the superior suggestions.

SUGGESTED REFERENCES

Arithmetic Teacher (February 1984), an issue on rational numbers.

ASHLOCK, ROBERT B., "Introducing Decimal Fractions with the Meterstick," *The Arithmetic Teacher* 23 (March 1976): 201–6.

BITTER, GARY G., MARY M. HATFIELD, and NANCY T. EDWARDS, *Mathematics Methods for the Elementary and Middle School: A Comprehensive Approach,* chap 11. Boston, Mass.: Allyn and Bacon, 1989.

BEHR, MERLYN J., AND THOMAS R. POST, "Teaching Rational Number and Decimal Concepts." In *Teaching Mathematics in Grades K-8: Research-Based Methods,* Thomas R. Post (ed.), Boston, Mass.: Allyn and Bacon, 1988.

Historical Topics for the Mathematics Classroom, Thirty-First Yearbook of the National Council of Teachers of Mathematics, chap. 2. Washington, D.C.: The Council, 1969.

THIESSEN, DIANE, MARGRET WILD, DONALD D. PAIGE, and DIANE L. BAUM, *Elementary Mathematical Methods,* 3rd. ed., chaps. 8, 9. New York: Macmillan, 1989.

WIEBE, JAME H., *Teaching Elementary Mathematics in a Technological Age,* chaps. 12, 14, Scottsdale, Ariz.: Gorsuch Scarisbrick, 1989.

chapter 12

APPLICATIONS: RATIO, PROPORTION, PERCENT, PROBABILITY, AND STATISTICS

OVERVIEW

The topics of ratio, proportion, percent, probability, and statistics could just as well have been fitted into the problem-solving chapter of this book. Why? They are all applications of mathematics, and they occur only in problem situations. It is impossible to be an informed consumer without a good working knowledge of these topics, and they are necessary for almost any mathematical application in science and social studies.

Every day the lives of both children and adults are affected by some application of these areas. Here are just a few decisions that are based on notions of probability:

1. Should I take an umbrella? What is the probability of rain?
2. Should I buy extra flight insurance?
3. How many plants should I put in the garden? What is my experience with growth? What is the probability of germination?
4. I have only a few minutes to get to the store before it closes. Is the probability that I will be in time high enough to make the trip? What were my experiences driving there at this time of day? etc.
5. A TV commercial states, "The probability is that one of every three persons listening to me will have a stalled car at least once this winter." Another says, "Statistics show that brand X is an effective aid in the prevention of tooth decay." What do they mean?

Probability theory can be a highly abstract mathematical topic, and statistical ideas go far beyond the scope of the elementary school curriculum. Thus, the study of these two topics should be of an exploratory nature, with emphasis on gaining insight, experimenting, and solving simple problems involving them rather than taking a formal look at probability and statistics. One of the continuing needs of the elementary mathematics program is the development of basic skills needed for life applications of mathematics and a basic understanding of these ideas. The level of the student will dictate the sophistication of computational skill obtainable in these areas.

In addition to the usefulness of these ideas themselves, they also (1) can be used to experiment, search for patterns, and generalize; (2) provide novel and interesting approaches to standard elementary school mathematics topics such as multiplication, fractions, ratio and proportion, and geometry; and (3) provide a means of practicing essential mathematical topics.

The National Council of Teachers of Mathematics Standards suggest close attention to probability and statistics. In this age of information there is an increasing need to understand and interpret data. Study a copy of *U.S.A. Today,* or study *The Wall Street Journal* and note the number of charts, graphs, and statistical information. In truth, the average person uses statistics much more in daily life than long division or computation with fractions.

The teacher should carefully consider the use of hand-held calculators in teaching these topics. It may well be that the study of ratio, proportion, and percent could involve a major emphasis on the calculator.

TEACHER LABORATORY

1. *Materials:* small nut cups, a bag of lima beans, and squared paper. How many lima beans does it take to fill one nut cup? to fill two nut cups? Make a chart that would help you find the number needed to fill 15 nut cups. Try to graph the information; use the X coordinates for nut cups, the Y coordinates for lima beans.
2. *Materials:* a major daily newspaper. Look through the paper carefully, and make a list of the number of times the word percent is used. What ideas for teaching percent can you get from a newspaper?
3. Some people in mathematics feel that percent should not be taught. Make a list of reasons for and against teaching percent.
4. Study the fourth-, fifth-, and sixth-grade material for any elementary school science program. How much use is made of ratio and proportion? Read this chapter and compare the approaches to teaching ratio and proportion.
5. After reading this chapter, develop a lesson on ratio and proportion based on elementary science.
6. After reading the chapter, use the newspaper to develop an introduction to percent.
7. Scale drawings could be used to teach ratio and proportion ideas. Develop a lesson by using scale drawings.

8. Make a guess as to what product would appear most often if you rolled a pair of dice and multiplied the two numbers that turned up. Now obtain a pair of dice and roll them 50 times. Keep a record of the products. What product occurred most often? Did your guess agree with the findings? Repeat the experiment. Did the same product occur most often?
9. By using the experiment just described, make up a game for children that would incorporate probability with multiplication practice. How could the experiment be used to practice addition facts?
10. If you made a voter survey in your community and were going to survey only 200 people (assuming over 1,000 in the area), how could you be reasonably sure that the 200 would be representative of the entire community? Make a list of procedures you would use. Compare your list with others.
11. Work with someone else and make up a simple survey instrument of three or four questions. Administer the survey and report your findings via a graph.

TAKE INVENTORY

1. Sketch the history of ratio in elementary school mathematics.
2. Develop a group-thinking lesson on ratio by using a topic in science as the motivation.
3. List four errors made in thinking about percent.
4. Give reasons for the use of percent.
5. Write a problem for each type of percent situation.
6. Illustrate the solution to each of the problems you devised in number 5 by means of each of the following approaches:
   a. Decimal approach
   b. Ratio approach
   c. Unitary-analysis approach
   d. Equation approach
7. Write a challenge problem using ratio and proportion.
8. Develop a lesson for a primary or an intermediate grade topic from the objectives on pp. 302–3.
9. List five errors made in statistical interpretations.
10. Give three suggestions for teaching charts and graphs.
11. Develop a game for teaching one idea about probability.
12. Suggest the role of the calculator in application work.

RATIO AND PROPORTION

Pairs of numbers may be used to indicate a relationship between the numbers. One of the most useful relations is the ratio of one number to another. The ratio of the number 3 to the number 4 is the quotient $3 \div 4$. In former times this relationship was written $3:4$ (three is to four). Today ratios are more often written in the form of a fractional numeral ($\frac{3}{4}$) or as an ordered pair $(3, 4)$.

The concept of sets provides a helpful method of studying ratio. The ratio is considered to express a numerical property that exists between two sets. For example, if apples are sold at two for 5¢, the relationship between sets of apples and pennies may be shown in this way:

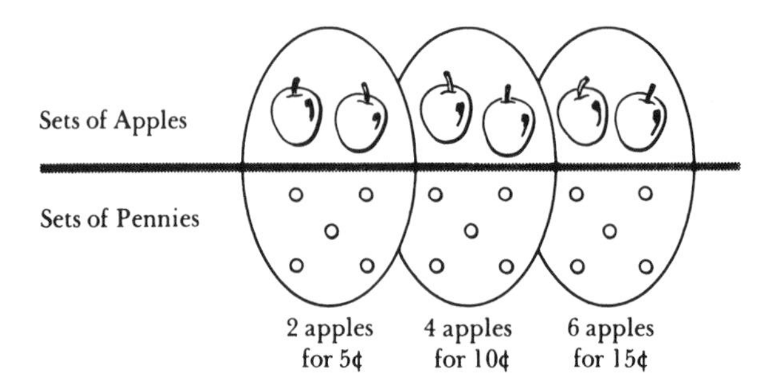

The greatest use of ratios in problem solving involves finding the equivalent ratio or proportion. The statement $\frac{3}{4} = \frac{6}{8}$ expresses a proportionality relationship and may be read "3 is to 4 as 6 is to 8." An infinite set of equivalent ratios such as $\{\frac{1}{2}, \frac{2}{4}, \frac{3}{6}, \frac{4}{8} \ldots \frac{n}{2n}\}$ can be generated. This set is representative of a proportionality relationship.

Illustrated procedures for teaching proportions are developed later in this chapter.

Historical Development

Ratio and proportion had their origin in antiquity. Egyptian and Babylonian documents containing ratios and proportions have been found that date back to the early periods of these civilizations. The Greeks made major advances in the use of ratio. Euclid treated ratios in developing a theory of music. He, along with a number of other Greek mathematicians, considered ratios as relations rather than as numbers.

Ratio was also used extensively by the Hindus and Arabs. For example, the Arab al-Khowarizini made extensive use of ratio in his algebra in A.D. 825, and the Moorish Arabs also used ratio extensively.[1] During the Middle Ages, ratio was taught in Italy and was employed in solving everyday business problems.

Ratio was emphasized in textbooks used in the United States until the end of the nineteenth century. Textbooks of the late 1880s contained highly complex proportionality problems, such as, "If in 8 days 15 sugar maples, each running 12 quarts of sap per day, make 10 boxes of sugar, each weighing 6 lbs, how many boxes weighing 10 lbs apiece will a maple grove containing 300 trees, make in 36 days, each tree running 16 quarts per day? Answer, 720 boxes."[2]

[1]Louis C. Karpinski, *The History of Arithmetic* (Chicago: Rand McNally, 1925), p. 139.

[2]James B. Thompson, *New Practical Arithmetic* (New York: Clark and Maynard, 1873), p. 318.

Because of the extremely long and complicated problems used in teaching ratio and proportion, and the belief that a linear equation could be used more easily, early-twentieth-century writers decried the heavy emphasis given to ratio and proportion. In 1909, David Eugene Smith attacked the wide use of ratio in the elementary school.[3] He suggested the use of a simple equation or unitary analysis. Unitary analysis involves finding the cost of one object and then finding the total. For example, "If two apples cost 6 cents, how much will five apples cost?"

Using Unitary Analysis	Using a Proportion
2 cost 6 cents.	$\frac{2}{6} = \frac{5}{N}$
1 costs $\frac{1}{2}$ of 6, or 3 cents.	$2N = 30$
5 will cost 5 × 3, or 15 cents.	$N = 15$

Thus, between 1910 and 1950, there was little formal study of ratio in the elementary school mathematics program. But, with the reform movement in mathematics during the late 1950s and the 1960s, the teaching of ratio again became an integral part of the elementary mathematics curriculum. The emphasis today on consumer education, basic skills, interdisciplinary teaching, and real-life problem solving dictates that ratio and proportion should play a part in the mathematics program of the upper grades.

Teaching Ratio and Proportion

Ratio involves matching sets in one-to-many, many-to-one, or many-to-many correspondence. Primary-grade teachers can begin to develop pupils' understanding of ratio by using sets of objects, the flannel board, and the chalkboard. Several possible situations and their representations are shown in Figure 12–1.

Children should be provided with many laboratory activities involving the use of objects and/or graphs to find the solution to problems of ratio and proportion. Several such situations are briefly described below.

The children were given four or five small nut cups and a bag of large lima beans. They were given the tasks of (1) finding how many beans it took to "fill" a nut cup, and (2) after agreeing on what they meant by "fill," finding the

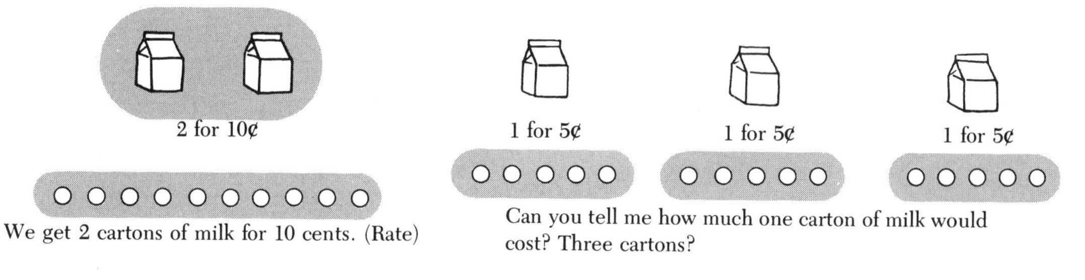

Figure 12–1 Depictions of Ratio

[3]David E. Smith, *The Teaching of Arithmetic* (Boston: Ginn, 1909), pp. 185–86.

number it would take to fill two nut cups, three nut cups, five nut cups, and so on. The following chart was given them to develop a pattern for solution:

Number of nut cups	1	2	3	4	5	6	7	8
Number of lima beans	7	14						

Then they were to find the information in as many ways as possible (working in groups of three).

JOE'S GROUP:

PHYLLIS'S GROUP: We used two number lines.

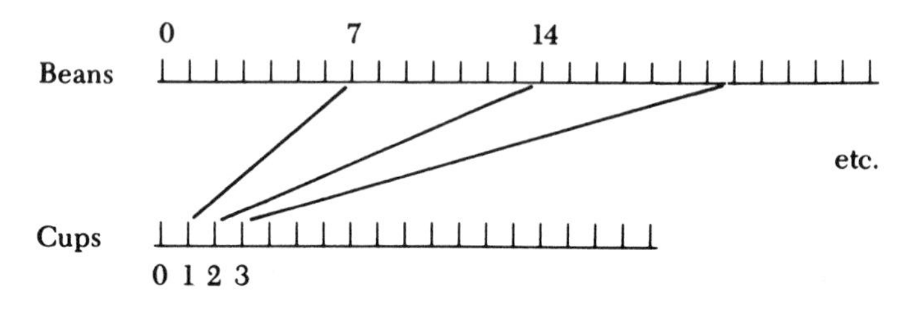

JANE'S GROUP: We used the number line to indicate 1 for 7. We marked off another 7 for two cups and still another for three cups. By this time, we had a pattern.

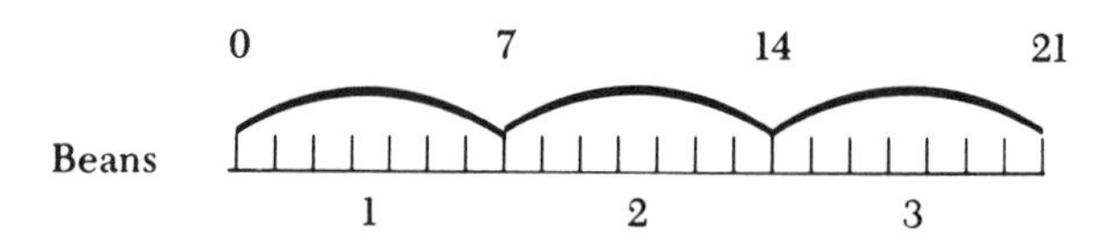

A teacher using these ideas may want to begin with a problem situation concerned with the planning for a party. The lima beans are about the same size as jelly beans, and the problem could involve the number of jelly beans needed to fill nut cups for the class and/or visitors.

Another teacher presented sticks of 10 cm, 20 cm, 30 cm, 40 cm, and 50 cm in length. Each group of four or five children was provided with a set and was given a sheet of paper like that shown on p. 293. They went out on the playground on a sunny day and recorded their data. Then the teacher suggested that they use their pattern to determine the height of several taller objects; for example, a basketball pole, a high fence pole, a tree, the school building.

HEIGHT OF STICK IN CENTIMETERS	LENGTH OF SHADOW IN CENTIMETERS
0	______
10	______
20	______
30	______
40	______
50	______

Another teacher used a study of levers from the science book:

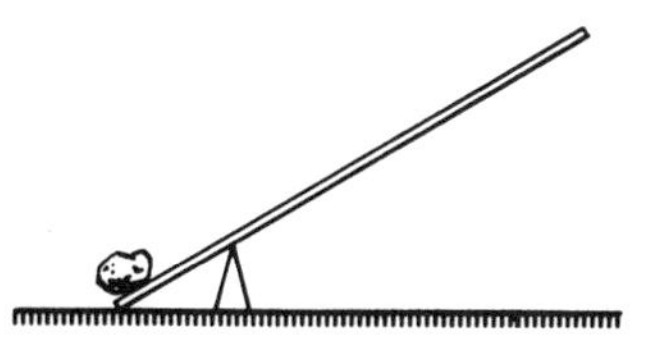

The Cross-products Approach

Knowing that the cross-products of two equivalent ratios are equal can be a valuable tool in problem solving. In looking for an unknown such as $\frac{2}{3} = \frac{N}{21}$, the cross-products are $3N = 42$. Pupils can then find a number for N by dividing both sides of the equation by 3, because this does not affect the equality. Thus, $\frac{3N}{3} = \frac{42}{3}$, $N = 14$.

The cross-products idea can be developed in the following manner. Pupils had been working on sets of equivalent number pairs:

$$\left\{\frac{1}{2}, \frac{2}{4}, \ldots, \frac{N}{2N}\right\} \qquad \left\{\frac{2}{3}, \frac{4}{\square}, \frac{6}{\square}, \frac{\square}{12}\right\}$$

$$\left\{\frac{5}{9}, \frac{10}{18}, \ldots, \frac{5N}{9N}\right\} \qquad \left\{\frac{3}{4}, ?, ?, ?, ?\right\}$$

They had found other members of the set by multiplying or dividing the members of the set by a form of the identity element, such as $\frac{3}{3}$ or $\frac{4}{4}$. Then the teacher said, "Let's study a few equivalent number pairs and see if we can find any relationship between them. Look at $\frac{1}{2} = \frac{2}{4}$, $\frac{3}{4} = \frac{6}{8}$, $\frac{2}{3} = \frac{4}{6}$. Do you notice any similarities? Work with these number pairs during your spare time and see what you can discover. Test your findings on other equivalent number pairs. We'll discuss your findings tomorrow."

The next day the members of the class discussed their findings. The majority of the pupils had found that the products of numerator (first pair) times denominator (second pair) and denominator (first pair) times numerator (second pair) were equal. They had verified their findings by testing a large number of instances. The teacher said, "You've discovered a valuable tool for working with number pairs. The equality that you've just mentioned is usually stated, 'The cross-products of two equivalent number pairs are equal.' What do we mean by *cross-products*?" The pupils discussed the meaning of cross-products; then, they were given several verbal problems with instructions to see if they could use the cross-product development as recorded below.

1. The pull of gravity on the moon is $\frac{1}{6}$ the pull of gravity on the earth. A track champion can high-jump 7 feet on the earth. What would be the height of a bar he or she could clear on the moon (all other things being equal)?

   Proportion: $\frac{1}{6} = \frac{7}{N}$ Cross products: $1 \times N = 6 \times 7$
   $N = 42$

2. A large gear turns three times while a small gear turns 15 times. How many times will the large gear turn if the small gear turns 60 times?

   Proportion: $\frac{3}{15} = \frac{N}{60}$

   Cross-products:
   $$3 \times 60 = 15 \times N$$
   $$15N = 180$$
   $$\frac{15N}{15} = \frac{180^*}{15}$$
   $$N = 12$$

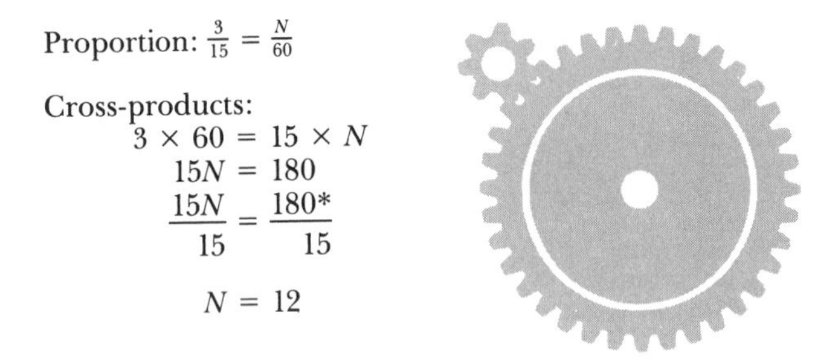

Analysis

Ratio and proportion can be valuable tools for verbal problem solving in the elementary school. The task for today's teachers is to establish a balance in teaching the topics, for it is possible to overdo ratio in the manner of the late 1800s. It should be used only as one possible tool in problem solving.

PERCENT

Percent has been taught as another means of viewing decimals. Actually, percents were used long before decimals were developed. The term *percent* is derived from the Latin *per centum,* meaning "by the hundred." Thus, the origin of percent and its major uses are more closely associated with rates and ratio than with decimals.

Many believe that the teaching of percent should cause little if any difficulty because of the close relationship of percents, ratios, and decimals. Those experienced in teaching the upper elementary grades often disagree with this conjecture. There are several reasons why the topic of percent often does confuse children and adults, among which are its languages.

The Precise Language of Percent

The language of percent is very precise. This precision is, in a sense, two-headed; it makes percent extremely useful, but on occasion it also causes confusion. Several common misconceptions found in the use of percent are analyzed:

Note: Both sides of the equation are divided by 15.

1. If a small-town weekly newspaper states, "Juvenile delinquency has risen by 50 percent this year," the population may become very upset. If, however, the paper states, "Last year there were two cases of delinquency; this year there were three," the reaction probably would not be the same. This difficulty occurs because when some rates are converted to hundredths, they may appear to represent a larger sample than is actually the case.
2. A statement, "Forty percent of the pupils in Washington Elementary School—160 out of the 400 pupils—gave to the cancer drive," may be correct. If a pupil then states, "Forty percent of our class gave to the cancer drive," he or she may be incorrect. A rate stated for an entire population may not be true for a sample of the population.
3. A store advertised, "Prices reduced 100 percent." Actually, the store had reduced its prices 50%; an item that originally sold for $10 was on sale for $5. The store based the 100% on the sale price when it should have been based on the original price.
4. When asked, "Would you be willing to lend money at .15 percent?" many people enthusiastically responded, "Yes." They believed the figure to represent $15 interest per $100, when it actually represented an interest rate of 15¢ per $100.

Foundation Work with Percent

Computation with percent is not usually developed until the fifth or sixth grade. However, an understanding of the theory of percent is necessary earlier to ensure proper interpretation of social studies, science materials, and situations outside of school. The meaning of percent can be introduced effectively at the fourth-grade level by using materials from science and social studies. One teacher introduced percent by saying, "In your social-studies book, you read today that 8 out of 10, or 80 percent, of the population of an African country live along the coast or rivers. What is meant by 80 percent?"

In class discussion, pupils judged that 80 percent must be equivalent to 8 per 10. Also, they felt that the term *cent* must have something to do with 100. Because they knew that 80 per 100 and 8 per 10 were equivalent ratios, they reasoned that 80 percent meant 80 per 100.

Several other examples from science and social studies were then used for interpretation: "The United States once produced 40 percent of the world's steel. Now the United States produces 17 percent of the world's steel." "Major auto-leasing firms expected a 15 to 20 percent gain in summer volume over last year's activity. Instead, they posted a 20 to 30 percent advance." "Foreigners now buy almost 40 percent of London's theater tickets."

From the discussion of these and other situations that use percent, the pupils decided that percent means per hundred or by the hundred. Then they checked reference books to verify their thinking, and they read up on the historical development of percent.

Since many percent ideas occur in daily life before decimals are introduced in the school curriculum, it is important to provide children with some way of visualizing the meaning of percent. One of the best devices is squared paper. Individual children can use 10-by-10 pieces of squared paper to indicate various percents. Through this device, the percent can also be interpreted as a fraction. For example,

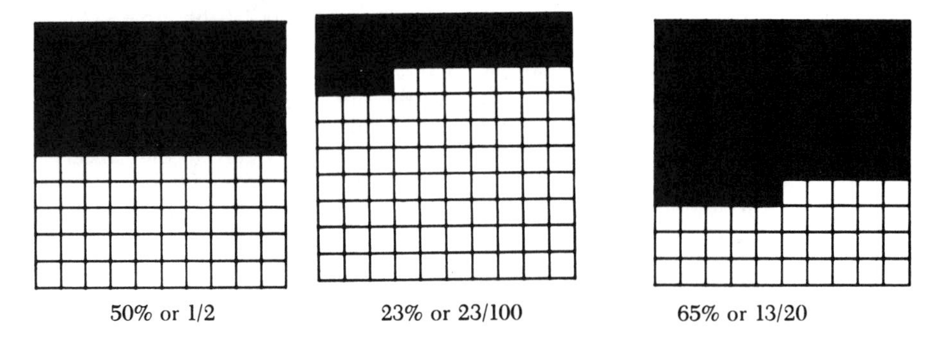

50% or 1/2   23% or 23/100   65% or 13/20

Problems Using Percent

The majority of problem situations that employ percent involve three problem structures.

1. The pupils may be asked to find the percent of a number; for example, "What is 25 percent of 64?"
2. The pupils may be asked to find what percent one number is of another; for example, "What percent of 64 is 16?"
3. The pupils may be asked to find the total (100 percent) when only a percent is known; for example, "Sixteen is 25 percent of what number?"

Several approaches can be used to solve percent problems that occur in the literature of elementary school mathematics. Among these are (1) the decimal approach, (2) the ratio approach, (3) the unitary-analysis approach, and (4) the equation approach. Pupils' thinking in the solution of each of the three problem structures is presented below for each of the approaches; then a teaching sequence is suggested.

The following problems are solved by using each approach:

1. During a sale, a 5 percent reduction is given on all merchandise. How much would a 40¢ item be reduced?
2. Joe won two games out of the five games he played. What percent of the games did he win?
3. Mr. Brown sold a garden cart for $150. This was 75 percent of the regular price. What was the regular price.

DECIMAL APPROACH

In using the decimal approach, the pupil is to think, "Percent means hundredths. I change a percent to a decimal and then work the problem."

Problem 1. *Pupil:* I am to find 5 percent of 40. I change 5 percent to .05 because percent means hundredths. Then I multiply .05 × 40 to obtain the answer of 2.

Problem 2. *Pupil:* I am trying to find what percent 2 is of 5. This is 2 to 5. I need to change this to hundredths. Percent means hundredths.

$$5\overline{)2.00}^{\;.40}$$

Therefore, .40, or 40 percent.

Problem 3. *Pupil:* Seventy-five percent means .75. To find the whole when I know a part, I must divide the number by the decimal.

$$.75\overline{)150.00}^{\;200}$$

RATIO APPROACH

All problems are solved by developing a pair of equivalent ratios, with a place holder representing the number sought. This approach employs the idea that percent means per hundred, implying a rate.

Problem 1. *Pupil:* Five percent is absent, which means 5 per 100 are absent. The ratio per 40 will be equivalent to the ratio per 100. Thus I set up equivalent ratios to solve the problem.

$$\frac{5}{100} = \frac{N}{40} \qquad \begin{aligned} 100N &= 200 \\ N &= 2 \end{aligned}$$

(For the early work, the student developed equivalent number pairs. Later, he or she may use the equality of the cross-products.)

Problem 2. *Pupil:* Joe won two games per five played. Now I must find out how many he would win per 100.

$$\frac{2}{5} = \frac{N}{100} \qquad \begin{aligned} 5N &= 200 \\ N &= 40 \end{aligned}$$

This means 40 per 100, or 40 percent.

Problem 3. *Pupil:* I'm to find the number of which $150 is 75 percent. Or, $150 is 75 percent of what number? I think 75 to 100 is equal to $150 to N.

$$\frac{75}{100} = \frac{150}{N} \qquad \begin{aligned} 75N &= 15{,}000 \\ N &= 200 \end{aligned}$$

To use the ratio approach, the pupils must understand ratio and proportion and be able to solve simple equations. Actually, these are not difficult to understand. Pupils should be solving simple equations and working with some ratios by the fourth-grade level. If they have not studied ratio by the time percent begins, it is quite possible to teach ratio and percent simultaneously. The use of ratio eliminates the need for previous study of decimal computation.

UNITARY-ANALYSIS APPROACH

The unitary-analysis approach, sometimes called the 1 percent method, was once very popular in schools in the United States. It has some advantages, particularly for the slower student, because it is easily understood and requires similar thinking for all three types of problems.

Problem 1. *Pupil:* I want to find 5 percent of 40. First I'll find what 1 percent would be. Since 100 percent of the amount is 40, I can find 1 percent by dividng 40 by 100.

$$100 \overline{)40.0} \quad = .4$$

One percent of 40 is .4. Five percent of 40 is $5 \times .4 = 2$.

Problem 2. *Pupil:* I'm finding what percent 2 is of 5.

5 games = 100% of the total
1 game = $\frac{100}{5}$ = 20% of the total
2 games = $2 \times 20\% = 40\%$

Problem 3. *Pupil:* I'm to find what number 150 is 75 percent of.

75% of the number is 150.

1% of the number $\left(75 \overline{)150} = 2\right)$ is 2.

100% of the number = $100 \times 2 = 200$

The solution of these problems by unitary analysis reveals several weaknesses of the method. Solution of type-1 problems often requires rather difficult division of decimals. Also, note the cumbersome development in problems 1 and 2. Probably the chief use of unitary analysis is in type-3 problems, because unitary analysis does add understanding to this problem structure.

EQUATION APPROACH

The equation approach makes use of the logical thinking required for unitary analysis. It also has some of the features of the formula. The writing of the mathematical sentence required in the equation approach ties it closely to the type of mathematical thinking wanted in problem solving.

Problem 1. *Pupil:* The essential mathematical statement is, Find 5 percent of 40, or, what number is equal to 5 percent of 40? Now I translate the statement into a mathematical sentence.

$N = 5\% \times 40$ (5% is equivalent to .05, or $\frac{5}{100}$)

$N = .05 \times 40$

$N = 2$

Problem 2. *Pupil:* The essential mathematical question is: Two is what percent of 5? I write this as a mathematical sentence.

$$2 = N \times 5$$

$$2 = 5N \text{ (I divide both sides of the equation by 5)}$$

$$\frac{2}{5} = N$$

$$\frac{2}{5} = .40 = 40\% = N$$

Problem 3. *Pupil:* I have to find the regular price when I know 75 percent of the regular price. A mathematical statement might be: 75 percent of the number equals 150. Now I write a mathematical sentence.

$$75\% \times N = 150 \text{ (75\% is equivalent to .75.)}$$

$$.75\,N = 150 \text{ (I divide both sides of the equation by .75.)}$$

$$N = \frac{150}{.75}$$

$$N = 200$$

The format of the equation approach becomes similar to that of the decimal approach, but the equation approach places greater emphasis on the structure of the problem and an understanding of the rationale.

Teaching Problems Involving Percent

Problems involving percent should be taught only after the children have had an opportunity to understand the ideas of rate and ratio and have some acquaintance with fractions and decimals. A problem such as, "Sarah decided to sell magazines; she keeps 30 percent of her collections as her salary. If she sells $50 worth of magazines, how much will she earn?" could be used to introduce percent problems. Either a lab or a Socratic-questioning approach can be used effectively.

After the children are able to handle problems of this type, the second type of situation can be introduced. An example from sports is usually effective; for instance, "The high point man for the Knicks made 15 of the 20 shots he took. What was his shooting percentage?"

The third type of problem is usually much more difficult for children and thus lends itself to a Socratic-questioning approach. Problems such as this will often require individual pupil attention by the teacher: "Nan saw a sign in the store window saying, 'Coats on sale for $65. This is only 75 percent of the regular price.' She wanted to know the regular price, to compare it with that of another store. What was the regular price?"

Analysis

The suggestions for teaching percent have focused on two important ideas. The first is that the use of the mathematical sentence (an equation, in this case) gives meaning to percent problems. In writing mathematical sentences, both the ratio equation and equations involving decimals were used. The second idea is the unitary-analysis approach, which was presented to give meaning to type-3 percent problems.

Because research results are still unclear on a single approach to problems involving percent, it is suggested that the teacher present to the pupils problems that use percent and then follow the lead they take (if it is mathematically correct). Such a procedure is in keeping with a discovery approach and also allows greater opportunity for success in handling individual differences.

Also, it should be remembered that percent is not a mathematical topic per se. It is an application of a particular type of notation system to quantity. In fact, many mathematicians would like to eliminate the use of percent and deal only with decimal or fractional notation. Thus, percent should be used in application situations. There is little if any justification for teaching percent in situations not tied to the "real world."

GOALS IN TEACHING PROBABILITY AND STATISTICS

What are the major topics in probability and statistics that should be taught in the elementary school? As yet, a body of subject matter in these areas has not emerged as dominant to the school mathematics program.

The National Council of Teachers of Mathematics Standards suggest the following topics for probability and statistics in the elementary program:[2]

For kindergarten through grade 4,

1. Collect, organize, and describe data.
2. Construct, read, and interpret displays of data.
3. Formulate and solve problems that involve collecting and analyzing data.
4. Explore concepts of chance.

For grades 5 through 8,

1. Systematically collect, organize, and describe data.
2. Construct, read, and interpret tables, charts, and graphs.
3. Make inferences and convincing arguments based on data analysis.
4. Develop an appreciation for statistical methods as powerful means for decision making.

[2]*Curriculum and Evaluation Standards for School Mathematics* Commission on Standards for School Mathematics, (Reston, Va.: National Council of Teachers of Mathematics, 1989).

5. Model situations by devising and carrying out experiments or simulations to determine probabilities.
6. Model situations by constructing a sample space to determine probabilities.
7. Appreciate the power of using a probability mode through comparison of experimental results with mathematical expectations.
8. Make predictions based on experimental or mathematical probabilities.
9. Develop an appreciation for the pervasive use of probability in the real world.

TEACHING CHARTS AND GRAPHS

The development of graphs can begin as early as the kindergarten level, with the children surveying their favorite TV programs and using colored blocks to record their findings. As the children progress up the ladder of mathematical understanding, the line-segment graph, pictograph, and circle graph can be introduced. Thus, various types of graphs should be developed, and the relationship between them should be understood.

In all cases, it is important that the first thing children learn is to develop the graph, not to read it. Current instructions often have this the other way around; but it is more natural and effective if a situation is first used in which children must work up a graph.

During the course of teaching graphs, the following generalizations should apply:

1. Pictographs, although they are the simplest type to read, give only an approximation of the data. Care must be taken to ensure that pictographs actually are representative. For instance, one picture can represent 10,000 cars, and two pictures of the same size can accurately represent twice as many. However, the use of a picture of one car to represent 10,000 cars and another that is twice as high to represent 20,000 cars is misleading, because the area of the second drawing is about four times that of the first.

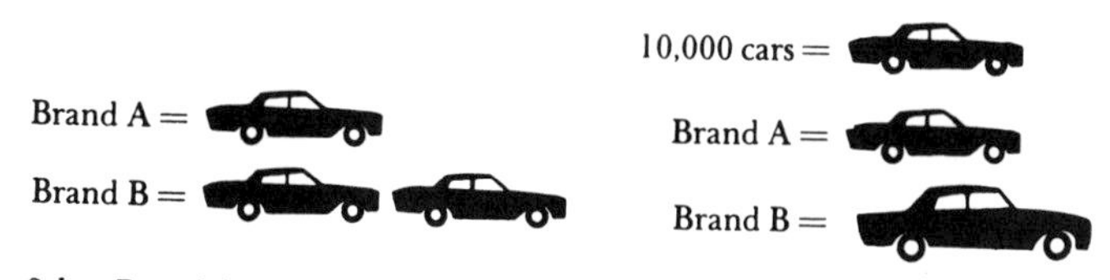

2. The bar graph, while not quite as easy to read as the pictograph, is easier to construct accurately. The bars in a graph should normally be of a constant width. Thus, the area of the bar is proportional to its frequency. Bar graphs can be misleading if units are compressed or extended and if zero is not the starting point.

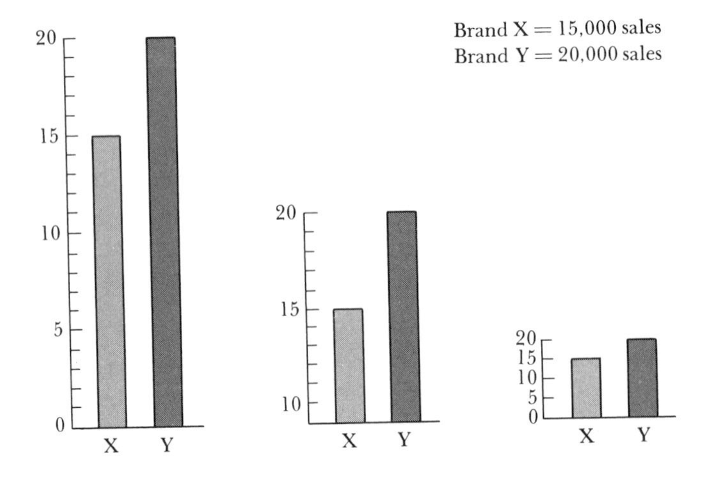

3. Line-segment graphs should be used only when there is a continued trend from one point to the next one. For example, the yearly per capita income in a country in 1960 was $2,000 and in 1980 was $2,500. If there had been a general trend of increase each year between 1950 and 1960, the line segment connecting the points for 1950 and 1960 would be helpful. If, however, the income had been $1,800 in 1975, the line segment connecting 1950 and 1960 would be misleading. Also, as is the case with bar graphs, line-segment graphs may be misleading if the scale is mixed or if the starting point is not zero. Line-segment graphs are often called *frequency polygons.*
4. A circle graph is particularly helpful if percentages totaling 100 are being graphed or when the graph deals with parts of a whole. A knowledge of angle measurement and decimal division is necessary for the construction of circle graphs.
5. On occasion it is desirable to use a combination of the formats described previously in constructing a graph. The general impression is conveyed by the pictograph. An accurate assessment can be made by the bar graph.

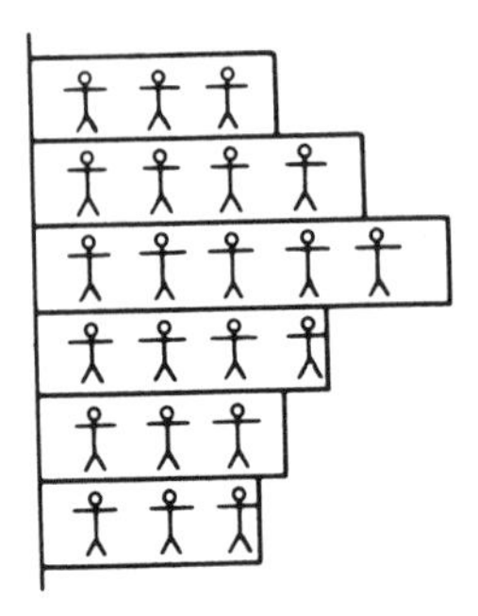

DESCRIPTIONS

Children and adults are often faced with the problem of finding a means to summarize the characteristics of data they have collected. Perhaps the data have been tallied or have been organized in a histogram. There are several ways of summarizing such findings, such as the mode, median, mean, and range. Children should have experience in using each and determining the appropriateness of each under various conditions.

Range

The range is one of the simplest means of looking at data. It is found by subtracting the smallest value observed from the largest. For example, (1) "We've made a height chart for our class. What is the height of the tallest? What is the height of the shortest? What is the difference in their height?" (Later, the term *range* is used.) (2) "What was the coldest temperature in the last 24 hours? What was the warmest temperature in the last 24 hours? What was the range in temperature?" (3) "You've checked the price of various items in the newspaper. What was the range in price of a pound of sugar?"

Calculators

Once the basic concepts of ratio, proportion, probability, and statistics have been developed, the calculator should take a major role in problem solving. When using the calculator, it is wise to have a class discussion concerning the meaning of the findings.

In one classroom the teacher said, "How could we use the calculator to find out what percent of the spelling words Jill wrote correctly? She was correct on 23 of the 25 words. What percent were correct?"

TIM: 92 percent, I divided 23 by 25 and got .92, that's 92 hundreds, percent is by the hundred, that's 92 percent.

JOYCE: I used the percent key. I wasn't sure what would happen when I put in 23 divided by 25 and pressed the percent key. My answer was 92—92 percent—the percent key gives us an answer as a percent rather than as a decimal.

The calculator can be used for (1) percents greater than 100, (2) percents less than 1, (3) testing estimates, and (4) working with sales tax.

Computers

Computers are ideal for the study of charts and graphs. Once children have developed the concepts, the computer provides speed and an attractive product. Programs such as "Exploring Table and Graphs," (Weekly Reader Software/Optimum Resources, Inc.) provide tutorials in one section and another section for children to construct their own graphs. Students construct tables on the screen and then create graphs based on the data. Creating bar, circle, and line graphs can provide a good basis for a discussion concerning the merits of each. It may be better to start out with the construction section and then use

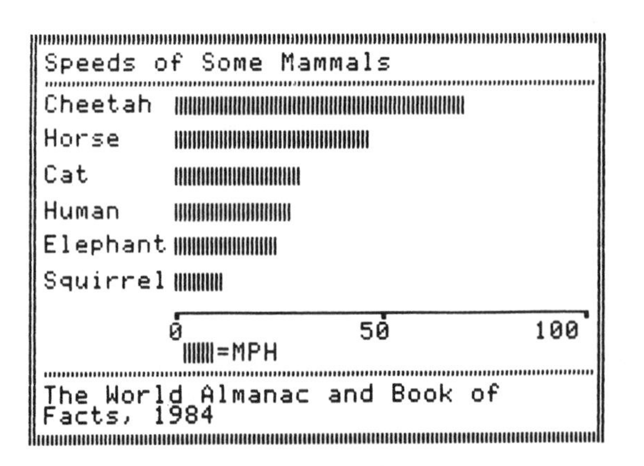

Making a Bar Graph with *Exploring Tables and Graphs*. (Weekly Reader Family Software/Optimum Resource, Inc.)

the tutorial only when necessary (in keeping with the discovery of oriented-developmental teaching).

Simulation programs, such as "Sell Lemonade," from the Minnesota Computer Consortium, give children the opportunity to experiment with selling under various conditions and provide data for the development of charts and graphs. This type of program is also excellent for problem solving, social studies, and statistics.

Measures of Central Tendency

THE MODE

The mode is probably the easiest of the measures of central tendency to introduce. It is simply the value of the category that occurs the greatest number of times. An easy way for children to observe the mode is to have them draw a histogram of their data. The "highest bar" on their graph is the mode. After children become familiar with the mode, they can easily record the value for the mode on their histogram. One example follows. The mode is 24.

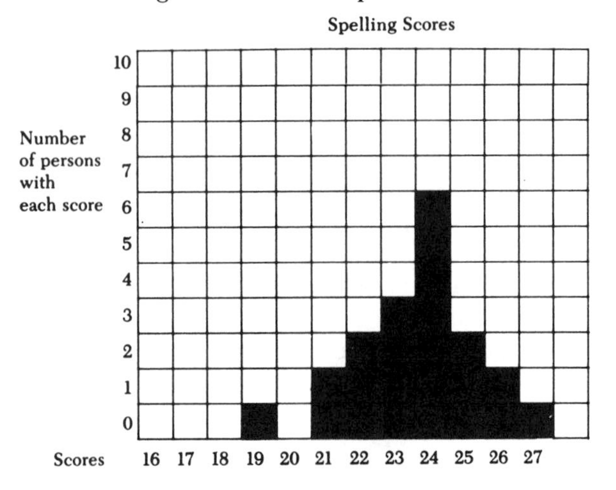

THE MEDIAN

The median, or "middle value," often represents the most typical measure. For example, Bill went fishing each day for a week. His catches were 0, 2, 3, 4, 8, 5, and 1. What was a typical day's catch? If children have experience with the mode, they can see that there is not a modal catch. Often children will suggest counting "halfway up" (after arranging the numbers in order, 0, 1, 2, 3, 4, 5, 8) to find a typical catch. In this case, the median would be 3. After many experiences finding the median, children can record both median and mode on their graphs. They soon discover that if there are an even number of observations, it is necessary to "split the difference." For example, with 1, 3, 5, 6, the median is 4.

THE MEAN

The arithmetic mean, often called the "average," is used to calculate batting averages, test averages, and the majority of measures of central tendencies reported in newspapers. Too often, the mean is thought of as simply adding up the scores and dividing by the number of scores. Although this does produce the average, children need experience at the manipulative level before arriving at the algorithm. One teacher used this lab sheet:

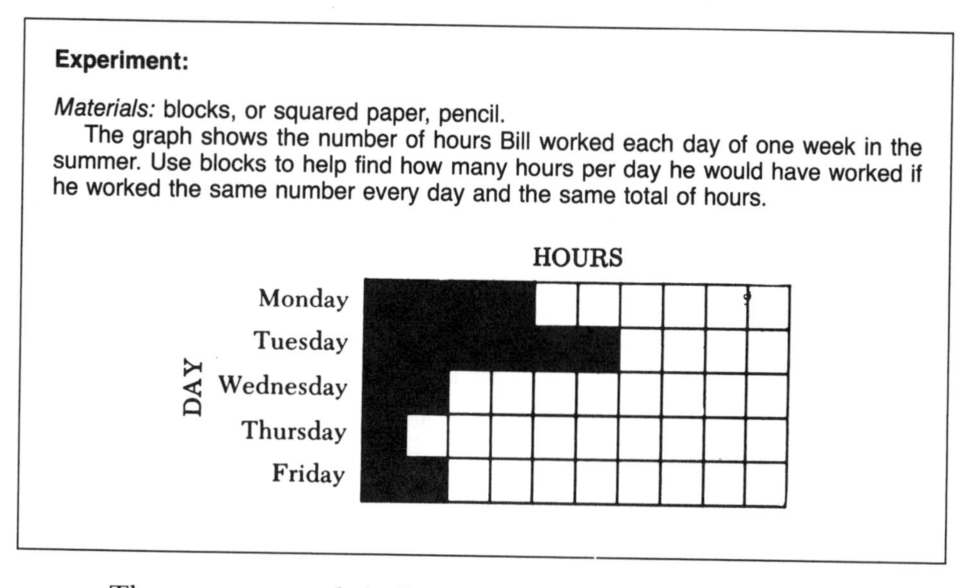

Experiment:

Materials: blocks, or squared paper, pencil.

The graph shows the number of hours Bill worked each day of one week in the summer. Use blocks to help find how many hours per day he would have worked if he worked the same number every day and the same total of hours.

There were several similar averaging problems on the lab sheet. The typical approach taken by the children follows:

CLAUDIA: I set out blocks to stand for 1 hour of work. Then I moved the blocks for the days when he worked more time to the days that he worked less time to make them all the same. At that rate, he would have worked 3 hours a day.

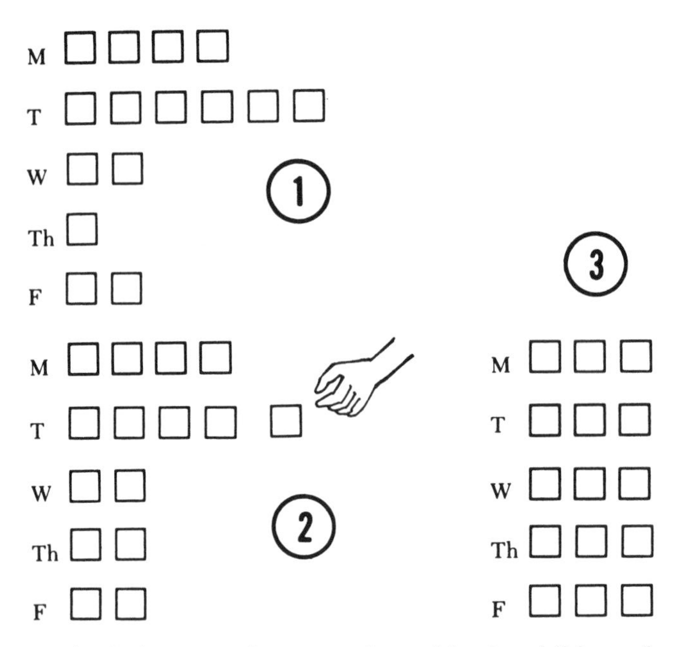

After many manipulative experiences such as this, the children should experiment to find a more rapid method of finding the mean.

Since variations in the value of the measures can affect the mean, children should have some opportunities to analyze differing situations that produce the same mean. For example, one teacher said, "Two star running backs each have a 5-yard rushing average. I'll show you the data on the runs they have made. Which one would you choose to run if you were third down and three yards to go at a very important point in the game? Why?"

Jones*	5	4	6	5	5	4	6	5	5	5	average 5 yards a carry
Smith*	10	2	3	0	15	0	0	5	10	5	average 5 yards a carry

The children said they would use Jones, since he could be depended on to pick up short yardage consistently. Smith would be used in a situation where they needed longer yardage.

The idea of deviation from the mean can be introduced in situations such as the football problem. The pupils list the set of distances run on each try and below each indicate the deviation from the mean. If a run is less than the mean, a negative sign is used (see table). Such an activity is a good application of integers, in addition to providing readiness for later study of the standard deviation.

SMITH										
Average	5	5	5	5	5	5	5	5	5	5
Run	10	2	3	0	15	0	0	5	10	5
Difference	+5	−3	−2	−5	+10	−5	−5	0	+5	0

*Use the names of two of their favorite backs. If possible, use actual information.

During the work on measures of central tendencies, the following generalizations should be developed and emphasized:

1. An incorrect selection of the measure of average used can often distort facts. For example, the mean yearly family income of a small community in Vermont is $30,000. But there is one very wealthy person in the community who earns several thousand dollars a day on investments. The median yearly family income of this community is $9,000. The modal yearly family income is $8,000. If the mean income were used in reporting, the type of community could be completely misgauged.
2. The mean provides some features that the median and the mode do not. The mean of a number of means can be taken by the add-and-divide method and be accurate. The mode of a group of modes or the median of a group of medians is not necessarily the actual mode or median of the entire group.
3. The mean is affected by extreme measures on either side of the scale, as the average income of the Vermont community shows.
4. An example in which the mode would be the most logical measure of central tendency would be a case in which a group of people were deciding on the time for holding a meeting. If six different times were proposed, taking the mean or median of these times would not be helpful, for it could be that very few of the people would be available at these times. More of them could be at the meeting at the modal time than at the mean or median time. In this case, the mode is by far the most useful measure.

Meeting Times Suggested	*Number of People Available*
5:00 P.M.	//
6:00 P.M.	//
7:00 P.M.	///
8:00 P.M.	////
9:00 P.M.	//////
10:00 P.M.	////
Mean = 8:00 P.M., median = 8:00 P.M., mode = 9:00 P.M.	

Also, the mode would be most useful in determining the average-size dress that a store should stock. If management were to decide to stock heavily the mean or median size, it could later find that very few women actually wear that size.

By using questions such as the ones that follow aid in determining the "best" measure for a particular instance:

1. What would you consider the most typical or average salary for newspaper deliveries?
2. Is it possible to have more than one mode? Give an example.
3. It is easier to find the median of 33 scores or 34 scores?
4. What measure would be best for telling the average height of houses in a community? of dining-room tables?
5. Could any of the three averages be as large as the largest measure? As small as the smallest measure?
6. If you were to predict the type of weather that Miami, Florida, will have tomorrow, would you use the mean, the median, or the mode?

A TEXTBOOK EXAMPLE

The accompanying lesson is a good example of how a sixth-grade textbook can be used effectively to present a useful application of statistics and give children an opportunity to make good use of a calculator.

Interpreting Data

Student book pages 306, 307

Purposes

1. To provide an opportunity to interpret, discuss, and make inferences from data.
2. To provide practice in approximation.

In this lesson (another pause in the 18-lesson sequence on functions and graphing), the students will study and interpret a chart giving information about library systems in the United States. The numbers are large and would be difficult to handle without a calculator. With a calculator, however, they become relatively easy to work with and it is possible to solve problems by asking questions that can be easily tested.

Note: If you have no calculators, or only 1 or 2, do page 306 as a whole-class activity.

Mental arithmetic

Multidigit arithmetic As in previous lessons, give multidigit problems that the students can solve mentally. Select problems according to the needs and ability of the class.

Student page 306

Go over the chart on page 306. Explain, if necessary, what each of the columns means.

Then show how one can make inferences and draw conclusions from these data without using a calculator—even though the numbers are large. For example, ask the students to make rough comparisons of Akron and Buffalo regarding population, number of libraries, number of books, and so on. Ask which system has more books per branch. Ask which population they think reads more. Continue such a discussion until the students begin to see how to understand and analyze such figures simply by inspecting them and asking relevant questions.

Next discuss the first question with all the students. Then have them work in small groups on the second and third questions. Have each group report its answers to the class.

316 Lesson 101

Answers to discussion questions (page 306)

First question. Answers will vary. One could argue, for example, that we would be most likely to find a specific book in Chicago, since its library system has the most books. But there are 48 branches in the Chicago library system, and if each branch had the same number of books, one would expect to find a little more than 100,000 books in each. In Buffalo, however, there are 2,701,424 books divided among only 17 branch libraries. If the books are evenly distributed among the branches, one might expect to find a little more than 150,000 books in each.

Second question. One could argue that the number of books circulated each year divided by the population is a good index of how much the population reads. Such figuring would make Buffalo, New York, a good candidate for the population that reads the most.

Third question. Expect a variety of good answers. One could argue that the operating cost per book circulated is a good index of efficiency. One might also want to consider the number of libraries, which might be an index of how available books are to the population.

From Stephen S. Willoughby, Carl Bereiter, Peter Hilton, and Joseph H. Rubinstein, *Real Math, Grade Six* (LaSalle, Ill.: Open Court Publishing, 1985), pp. 306–7.

STATISTICS

Developing Critical Analysis

Present-day reporting of statistics is often sketchy, and in some cases the devices used to obtain the statistics are questionable. Analysis of newspaper reports, advertising, and magazine articles can be helpful in developing a spirit of "proceed with caution" among elementary school pupils. When a report states that three out of four doctors interviewed favored a certain medicinal product, pupils should think, "I wonder how many doctors they asked, and how they decided which doctors to ask." It might be possible that they surveyed only the staff doctors working for the company. Also, did they survey medical doctors or Ph.D.s? A historian may not be a good authority on the quality of medicinal products.

Analysis of statistical reports is a fascinating project for elementary school pupils. They enjoy looking at reports and raising questions. This is good; however, the teacher should take care that pupils do not become so critical that they no longer consider any information at all accurate. A good start for looking at statistical misinformation is a paperback book by Darrell Huff.[5] The teacher can find a number of examples of common statistical errors. Common errors that can be detected by elementary school pupils include

1. A shift in definitions. For example, at one point, the writers may use the *mean* as the average and, at another point, the *median*.

Interpreting Data: Library Statistics

SELECTED LIBRARY SYSTEMS IN THE UNITED STATES*

City	Population	Number of Libraries	Total Number of Books	Average Number of Books Circulated Each Year	Annual Operating Costs
Akron, Ohio	275,425	17	777,285	1,945,499	$ 2,255,436
Buffalo, New York	462,768	17	2,701,424	5,817,993	$ 7,446,232
Chicago, Illinois	3,366,957	48	5,069,533	8,969,194	$15,661,185
Detroit, Michigan	1,513,601	29	2,210,277	2,472,750	$10,138,523
Miami, Florida	334,859	17	780,190	2,358,834	$ 4,075,304
Philadelphia, Pennsylvania	1,949,996	45	2,880,941	5,410,818	$10,828,852
St. Louis, Missouri	622,236	19	1,295,726	2,772,464	$ 5,333,950
San Francisco, California	715,674	26	1,443,848	3,195,475	$ 5,501,847
Washington, D.C.	756,510	19	2,133,102	2,050,690	$ 7,613,800

*Adapted from *The World Almanac & Book of Facts*, 1976 edition, copyright © Newspaper Enterprise Association, Inc. 1975, New York, New York 10017.

This chart gives information about selected library systems in the United States.

First study the chart. Then use the information in the chart to answer and discuss the questions below. Use a calculator. Try to support your answers.

- Which city library is most likely to have a specific book? Answers are in margin.
- Do you think that, on the average, the people in some cities read more than people in other cities? Which city do you think has a population that reads a lot?
- Which library system do you think is the most efficient? Why?

Now use the information in the chart to make up questions. Discuss them with a friend.

306

[5]Darrell Huff, *How to Lie with Statistics* (New York: Norton, 1954); see also Stephen K. Campbell, *Flaws and Fallacies in Statistical Thinking* (Englewood Cliffs, N.J.: Prentice Hall, 1974).

2. Inappropriate comparisons. For example, a small manufacturer may report, "We have the fastest growing sales record in the industry." This might be based on the fact that in one year, the company sold 250 machines, and during the next year, it sold 500 machines. In the meantime, a large producer of the machine may have increased its sales from 250,000 to 450,000. The first company has increased sales by a greater percentage than the second company, but its claim does not mean much.
3. Inaccurate measurement. In any statistical treatment, an inaccurate measurement may cause enough difference to affect the results greatly.
4. An inappropriate method of selecting a sample. [See section on sampling.]
5. Technical errors.
6. Misleading charts. [See section on charts and graphs.][6]

PROBABILITY

Bettor Without Odds and With Odds

The teacher gave each group of four children an opague paper bag in which were 10 yellow marbles and 20 red marbles (different-colored lima beans could have been used). The teacher said, "We're going to play a guessing game for a while. Each of you get out a sheet of paper to keep track of your right and wrong guesses. In the bag there are some marbles. Some are yellow and some are red. Guess a color and, without looking, take out a marble. Mark your guess right or wrong on the paper. Put the marble back in the bag and have another child in your group follow the same procedure. You can use a chart like this [drawn on the chalkboard]."

R = Red Y = Yellow	Jill
Guess	Color Drawn
1. Y 2. R 3. R	R R Y

After the children had played the guessing game for about 10 minutes, the teacher asked, "How did your guessing come out? What color did you guess right most often?" The children said that after a short time, they had always guessed red, since they seemed to draw a red marble much more often than a yellow marble. The teacher suggested drawing a graph of their group experiences to get an idea of the mix of yellow and red marbles.

Each of the graphs formed a similar pattern. The teacher then said, "Before we played this game, I put 30 marbles in each of the paper bags. We're

[6]Richard S. Pieters and John J. Kinsella, "Statistics." In *Growth of Mathematical Ideas K–12, Twenty-Fourth Yearbook* (Washington, D.C.: National Council of Teachers of Mathematics, 1959), p. 277.

going to count them to see how many are yellow and how many are red. But first I'd like to have you guess the number of each color that are in the bags."

Several different guesses were made, but the majority of the children guessed 10 yellow and 20 red.

The next day, the small groups used the same set of marbles in the bags. The teacher presented these rules for the guessing game: "There are twice as many red marbles as there are yellow ones. So let's score points somewhat differently. If you score two points for every correct guess of yellow, how many points do you think you should score for every correct guess of red?"

After some discussion and argument, the children arrived at the notion that since there were twice as many red marbles, they should probably only score one point. The game was played in a way similar to that of the first day.

There are many activities and laboratories that can be developed as readiness for probability ideas. Thus, the children should have many experiences similar to the one just discussed.

Later, the teacher gave the children a bag with 100 lima beans: 25 with a red dot, 50 without any marking, and 25 with a blue dot. The children worked in groups of three, with instructions to take turns drawing 10 lima beans, record their draw on a graph, return the beans to the sack, and repeat the procedure. In addition to the individual graphs of the draws, each group was to make a continuous graph (a long-run relative-frequency distribution).

Developing Some Basic Probability Ideas

After the children had conducted a wide variety of experiments involving guessing the outcome of events, the teacher used the following discussion to pinpoint some of the ideas involved in the experiments and games. (Note: More than one day should be used to present the material that follows; also, there should be some experimentation at various points in the development.)

The teacher stated, "Today we're going to be drawing a name out of the hat to see who will be the leader of the lunch line. How many do we have here today?—Thirty-three. Each person's name will appear once. What is the probability of Hank's getting to be first in line?"

The pupils responded that they felt that he had 1 chance out of 33, or $\frac{1}{33}$. The class members admitted that they were somewhat hazy as to the meaning of the question, "What is the probability of?" To clarify this concept to a certain extent but to avoid treating the definition formally, the teacher asked several questions. Often a debate as to the answer followed. The questions were the following:

1. You said that the probability of Hank's getting to go first was $\frac{1}{33}$. Does this mean that were I to draw a name from the hat 33 times (we put the name back in after every draw), Hank would be sure to win at least once?—No.
2. Would he probably win at least once?—Yes.
3. Were we to draw 66 names (replacing the name after each draw), would Hank win 2 times?—Not necessarily. He might, but he might not.

4. Were we to draw 660 names out of the hat, would Hank ever have to win?—No, but he probably would.
5. From 1, 3, 5, 7, 9, 11, 15, picking a number at random, what is the probability of choosing a number less than 8? Of 5 or more?—$\frac{4}{7}$; $\frac{5}{7}$.

From this discussion the class may move to a topic such as coin tossing or opinion sampling. Introductory remarks and possible questions and answers are given in the material that follows.

The teacher said, "We can get a better idea of probability if we try to answer a few more questions and then do some experiments."

1. One of you registers six times for a drawing on a bicycle. There are 2,567 tickets in the drawing. What is the probability that you will win? Answer: $\frac{6}{2,567}$.
2. There are 16 girls and 17 boys in our class. What would be the probability of drawing a girl's name from a hat containing the names of all class members? Answer: $\frac{16}{33}$.
3. Bill knew that the answer to a social-studies question was either Washington or Lincoln. He decided to pick one of the names at random. He said, "The odds are 50 to 50 that I will be right." What was the probability he would be right? Answer: $\frac{1}{2}$. Discussion should bring out the idea that the *odds* of success are the ratio of chances of success to chances of failure—in Bill's case, 1 to 1. However, the *probability* of success is

$$\frac{\text{Chances of success}}{\text{chances of failure} + \text{chances of success}}$$

After similar questions, the teacher concluded the lesson. The next day, each pupil was given a penny and asked to state the number of heads they would expect in 100 tosses. The children responded that they would expect about 50. Several groups then tossed the coin 100 times and recorded their findings. The outcome was close to 50 in many cases, but not 50 in any of the cases. The teacher used questions to emphasize that although the probability of heads was 50/50, heads would not necessarily come up half the time. Children then used ten coins to record the result of 1,000 throws. They found that as the number of throws increased, the number of heads thrown more closely approximated the predicted proportion.

Sampling

One day, the teacher brought in a fishbowl filled with red and blue marbles and presented this situation: "At election time, a prediction of the election results is often desired. Let's assume that the blue marbles represent people who are for one candidate and the red ones represent those who are for the other. How could we get an idea of the election outcome without counting all the marbles?"

Pupils suggested that a sample of the marbles could be taken, and they discussed an appropriate method for doing this. The discussion emphasized the term *random sample*. (A random sample is one drawn in such a way that every possible combination of the given size has an equal chance of being selected.)

A paddle constructed of plywood (see illustration) was used to dip into the bowl to select a sample.

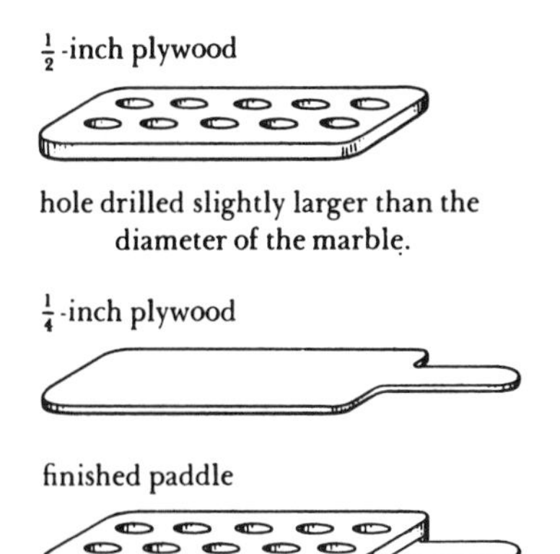

Five samples were drawn, and the following results were obtained. (Each sample was returned to the bowl before drawing the next. If this is not done, the result is inaccurate.)

Red	Blue
8	2
7	3
6	4
6	4
7	3
34	16

The pupils estimated that the candidate represented by the red marbles would probably win by something like a 2-to-1 margin. (There were actually 800 red marbles and 400 blue marbles in the fish bowl.)

At this point the teacher directed questions to the class concerning the basis of obtaining a good sample. The sampling of people in political and other surveys was emphasized, and the need for a *representative sample* was stressed. (A representative sample matches the population from which the sample is drawn. For example, it would be poor sampling procedure to take an entire national survey from among only people living in small towns.) Some of the generalizations reached were these:

1. Each individual (or object, etc.) should have some known probability of being selected; for example, $\frac{1}{23,000}$ in a sample of voters in a community. The choice of one is not dependent on the choice of another.
2. The sample should be taken by some automatic or prescribed means. It should avoid bias toward or against any portion of the population.
3. There should be some kind of randomness in the selection.

From this discussion, the class may move to a topic such as coin tossing or population of children in the school. When each group had surveyed its sample, the findings were graphed. Then the graphs were compared.

TRACER TECHNIQUE

An interesting laboratory can be developed by using the tracer technique. The idea is often used in checking wildlife populations, insect life, and so on. The technique is to introduce a number of "labeled" animals into a population of animals that cannot easily be counted. For example, 30 banded cardinals were released in an area in which the approximate population of cardinals was of interest to scientists. Then observers kept track of the proportion of cardinals banded. Thus, if after many observations the ratio of banded to nonbanded cardinals was 10 to 100, the scientists would approximate the cardinal population to be about 330 (300 nonbanded and 30 banded).

In the classroom, the teacher used a larger bag with 200 blue marbles, and the children took turns drawing a marble and replacing it. After a number of draws and the making of a graph of the draws, the children felt reasonably sure that all the marbles were blue. The teacher then asked, "How could we find out about how many blue marbles there are without counting them all?" After several suggestions that did not prove productive, the teacher related the technique used with the cardinals. Then several of the children suggested putting 20 red marbles in the bag, shaking it up, and then drawing samples of 10 marbles 10 times. Other children noted that it would be very important to replace the marbles after each draw and to be very sure that the marbles were completely mixed up.

The children made the table shown from the results of drawings. Then they found the average number of red to blue marbles and predicted that there were about 200 blue marbles in the bag, after which they counted the marbles to check their result.

		MARBLE COLOR	
		Red	*Blue*
	1	0	10
	2	1	9
	3	0	10
	4	2	8
	5	1	9
	6	0	10
	7	2	8
	8	1	9
	9	0	10
	10	2	8
Total		9	91

At the end of this activity, groups of four children were each given 200 marbles (corn or lima beans of different colors could be used) and allowed to place any multiple of 25 marbles in the bag. Each group then traded bags with another group and used the tracer technique to estimate the number of marbles in the bag. Later, similar activities were conducted.

OTHER EXPLORATIONS

The following ideas are given for the development of activities dealing with probability and statistics. What teaching strategies could be used with each? What ideas does each suggestion develop?

1. Make a survey of vehicles (cars, trucks, buses, etc.). Possible information for graphing the data: make, year, color, time of day for survey, location of street for survey, day of week for survey, length of time for count, weather at time of count, direction of traffic flow, streets two-way or one-way, etc.
2. Find the frequency of letters in use in the English language. Questions to be considered: What books should be surveyed? How many letters would make a good sample? What type of table would best describe the information? How could this information be used to develop simple codes? How could it be used to break a code? (*Note:* See *The Gold Bug* by Edgar Allan Poe for an interesting detective story using this technique.)
3. Conduct the same type of survey on the length of English words.
4. How many different phone numbers can be generated from the typical seven-digit sequence? Why are there more area codes in highly populated areas?
5. Study the newspapers for a week. What surveys are reported? How much information is given on each? What are the strengths and weaknesses of each survey?
6. Develop the idea that there may be physical properties of objects that cause bias. Use variously biased spinners and compare the actual outcome with predicted outcomes.
7. A device known as Pascal's triangle is sometimes useful in determining the different outcomes from probability experiments.

Computers

There are a number of computer-assisted instruction programs that simulate probability situations, generate random happenings, or present problem situations related to probability. See the references in "Selected References" at the end of the chapter, specifically Clements (1989).

ANALYSIS

Once pupils have mastered the basic ideas of sampling, graphing, averages, and so on, a variety of statistical projects may be undertaken. A survey of the TV interests of an elementary school was one such project. The following steps were used in the survey:

1. The class agreed upon the nature of the problem: "What are the television viewing interests of the pupils in our school?"
2. The class discussed the selection of a "representative sample" of the school. (It was agreed that surveying every student would be too time-consuming. Also, the pupils wanted to test sampling techniques.) A sample was selected.
3. An appropriate method of gathering data was decided on.
4. The data were gathered.
5. Interpretations were made from the data.

With activities such as those described in this chapter, extreme care should be taken to make the pupils aware of the difficulties that arise in statistical work and the inaccuracies that may occur. For example, in one fifth grade, the pupils thought they could report the reading interests of fifth graders in general (throughout the United States) from a sample taken in one community of 1,200 people. Such errors in thinking should be corrected. In fact, it is sometimes wise to let pupils jump to false conclusions and then by discussion have them discover the fallacies in their thinking.

Many worthwhile learning experiences can be developed by exploring the ideas of probability and statistics. The teacher should be alert to situations in the sciences and social studies that can be interpreted more effectively with probability statements and simple statistics.

Since graphs, statistics, and probability are a part of the daily life of all children and adults, care must be taken to include a balanced amount of them in all basic-skill and essential evaluation measures.

KEEPING SHARP

Self-Test: True/False

_______ 1. There are many applications of statistical inference outside the classroom.

_______ 2. The study of probability and statistics during the elementary school years should be concerned with exploring problems and experimenting with predictability of simple events (coin toss, etc.) rather than formal work on probability and statistics.

_______ 3. Data collecting and sorting can lead to graphing and modeling activities even in the primary grades.

_______ 4. Prediction is an important use of statistics.

_______ 5. Children must know how to read a graph before they are taught to construct their own graphs.

_______ 6. When comparing bar graphs representing the number of immigrants by country of origin for each of several years, the relative size of the units and the starting point become unimportant.

_______ 7. Early experiences in finding the mean can be developed by using manipulatives before computation is used.

_______ 8. Children should be sufficiently familiar with the mean, median, and mode to recognize and explain their advantages and limitations.

_______ 9. Analysis of advertising in newspapers and magazines is best left to the secondary school, because it can serve little purpose in elementary math.

_______ 10. Critical-thinking techniques should be developed from the earliest experiences with statistics.

_______ 11. Very young children can understand probability when they are encouraged to experiment with concrete materials and make predictions.

_______ 12. Good teaching strategy includes numerous laboratory experiences using the social situations in which ratios are used.

_______ 13. Using the cross-products approach is helpful in problem solving.

_______ 14. The precise language of percent can cause confusion.

_______ 15. The use of squared paper is not a recommended strategy for helping pupils visualize percents.

_______ 16. Elementary school mathematics texts commonly suggest using decimal, ratio, formula, and equation approaches as well as unitary analysis in solving percent problems.

_______ 17. From time to time, pupils should be encouraged to solve the same problem in different ways.

_______ 18. Percent is the application of a particular notation system to quantity rather than a mathematical topic per se.

Vocabulary

ratio	tracer techniques	range
proportion	probability table	mode
percent	unitary-analysis approach	mean
random sample	formula approach	chance

THINK ABOUT

1. Analyze the approach to ratio and proportion in two current textbook series. Is this treatment adequate to develop children's understanding and use?
2. Collect examples of the use of ratio in advertising. Relate these to common misunderstandings and discuss the sources of these problems.
3. The word *percentage*, as used in arithmetic books, in educational literature, and in everyday business transactions, has three distinctly different meanings. What are they, and do these different meanings for the same word pose an instructional problem?
4. How might the use of ancient ways of figuring taxes, tithes, and tolls be of value in teaching percents?
5. What feature of the unitary-analysis method of solving percentage problems makes it attractive to those who wish to use the developmental, or figure-it-out-yourself, approach to the teaching of percents?
6. Plan a series of laboratory experiences for third-graders on the probability of moves in a *Monopoly*® game when dice are used or when a spinner with equal segments is used.
7. Explain both the activities and the expected outcomes for a unit on graphing in kindergarten or first grade.

8. Working with two others, develop a list of probability and statistics lessons, laboratories, or games that you could use to develop other mathematical ideas, such as fractions, geometry, and multiplication.
9. Collect examples of advertising that show one or more of the common errors discussed under "Developing Critical Analysis." In each case, what is the reader led to believe as a result of the error, and how would the ad be affected by correcting the error?
10. Get three or four shoeboxes. Begin collecting inexpensive materials you could use to develop probability in the grade you plan to teach. What will you include? How much will you need for whole-group lessons? for individual experiments? Will it get "used up" or will it last through numerous sessions? What materials will you need for recording information and results? Can these be stored with the manipulative supplies?

SUGGESTED REFERENCES

BARATTA-LORTON, ROBERT, *Mathematics: A Way of Thinking*, Menlo Park, Calif.: Addison-Wesley, 1977.

BITTER, GARY G., MARY M. HATFIELD, and NANCY T. EDWARDS, *Mathematics Methods for the Elementary and Middle School*, chap. 11, 12. Boston, Mass.: Allyn and Bacon, 1989.

CAMPBELL, STEPHEN K., *Flaws and Fallacies in Statistical Thinking*. Englewood Cliffs, N.J.: Prentice Hall, 1974.

CLEMENTS, DOUGLAS H., *Computers in Elementary Mathematics Education*, chaps. 8 and 10. Englewood Cliffs, N.J.: Prentice Hall, 1989.

COBURN, TERRENCE, *How to Teach Mathematics Using a Calculator*, Reston, Va.: National Council of Teachers of Mathematics, 1987.

Mathematics for the Middle Grades (5–9), 1982 Yearbook, Reston, Va.: National Council of Teachers of Mathematics, 1982.

Teaching Probability and Statistics, 1981 Yearbook, Reston, Va.: National Council of Teachers of Mathematics, 1981.

THIESSEN, DIANE, MARGARET WILD, DONALD PAIGE, and DIANE L. BAUM, *Elementary Mathematics Methods*, chaps. 12, 13, 15. New York: Macmillan, 1989.

VAN ENGEN, HENRY, and DOUGLAS GROUWS, " Relationships, Number Sentences and Other Topics." In *Mathematics Learning in Early Childhood Education*, Reston, Va.: National Council of Teachers of Mathematics, 1975.

WIEBE, JAMES H., *Teaching Elementary Mathematics in a Technological Age*, chaps. 14, 18. Scottsdale, Ariz.: Gorsuch Scarisbrick, 1988.

chapter 13

GEOMETRY

OVERVIEW

> "Geometry is grasping . . . that space in which the child lives, breathes and moves. The space that the child must learn to know, explore, conquer, in order to live, breathe and move better in it."[1]

The amount of geometry contained in the elementary school mathematics textbook series steadily increased between 1960 and 1975. Since that time the amount has remained rather stable, but the caliber of the material has greatly improved.

Even with the increase in the teaching of geometry, however, teachers tend to move geometry aside for the more important "basics." This is unfortunate, since the world is a maze of geometry. From the time children in their cribs first handle objects, they are interested in and concerned with shape, size, and location.

[1]Freudenthal, H., *Mathematics as an Educational Task* (Dordecht, The Netherlands: D. Reidel, 1973).

Tangrams

Children enjoy all types of puzzles and puzzle games. Tangrams can be used for enrichment, for free-time study of geometric relations, or just for fun. Children can find multiple solutions to some tangram puzzles, and they gain insight into geometric relations by building and laying out designs. Also, there are many perimeter-area ideas imbedded in the activities with tangrams.

Make a set of seven shapes like those in white. Try to make each of the figures. (Each uses all seven pieces.)

TAKE INVENTORY

Can You:

1. Give at least three reasons for teaching geometry in elementary school?
2. Plan a lesson for introducing a geometric term using discovery-oriented procedures?
3. Suggest uses that could be made of LOGO to teach geometry to elementary school students?
4. Make a table of properties of common geometric figures?
5. Develop a discovery lesson using a topic in geometric construction?
6. Develop activities for contrasting three-dimensional object and two-dimensional objects?
7. Develop lessons by using slides, flips, and turns?

Approaching the Teaching of Geometry

The following suggestions are designed to give the reader a perspective on teaching elementary school geometry:

1. Topics in geometry should be introduced through situations that present a purpose for studying them. Rather than stating, "Today we are going to learn to bisect lines," present a situation in which pupils have a reason for bisecting lines.
2. A study of geometry in the elementary school should be based on exploration. Emphasis should be placed on questions such as, "What new ideas did you get from today's work?" "What likenesses, differences, and patterns did you see in the geometric figures?" "Why do you think mathematicians consider dots and points as being quite different?" Less emphasis should be placed on questions such as, "Why should you be very careful to sharpen your pencil before working with geometry?" "How many sides does a hexagon have?" and other questions that require a rote response. It should be noted that mathematical abstractions and deductions follow guesses, approximations, investigations, and corrections.
3. Topics in geometry should be taught in short- or medium-length units rather than in isolated, single-day lessons.
4. There should be a *sequential development* of portions of topics in geometry at each grade level. Geometry should be a part of the entire elementary school mathematics curriculum.
5. Imprecise vocabulary should be carefully avoided. However, it is sometimes better to place less emphasis on the early identification of the name of a geometric idea or concept. Too often in the past, pupils had a good memorized mathematical vocabulary and a poor understanding of the mathematical ideas. Vocabulary should follow the development of the geometric idea.
6. Since one of the major uses of geometry is connected with measurement, there should be a concerted effort by the teacher to integrate geometric ideas and measurement whenever reasonable.
7. The use of LOGO can play an important role in the geometry program. Also, there are a number of excellent computer-assisted instruction programs that should be considered a part of the program.

The serious student of geometry teaching should become familiar with the van Hiele model of development of geometric thought. Beginning in the middle 1950s, Dina and Pierre van Hiele studied children's geometric thought. Dina died shortly after her thesis work (1957), and Pierre continued the work. They suggest that there are four levels of geometric thought, (1) visualization, (2) analysis, (3) informal deduction, and (4) deduction, through which children pass through more by instruction rather than maturation.[2]

STUDYING GEOMETRIC SHAPES

Spheres, Cylinders, and Rectangular Solids

The most common experiences of pupils with geometry are those involving three-dimensional rather than two-dimensional shapes.[3] At a very early level,

[2]A number of suggestions for using their work appear in *Learning and Teaching Geometry K–12, 1987 Yearbook* (Reston, Va.: National Council of Teachers of Mathematics, 1987).

[3]A portion of this section is based on Ruth Hutcheson and George Immerzeel, *Non-Metric Geometry* (Cedar Falls, Ia.: Malcolm Price Laboratory School, 1989).

situations that further ideas about the nature of three-dimensional geometry should be planned.

One teacher said to the class, "I watched a little boy sort his toys and put them in boxes according to shape. However, he made a number of errors. Here are some objects similar in shape to the ones he was sorting out. Let's see if we can sort them according to shape." On a large table, the teacher had placed balls (spheres) of various sizes, several building blocks and boxes containing games (rectangular solids and prisms), and cylindrical building materials, pickup sticks, and containers for Tinker Toys® and Lincoln Logs® (cylinders).

The pupils were asked to separate the objects according to shape and to explain why they considered certain objects to be of the same shape.

After they had done this, the teacher said, "Phyllis and Ray were at the grocery store. They couldn't agree as to the best way to describe the three shapes that I've brought along. How could you describe them? First, how are they alike?"

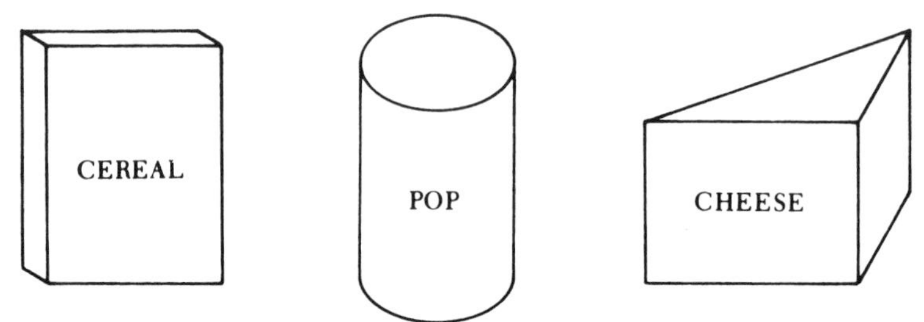

Some of the children said that the cereal box and the cheese had corners and edges and had some sides that were rectangular. The discussion continued, and as pupils introduced vocabulary words such as *triangle, circle, square, solid, line,* and *point,* these were informally discussed.

The next day, pupils discussed ways in which the objects were different. Toward the end of the period, each pupil was given a wooden model of each object and asked to develop a summary of the characteristics of the objects. They made use of a table such as the one shown to summarize their findings. When pupils disagreed on a portion of the table, they were asked to demonstrate how they had found an answer.

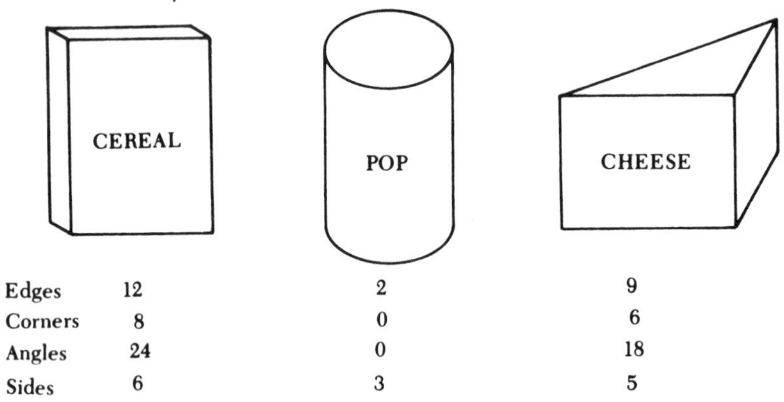

	CEREAL	POP	CHEESE
Edges	12	2	9
Corners	8	0	6
Angles	24	0	18
Sides	6	3	5

Similar methods can be used to introduce other geometric shapes. The emphasis should center on the identification of their characteristics. Correct geometric vocabulary can be developed, but extreme care should be taken that the study of geometry is not just the naming of the shape itself.

The study of three-dimensional objects can move smoothly to the identification of *line segments* as the meeting of two faces and *points* as the location at which three sides (surfaces) meet.

Points, Curves, Lines, and Planes

Geometry is based on several important constructs. First, a *point* is thought of as being so small that it has no size. It can be used to model an exact location in space. A *curve* may be thought of as a model of a set of points represented by a pencil drawing made without lifting the pencil off the sheet of paper. *Lines* are a particular type of curve. *Line* means "straight line" and extends in both directions in space without limit. When a line has a measurable length, it is called a *line segment. Planes* are also particular sets of points. A geometric plane is a flat surface that extends without end. *Space* is considered to be an unlimited number of points; points in space can be thought of as described by their positions.

Pupils' understanding of these geometric concepts should be gradually developed throughout the elementary school years. Several approaches to developing this understanding are presented below.

POINTS

The idea of a point can be developed through the need to identify an exact location. One primary teacher began with a discussion of a candy developed by Willy Wonka[4] that was never used up. The ball of candy kept getting smaller and smaller each day but would never disappear completely. The teacher asked the children what they would do with the candy at night if it were so small that they could no longer see it. After some discussion, one child suggested making a dot on a piece of paper and setting the "invisible" piece of candy on that dot.

The teacher made use of this idea to further develop the concept of a point and to introduce the word *point.*

Other situations can be used, such as, "On a recent trip, I saw the point at which three states meet. Who owns the exact location at which these states meet? Can it have dimension? Would this exact location (that we can't see) be a dot or a point? What is the difference?" "I've drawn a circle cut into fourths. Notice that each of the fourths is a different color. What is the color of the point at the exact center of the circle? Does it have dimension?" "What do we mean by a point at which a space capsule reenters the earth's atmosphere?" "A friend of mine said, 'I know a very good spot for fishing. I'll take you right to the point where we should fish.' What did he mean?"

[4]See Roald Dahl, *Charlie and the Chocolate Factory* (New York: Alfred Knopf, 1964).

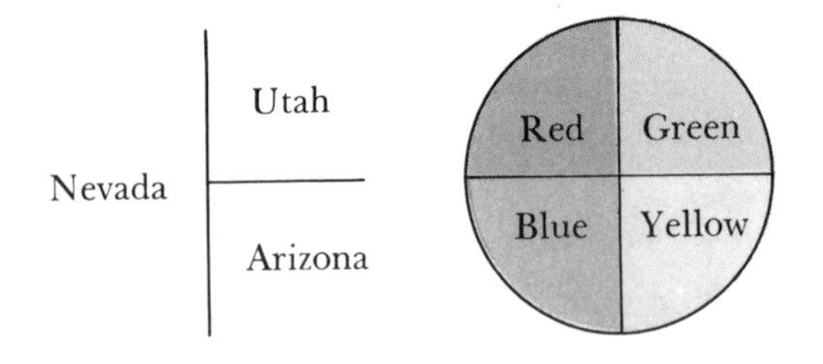

Children were also directed back to the ideas that had developed working with blocks and were led to recall that when three surfaces meet, they form a point.

CURVES

The topic of curves was developed through the following situation: "I've reproduced a Cub Scout map of a trip to the woods. What would we call the dotted line?" Pupils responded that it would be the trail or the path that Mark took. Then the teacher said, "Right, we could call this Mark's path. Also, we can identify the mathematical idea given by this path as a curve."

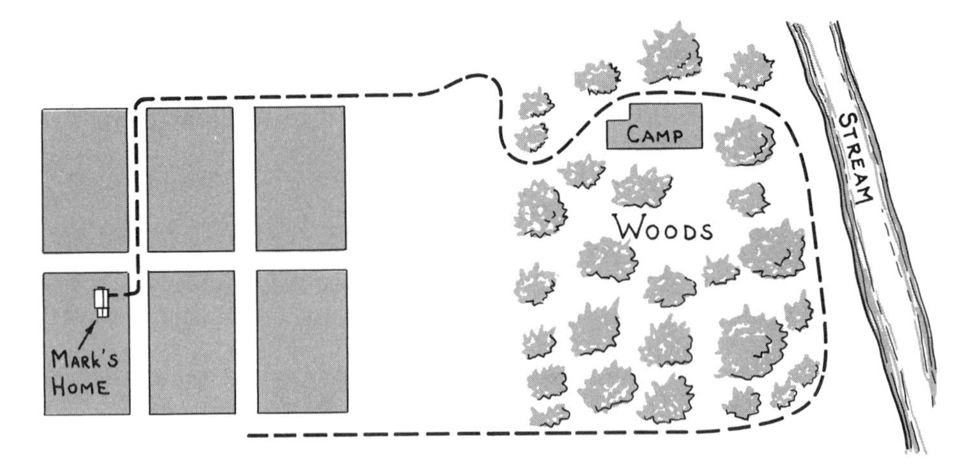

Pupils then remarked that not all of the path was curved; some of it was made up of straight lines. (In this discussion, only curves that lie in the same plane are considered.) The teacher asked, "Is a straight line a curve? Is a curve a line?" This type of questioning gave rise to some confusion, which provided a need to check reference sources concerning the mathematical meaning of *curve*. From the sources, the pupils learned that all lines and line segments are curves, and that a curve is made up of an infinite set of points.

The teacher then asked for other representations of curves and inquired if the path around the lake shown below is a representation of a curve. Again the class found it necessary to check references to find the answer. They found that the path around the lake is a representation of a closed curve. They also found that geometric shapes such as circles and rectangles are special types of closed curves.

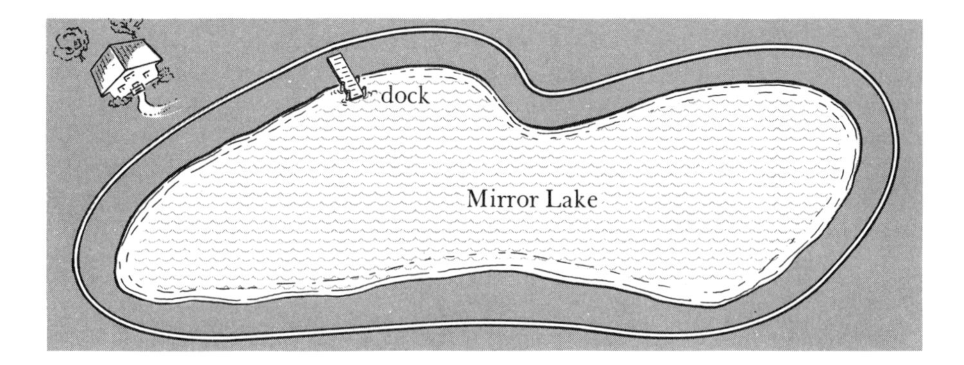

As the study of curves progressed, the teacher made use of laboratory activity cards using treasure maps to develop the idea of points on the interior, the exterior, and the boundary of a closed curve. The idea of a simple closed curve (a closed curve that does not intersect itself) is of great importance in geometry, because these curves are the boundaries of plane regions such as triangles and rectangles.

LINES

Discussion of the idea of lines might be developed by beginning with a statement such as, "We talked about a point at which several states meet. What would we call the border between two adjoining states? Which state would own this line? Does a line have dimension? Can I really draw a line on the board? How long should it be? If a line has two end points, it is called a line segment. [See diagram.] Is the border between the two states a line or a line segment? How could we represent a line to show that it goes on without end?"

A ———————— B ←————————→
Line Segment Line

"Nancy and Jane share a room. They draw a representation of a line down the middle of the room to divide it. How wide should this strip be if each girl is to have maximum space?"

To give the idea of a line segment, one teacher used the idea of a magic peppermint stick that was never used up but got smaller and smaller so that it eventually could not be seen. The children noted that they would have to make a drawing of a line segment on paper to keep track of the peppermint stick after it was used for a time.

PLANES

The topic of planes was introduced by placing a plastic rectangular solid and a plastic rectangular prism on a table and labeling the end surfaces. The teacher began the discussion by saying, "Ken said that surface A and surface B were just alike; Nancy didn't think they were. See if you can resolve their argument by thinking about the two surfaces. First, how is surface A like surface B?" Pupils said that both were flat and that both were the same distance above the table top. Then they were asked to give possible names for smooth surfaces such as A and B. Many suggestions were given; however, the teacher capitalized on the term *plain,* which they had been discussing in geography, to develop the term *plane.*

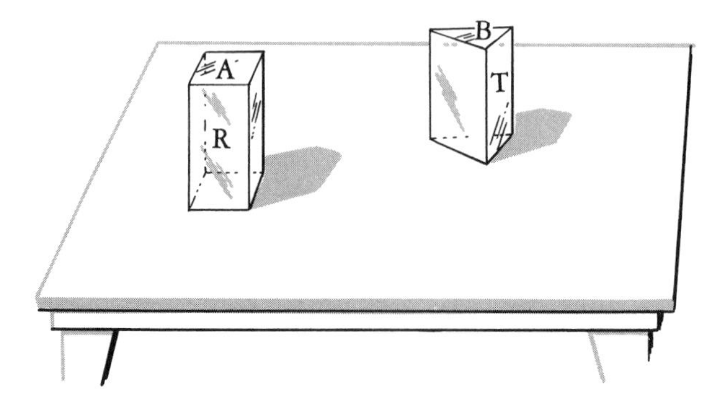

Next, pupils were asked, "Are A and B in the same plane?" Many pupils were not sure. Through questions such as, "Will a jet fly above plane A?" "What if surfaces A and B went on forever? What could you say about them?" pupils developed the concept of a plane as a flat surface that extended in every direction "forever."

The lesson continued with a discussion of only the plane determined by A. The teacher asked, "Will the peak of Nittany Mountain be above plane A?" (Yes) "If you were in Pittsburgh, could you find a point on plane A?" (Maybe; it would be there, but it would be hard to identify.)

Further discussion can lead pupils to visualize two planes that intersect and to identify the points of intersection of the two planes as a line. Then, the difference between the ideas about planes that the pupils have been discussing and physical-world examples that are like planes can be discussed. For example, doors, windows, and sections of a wall are like planes; however, they are three-dimensional and don't "go on and on."

RAYS

The concept of a ray can be developed by questions such as, "How would the representation of a rocket sent from earth into space appear on the chalkboard?" "What would the representation of the beam of a flashlight look like?" "Would a lighthouse beacon be a representation of a line? of a line segment? How is it different from either a line or a line segment?" "What do we call the beam of light from a lighthouse or a flashlight?" Pupils then combine their understanding of representing lines with their understanding of representing line segments to represent rays. Also, they associate an infinite set of points with the ray and know that a ray has one end point.

Further discussion can be used to develop these generalizations about the properties of lines and planes: (1) exactly one line joins any two points in space; (2) if two different points of a plane lie on a line, that line lies in the plane; (3) three points that are not on the same line are in only one plane;

(4) if two different planes intersect, their intersection is a line; and (5) when two different lines intersect, their intersection is a point.

Circles, Rectangles, and Triangles

A teacher brought several pictures of rectangular, circular, and triangular forms to class and discussed with the students the shape that would frame each picture. Pupils identified the shapes of these objects from home experiences. The teacher then brought out a rectangular picture frame and asked, "What do we say the shape of this picture frame is?" When it was identified as rectangular in shape, the teacher continued by saying, "Yes, this picture frame is a model of a rectangle. How can you tell an object that is the shape of a rectangle?" Pupils noted that opposite sides were the same length and that the rectangle had "square corners."

The next questions were designed to identify the difference between a rectangle and a rectangular region. The teacher asked, "How is the frame different from the picture?" Pupils suggested that they were the same shape but that the picture frame had nothing inside. The teacher commented, "We can describe the picture frame and the picture by saying that the picture frame is a model of a rectangle and the picture is a model of a rectangular region. What do you think I mean by a rectangular region? Can you give me examples of representations of rectangles and rectangular regions? Look around the room." Pupils identified what follows as models of rectangles: the window frame, other picture frames, and the door molding. They identified window panes, doors, and pieces of tablet paper as models of rectangular regions.

Similar discussions of triangles and triangular regions and of circles and circular regions can then be developed. In addition to circular and triangular picture frames, models can be made from pipe cleaners, wire, and heavy paper. Darning hoops and bicycle tires make good models for a circle, and metal musical triangles can be effectively used as models for a triangle.

When the teacher feels that pupils have gained some insight into the difference between a geometric shape and its region, he or she can make effective use of worksheets in which pupils are asked (1) to identify triangles, circles, and rectangles; (2) to be ready to describe their characteristics; and (3) to outline the model of the geometric shape and color the region. A bulletin board using various geometric shapes is also useful.

Other Geometric Shapes

As pupils make mathematical progress, a wide variety of geometric shapes should be identified. In addition, there should be an increased emphasis on classifying these shapes. The set of developmental questions concerning quadrilaterals that follows is representative of this type of classifying.

TEACHER: You have been working with a number of geometric shapes that contain four sides. Name all that you can.

PUPIL:	Squares, rectangles, parallelograms, quadrilaterals, trapezoids.
TEACHER:	Which of these shapes is the most general? That is, which of these has all the other shapes as subsets?
PUPIL:	Quadrilateral.
TEACHER:	Try to use the Venn diagram to show the relationship. [The following diagram can then be developed.]

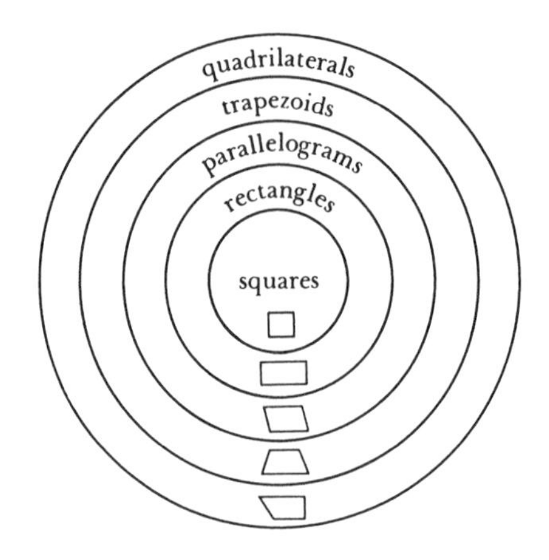

One of the most effective means of familiarizing children with the great variety of geometric shapes is to conduct several "shape" scavenger hunts. One teacher partitioned the class into groups of three and gave them the following instructions: "Today we're going to have several scavenger hunts. First we'll spend 20 minutes hunting for shapes. Your group is to quietly find all the different shapes in the room, and then list all the objects that are that shape. If you don't know the name of the shape, make a drawing of that shape."

When the children had finished, some time was spent determining the names for some of the shapes and developing lists of all the objects with a particular shape. Following this, the composite findings were listed on the board, and a graph was made to indicate the comparison of the number of various shapes. The next day, most of the time was spent in discussion of likenesses and differences between the shapes.

Several days later, the teacher again formed scavenger-hunt groups and said, "After school today, see how many different shapes you can find; let's see if we can find any new ones. You can get your parents to help if you wish." This resulted in the children's identifying a large number of shapes, which were categorized and compared.

Children learn names and characteristics of shapes in this type of activity in much less time than with workbooks. For one thing, three-dimensional shapes that are "booked" are difficult for many children to visualize. Also, children enjoy the search, the classification, and the comparison when they are engaged in the activity with their friends.

MOTION GEOMETRY

Euclidean geometry is sometimes called "rigid motion geometry." That is, it consists of the study of properties of figures that do not change when we move them in space without altering the angles or the distances in the figures. These ideas can probably be best explored with children by developing a series of movements with figures and studying these. Figure 13–1 contains basic ideas of what might be called motion geometry (transformational geometry).

There are many situations that lead children into checking out slides (translations), turns (rotations), and flips (reflections). Several are briefly described in this section.

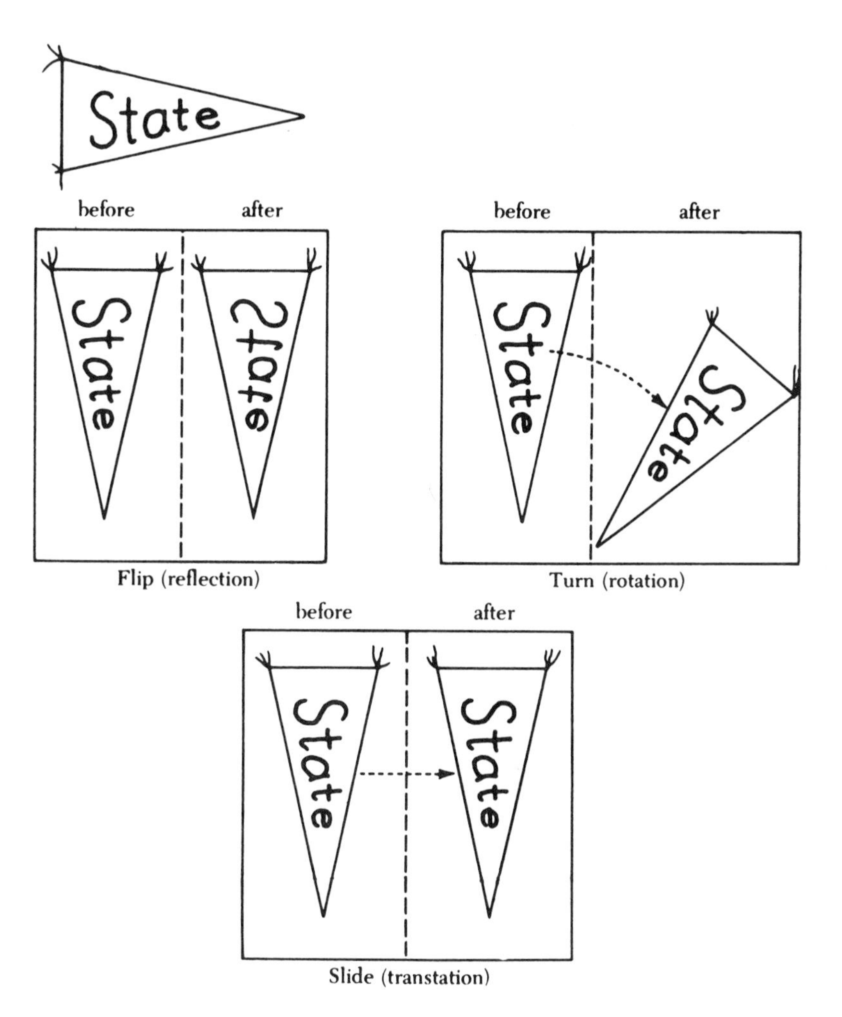

Figure 13–1 Motion Geometry

Slides

Make a TV or movie by using a shoe box with sticks as rollers and adding-machine tape as film. What happens to the figures as they move along? By using tracing paper, make slide images of skiers on a hill; a diver at various stages of a "perfect" swan dive; and an airplane taking off, in the air, and landing. Using coordinates (see the coordinate geometry section later in this chapter), show a figure on coordinates, such as a small house, and give the number pair move for each vertex of the house. Slide it to cover the slide-image house; and use templates of children, cars, and places. Slides can also be developed by using LOGO or by using a computer-assisted instruction program such as *The Factory* (by Sunburst Communications).

Turns

What happens to the hands on a clock? Use a game-type spinner. What happens when you move from 5 to 8? Use a "pop-up" puppet (see drawing).

Flips

What do you have to do to position your airplane template to draw a plane going in the opposite direction? Use symmetry-type folding of hearts and the like. Show the plane flying upside down.

To combine slides, turns, and flips, a variety of treasure-hunt activities can be used. Tracing paper will prove helpful. See p. 333.

After children have had a good deal of experience with these movements, they can work on analysis of the effect of the movements on a figure. They should check out the effect on the size of the figure, shape of the figure, relationship of points on the figure, straight lines, parallel lines, convex lines, angles, distances on the figure, and location of the figure. What would they note?

A Computer Approach to Motion Geometry The computer problem-solving program, *The Factory* is excellent for the development of motion-geometry ideas. The program works well with the teacher's leading a class discussion concerning, "What do we have to do to make this machine?" I suggest

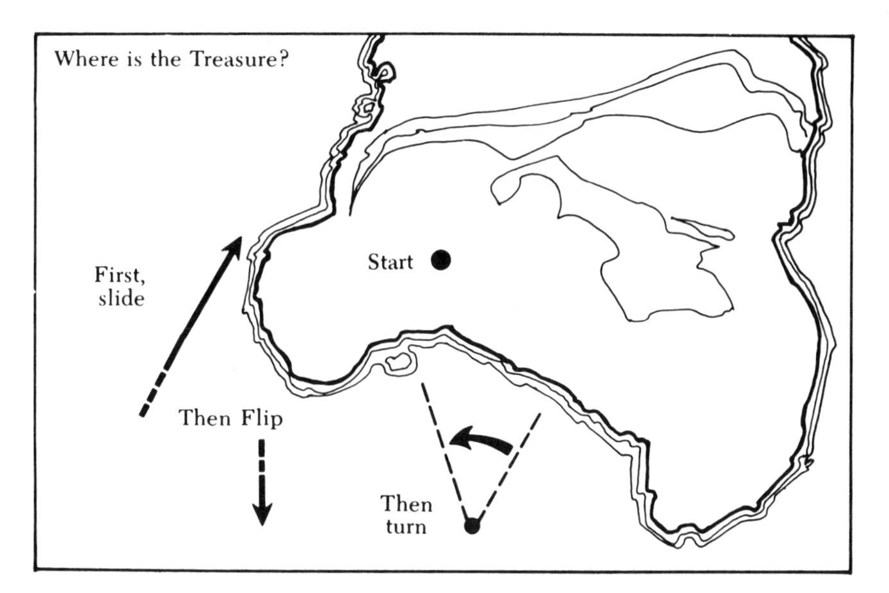

you begin with the portion called "make a product." In this portion of the program, the students are asked to look at a product such as the disc shown here.

Then the students decide how to punch, rotate, and stripe a blank disc to get the products. See below:

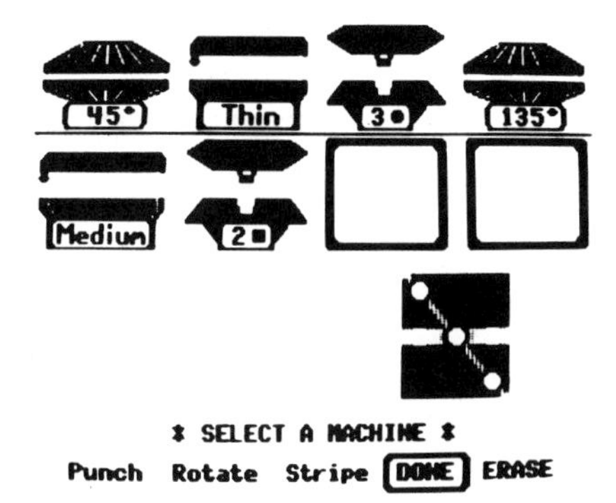

In *The Factory,* students rotated the product 45 degrees, made a thin stripe, punched three round holes, rotated another 135 degrees, made a medium stripe, and punched two square holes (Sunburst Communications).

The entire program is involved with slides, turns, and flips in a problem-solving manner, which allows for three levels of difficulty. This is one of the best examples of a superior computer-assisted instruction program available today. At a higher level of motions you can try *The Super Factory,* which extends motion geometry into three dimensions. The student builds a cube, rotates the cube, and places various designs on it. Also, students are shown a cube rotated to show all sides. They are then challenged to recreate the cube on a planning sheet and display it next to the model. This program is extremely challenging and is excellent for trial and error problem solving.

COORDINATE GEOMETRY

In previous chapters, coordinates were used in relation to number and in problem solving. In fact, the use of coordinates in the elementary school mathematics curriculum is almost endless.

Coordinates can be introduced at the primary-grade level by means of a tic-tac-toe game using teams. If the class is used to working in groups, the teacher can introduce the idea to a group of six to eight children and then continue working with other groups.

One teacher selected six children and made two teams of three. The teacher said, "How many of you have ever played tic-tac-toe? [All had.] Today we are going to play a variation of that game. Look at the squared paper I have on the easel. [A large sheet with two-inch squares.] In this game, you have to get four in a row rather than three in a row, and I will mark the place where two lines meet. If you give me a pair of numbers, I'll mark an X or an O. Jane, you may begin. Your team will be the X team." This dialogue followed:

JANE: Two, four.
TEACHER: OK, I've marked it; now Bill for an O.
BILL: Five, three.
TEACHER: Fine. Now Joe for an X.
JOE: Two, five.

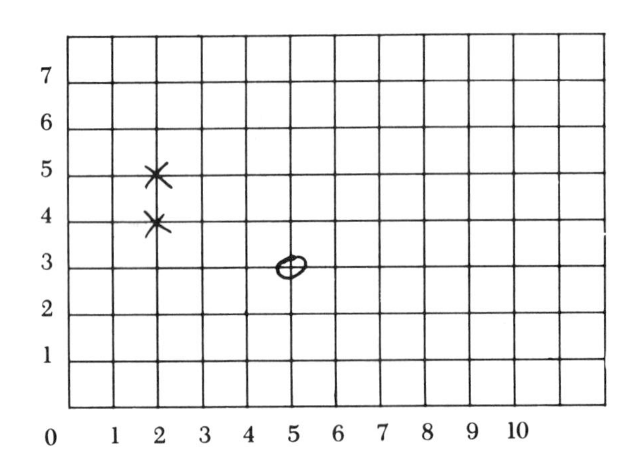

The game continued until a winner was determined.

Teachers using this approach usually find that most children quickly learn to find points on the coordinates. They also recognize quickly that the first number in the pair is the horizontal move and the second the vertical.

Later, the teacher can go on to the other three coordinates. I propose that they not be labeled or used until one or more of the children suggest a correct number pair, such as $(-3, 4.)$

In addition to a game such as tic-tac-toe, road maps and city maps can be used for early working with locations on a coordinate grid. On the road map, coordinates such as A7 can be given and the children asked to tell the name of the city located at that point. Also, one of the best devices for developing an understanding of the coordinate plane is to use the children's game, *Battleship.*® Another activity that will interest children is plotting pictures that classmates have designed. All four sections of the coordinate plane can be used.

If a microcomputer is available, you can make use of one of a variety of coordinate geometry games. One such game is "Hurkle." "Hurkle," a cartoon-type man, jumps from an airplane and hides on a coordinate plane. Class members make guesses on his location by giving number pairs. After a number pair is typed into the computer, there is an indication on the screen of the relationship between their guess and the location of Hurkle. For example, if Hurkle is located at 3, 5 and the class guesses 5, 7, the directions south and west would appear on the screen. Besides learning the location of points on the coordinate plane, the children learn problem-solving strategies. I have used Hurkle with children from 6 to 12 years old. After the first game, the older children quickly begin the questions by giving a location at the center of the coordinate plane.

Congruence

One of the basic ideas of geometry is the congruence of geometric figures. Two figures may be thought of as being *congruent* when the model of one figure will fit exactly on the model of the other. We could also say two figures are congruent if the first figure "maps" onto the second figure.

To introduce the idea of congruence, one teacher developed a pattern set of cartoons for a science report dealing with ecology. A portion of the sheet follows. The children made various cartoons by using the sheets, and then the teacher conducted a discussion on which pictures were congruent. The children found that placing one figure on top of another was a good way to see if they matched.

ALTERNATE APPROACH

Another teacher began with this: "Modern companies often employ people to check the quality of the product. This is called quality control. See if you can identify the major quality-control check for each of the products I mention."

Pupils were then given a worksheet on which to identify geometric forms that were of the same size and shape. The teacher asked, "Do any of you know what geometric forms that fit exactly over one another are called?—No?—Turn to your textbooks and find the term that is used." Pupils identified the vocabulary word *congruent* and checked their description of congruence with that given in the book.

Symmetry

Another important geometric idea is that of symmetry. A common notion of symmetry is "things that balance." One of its chief uses for children is in cutting out valentines. In this case, the two halves have *symmetry about a line.* There are other cases of symmetry: *symmetry about a point,* such as that of circles and some flowers, and *symmetry about a plane,* such as that of a sphere. The suggestions that follow focus primarily on line symmetry.

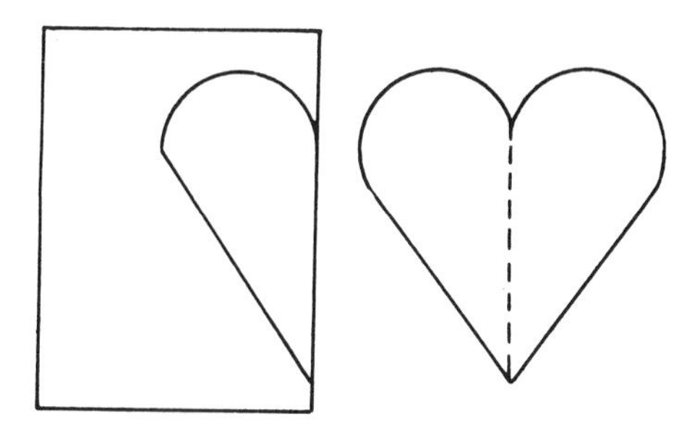

Several activities that can develop ideas of symmetry with children follow:

1. Trying various types of pattern folding, such as making valentines, "snowflakes," paper flowers, etc.
2. Experimenting with figures drawn or printed on paper. The child tries various folds to determine whether or not there is a line or lines of symmetry.
3. Using mirrors and a variety of pictures to search for patterns of symmetry.[5]
4. Analyzing common figures for symmetry. For example, which of the letters that follow are symmetrical?

A B C D E F G H I J K L

M N P Q R T U V W X Y Z

5. Using compass and straightedge to construct various figures that have point or line symmetry.

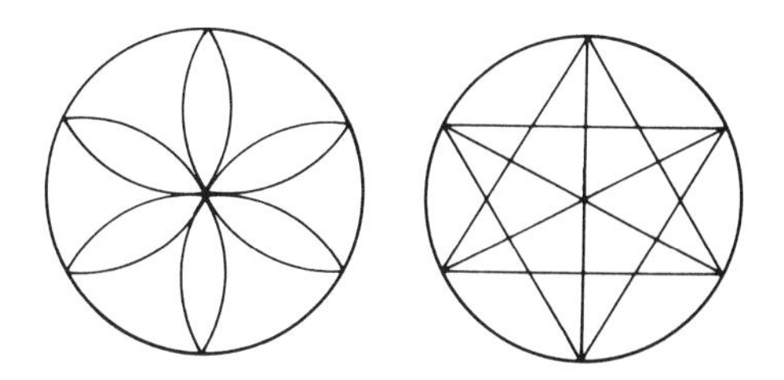

6. Trying "inkblot" ideas. Put a drop of ink on a piece of paper and fold the paper in half.
7. A "mira" can be made from two pieces of clear Plexiglas® and one piece of colored Plexiglas.® The construction is shown below. A figure drawn on paper will be reflected on the colored Plexiglas® and can be traced on the other side of the paper. This device can also be used with symmetry and bisection.

[5] See "Elementary Science Study," *Teacher Guide for Mirror Cards* (St. Louis: Webster Division, McGraw-Hill, 1968). The manual presents a variety of activities for symmetry and congruence.

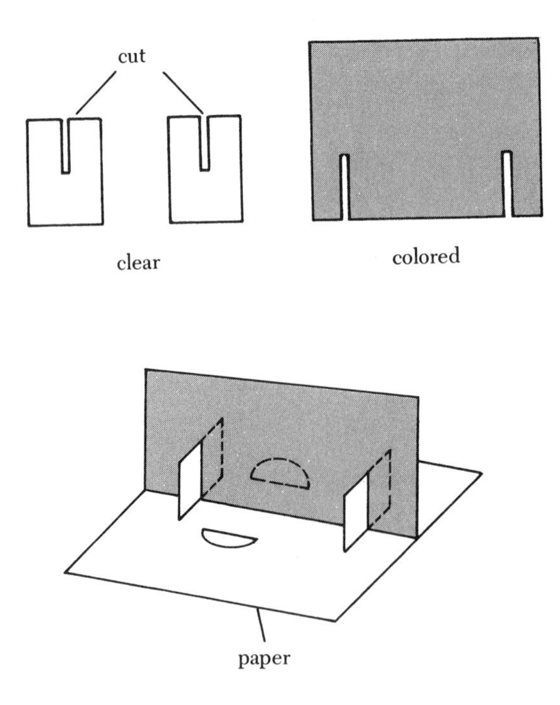

GEOMETRIC CONSTRUCTION

Use of a Compass

A variety of situations can be provided in which a compass and a straightedge are used for solving problems or developing materials. The vignette that follows shows teaching strategies for using a compass and bisecting an angle. Several other possible geometric constructions are then listed.

The following situations were used by a teacher to introduce the use of the compass. "A group of students at Longfellow School needed 45 circles that measured 8 inches across to decorate the stage for a school play. How could they make these circles?"

Class members made suggestions such as, "They could get one and use it as a pattern." "They could use an 8-inch pie pan as a pattern." "Make use of a compass to draw them." The various proposals were discussed. The class decided that (1) if a good pattern of the right size was available, it could be used; (2) an 8-inch pie or cake pan was 8 inches on the inside and thus could not be used for the circle; and (3) in general, the compass would probably be the best choice.

The teacher provided each member of the class with a compass, paper, a piece of cardboard, to prevent the compass point from marking the desk, and a set of instructions for using a compass. The teacher said, "Other students have found these directions for the use of a compass to be handy. See if you can think of other suggestions. Now see if you can draw a circle that is 8 inches

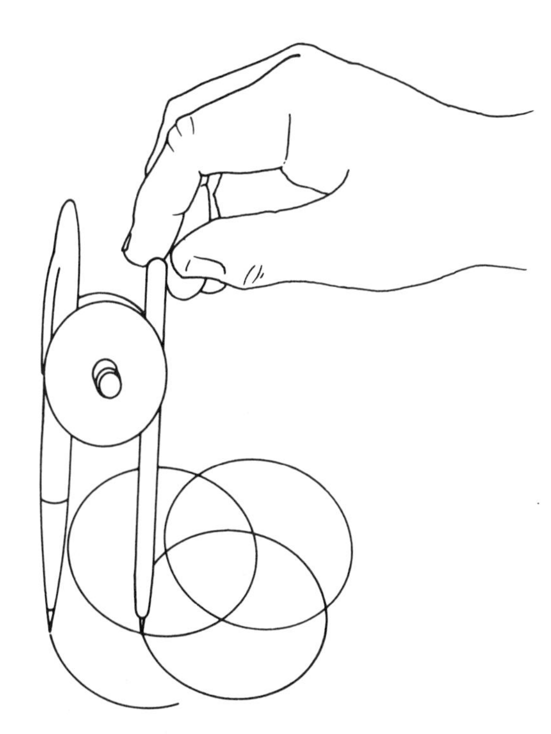

Compass available from *Creative Publications,* 3977 East Bayshore Road, P.O. Box 10328, Palo Alto, CA 94303.

across. Experiment to find a means of getting a circle as close to 8 inches across as possible. Incidentally, we call the distance across a circle that goes through the center of the circle a *diameter.*"

The following suggestions were reproduced on the paper:

1. Place the steel point of the compass on the point you wish to be the center of the circle.
2. Hold the compass at the top with one hand.
3. Let the compass swing around on the steel point so that the pencil draws a complete circle in one sweep.

LOGO

LOGO has already been introduced in Chapter 5. Children can quickly discover methods of drawing geometric shapes using LOGO. Because the turtle turns right and left in degrees, children quickly find that they are associating shapes with degrees. With as little information as this, the children can experiment to draw rectangles, triangles of various types, hexagons, and even circles. With children of fourth-grade age and above, LOGO is learned at a tremendously rapid pace, and, more important from the stand-point of geometric instruction, the children are able to learn with little pressure and have high retention.

The accompanying art is a LOGO geometric construction created by a fourth grader in his third hour of "playing" with LOGO.

Geometry, as studied using LOGO, is thoughtful rather than factual. It promotes relational understandings and, thus, is much more in keeping with the geometry program I have suggested than is reflected in current elementary school textbooks.

KEEPING SHARP

Self-Test: True/False

_______ 1. Geometry topics should be taught in isolated, one-day lessons.

_______ 2. A point is a model of an exact location in space.

_______ 3. Curves are always bent.

_______ 4. Laboratory activities for a unit on curves and points could include reading treasure maps.

_______ 5. It is not necessary to distinguish among points on the interior, exterior and boundary of a closed curve.

_______ 6. Simple closed curves form the boundaries of plane regions (e.g., squares, triangles, etc.).

_______ 7. Squares are a special case of rectangles.

_______ 8. Two figures are similar if, and only if, the model of one exactly fits on the model of the other.

_______ 9. Precise vocabulary should be well developed before the geometric ideas represented are studied.

_______ 10. Sorting activities that encourage children to explain their choices can help clarify geometric ideas.

_______ 11. Three points not on the same line are in only one plane. An application of this property is the use of a tripod to keep a camera steady.

_______ 12. Geometric constructions using compass and straightedge are not appropriate for elementary school pupils.

_______ 13. For every object drawn on paper, there is only one line of symmetry.

_______ 14. Projecting a map from a sphere to a sheet of paper in no way alters the shape of the figure projected.

Vocabulary

transformation	coordinates	symmetry	point
diameter	translations	trapezoid	curve
congruence	similarity	equivalence	arc
rhombus	bisect	ray	quadrilaterals

THINK ABOUT

1. What proportion of the elementary mathematics program should be geometry? Why?
2. Study motion geometry. What are some everyday uses?
3. How can geometry and art be taught together?
4. What is the effect of geometry on science?
5. Develop a lesson on geometric construction.
6. Review 5–10 computer programs on elementary geometry.
7. Study the geometry in two or three elementary mathematics series. How does it compare with the National Council of Teachers of Mathematics Standards and the suggestions presented in the chapter?
8. The use of LOGO can play an important part in the geometry program.

SUGGESTED REFERENCES

BIDWELL, JAMES H., "Using Reflections to Find Symmetric and Asymmetric Patterns," *The Arithmetic Teacher,* 34 (March 1987): 10–15.

CLEMENTS, DOUGLAS H., *Computers in Elementary Mathematics Education,* chap. 9. Englewood Cliffs, N.J.: Prentice Hall, 1989.

CROWLEY, MARY L., "The van Hiele Model of the Development of Geometric Thought." In *Learning and Teaching Geometry, K–12,* Mary M. Lindquist, (ed.), Reston, Va.: National Council of Teachers of Mathematics, 1987.

HOFFER, ALAN (ed.), *Geometry and Visualization,* Palo Alto, Calif.: Creative Publications, 1978.

HOFFER, ALAN R., "Geometry and Visual Thought." In *Teaching Mathematics in Grades K–8: Research-Based Methods,* Thomas R. Post (ed.), Boston, Mass.: Allyn and Bacon, 1988.

Learning and Teaching Geometry K–12, 1987 Yearbook, Reston, Va.: National Council of Teachers of Mathematics, 1987.

PIAGET, J. and B. INHELDER, *The Child's Conception of Space,* New York: Harper and Row, 1968.

POHL, VICTORIA, *How to Enrich Geometry Using String Designs,* Reston, Va.: National Council of Teachers of Mathematics, 1986.

chapter 14

MEASUREMENT

OVERVIEW

Measurement is one of the topics in the elementary school mathematics program that closely relates to the everyday life of all children. If you listen carefully to the questions and comments that children make to their peers or to adults, you are certain to hear such comments and questions as

"I'm bigger than you are." "No you're not, I'm taller."
"I weigh more."
"When I get to be big, I'm going to get a car like that one."
"How much longer will it be until we get to Grandma's?"
"What time are the cartoons on TV?"
"How high am I now?"

Since antiquity, we have been interested in measuring. Many phases of mathematics have developed because of this interest, and any educated citizen today makes use of some type of measurement in every kind of endeavor.

Today's space scientist refers to measures such as *light-years, parsecs,* and *angstrom units*. Textbooks of the early 1800s contained terms such as *perch, ell-Flemish,* and *hogshead*. Thus, although the units have sometimes changed, the need for measurement continues to exist.

The National Council of Teachers of Mathematics Standards suggest that the following concepts be developed during the study of measurement in the elementary school:[1]

For kindergarten through grade 4,

1. Understand the attributes of length, capacity, weight, area, volume, time, temperature, and angle.
2. Develop the process of measuring and concepts related to units of measurement.
3. Make and use estimates of measurement.
4. Make and use measurements in problem and everyday situations.

For grade 5–8,

1. Extend understanding of the process of measurement.
2. Estimate, make, and use measurements to describe and compare phenomena.
3. Select appropriate units and tools to measure to the level of accuracy required in a particular situation.
4. Understand the structure and use of systems of measurement.
5. Extend understanding of the concepts of perimeter, area, volume, angle, measure, capacity, and weight/mass.
6. Develop the concepts of rates and other derived and indirect measurements.
7. Develop formulas and procedures for determining measures to solve problems.

Even though measurement is recognized as being of great importance, it typically does not receive its fair share of the elementary school mathematics program. This is probably because it is difficult to develop good textbook materials, and superior teaching of measure requires special equipment. As with problem solving, the development of a good measurement program rests heavily on the individual teacher.

TEACHER LABORATORY

1. Experiment: Use a large rubber band, paper clips, and a felt pen. Make a "spring" scale that you could use to weigh objects. Use washers or pennies to act as units for marking your scale. How many washers or pennies does a pencil weigh?
2. Use a pegboard beam balance. Develop a plan for weighing objects up to a kilogram in weight. What are some possibilities for your units of weight?
3. Below are the instructions for making a *hypsometer* (a height-measuring device). Measure the height of two buildings by using the device. Find an intermediate-grade child and work with him or her to make a hypsometer.

   Tape a sheet of cross-section paper on stiff cardboard and fasten a plastic straw along one edge of the paper as shown in the sketch. Hang a weight on a piece of thread from A.

[1]Commission on Standards for School Mathematics, *Curriculum and Evaluation Standards for School Mathematics* (Reston, Va.: National Council of Teachers of Mathematics, 1989).

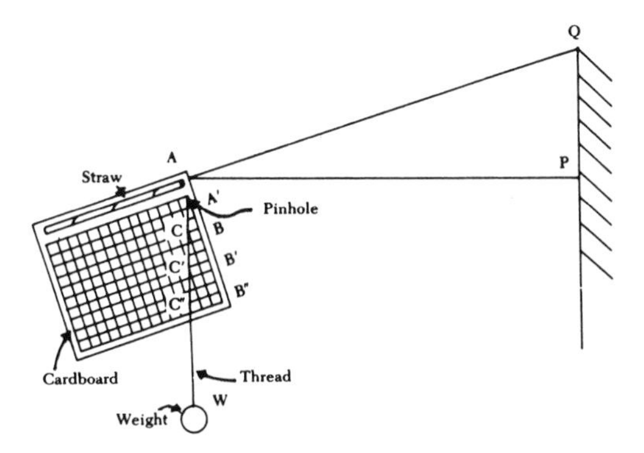

The string AW establishes a vertical line. How would you hold your hypsometer so that you know the straw is level? This gives you a way of locating a point P on the building level with point A. Measure the distance AP.

By sighting at point Q, you automatically get a collection of triangles A′BC, A′B′C′, A′B″C″ on your hypsometer that are all similar to triangle APQ. Choose a suitable scale and find PQ, the height of the buildings above your eye level.

4. Experiment: Use squared paper. Find the largest polygon in area that you can draw with a perimeter of 24 units. Find the smallest. Use squared paper or blocks and find all the possible perimeters of 36 squares. Find all the areas with a perimeter of 36. Make a table of your findings.

TAKE INVENTORY

Can You:

1. List at least six basic principles of measurement?
2. List the basic units in the metric system and give a suggestion for teaching each?
3. Develop a historical lesson on measurement?
4. Explain the basic metric relationships and use the system?
5. Write a sequence for teaching "telling time"?
6. Explain area relationships, including π, and give suggestions for teaching them?
7. List eight activities for teaching measurement?

Measurement Generalizations

Two types of measurement can be readily discerned. The first type involves use of discrete variables (those that can be counted), such as the size of families and the number of postage stamps. The second type involves the use of continuous variables, such as heights, weights, time, temperature, and blood pressure.

Continuous variables may be viewed as represented on an uninterrupted scale. The material in this chapter emphasizes continuous measurement. Measurement is often not taught because it does not lend itself to textbook instruction.

In its attempt to measure such things as time, volume, area, temperature, energy, and so on, we have arrived at the following generalizations, which should be considered when developing possible learning experiences for children.

1. Measurement is best learned by measuring, not by reading about measuring. Approximating measurement should be stressed first, before actual measurement.
2. Have children measure things of interest. Remember that measurement does not lend itself to textbook instruction.
3. Normally, the sequence of events in the study of a type of measurement should be making gross comparisons of objects or sets (this stick is longer than that stick, this ball is heavier than the apple); using units of measure devised by the pupils (the paper is 6 pencils long); and using standard units of measure (the stick is 7 centimeters long, the pencil is 5 inches long).
4. Measurements of continuous data, no matter how precise they may be, are approximations. Most measures can be refined enough to be precise for practical purposes, but the general statement, "*Every time you measure, you're wrong,*" merits consideration. Even though a scientific instrument measures to the nearest millionth of a meter, there is still some error. Thus, statements such as "He's exactly 164 cm tall" should be avoided.
5. Some things, such as the length of a room, can be measured directly. Other things, such as the distance to a star, must be measured indirectly.
6. The metric system and the English system should be taught simultaneously from the time measurement with standard units begins. However, little if any stress should be placed on converting from one system to the other.
7. Precision in measurement should be attained gradually. Children should refine their measuring from a need to increase their precision.
8. A "measurement sense" is essential. A "feel for" the units, combined with a knowledge of the basic structure, is a requisite for mastery of measurement.

REFERENCE MEASURES

Children often develop a knowledge of standard measure without really understanding actual measurement situations. Thus, a third grader who knows that there are 1,000 meters in a kilometer and the approximate distance of a kilometer may say, after 3 hours of car travel on a 600-kilometer trip, "Are we almost there?"

It is desirable to develop a list of common reference measures that can be used to develop the child's "measurement sense." The list below is a start in that direction (adjustment should be made for locality). Also, there should be distances and weights that are specific to the particular school building and neighborhood used. A typical set of reference measures follows.

LINEAR MEASURE

1 meter	Distance of chalkboard from floor
	Height of a large couch or wingback chair
2 meters	Height of a very tall man
	A little less than the height of a doorway
1 centimeter	Width of a paper clip
	Thickness of a little finger (pinky)
	Head of a thumbtack
10 centimeters	Width of a hand (including thumb)
1 millimeter	Thickness of one strand of a paper clip
	Thickness of a dime
	Thickness of heavy tagboard
1 kilometer	About 11 football fields long
10 kilometers	Distance from SUNYAB Main Campus to the airport
100 kilometers	Distance from Buffalo to Rochester between the New York Thruway toll booths

VOLUME MEASURE

5 milliliters	A teaspoonful of cough syrup
350 milliliters	Capacity of a regular-size soft-drink can
1 liter	Capacity of a canning jar filled to the top
100 liters	Capacity of a large car's gasoline tank

WEIGHT

1 gram	A medium-size paper clip
	A jelly bean
	A 9-by-5 cm section of a manila folder
2 grams	A sugar cube
5 grams	A nickel
10 grams	A whole piece of chalk
300 grams	A can of Campbell's soup
1 kilogram	Six medium-size bananas
	A small (350-page) volume of an encyclopedia
100 kilograms	A professional football player

TEMPERATURE, CELSIUS

0°C	Water freezes
10°C	Cool but pleasant spring day
20°C	Normal room temperature
30°C	Summer day
37°C	Body temperature
100°C	Water boils

TIME

5 minutes	The time it takes to get ready for lunch
1 minute	The time it takes to count each pupil in the room twice (if there are about 30 pupils)
1 hour	TV program
$\frac{1}{2}$ hour	TV program

GROSS METRIC-ENGLISH COMPARISONS

A meter is a little more than a yard.
A liter is a little more than a quart.
A kilogram is a little more than two pounds.

THE BEGINNINGS OF THE METRIC SYSTEM

BASIC METRIC UNITS

Metre or meter (length)
Litre or liter (liquid)
Gram (dry weight)

The ancient systems of measure were not uniform from country to country, nor were they planned in a fashion that allowed for simple conversion from one unit to the next larger or smaller unit.

Probably the leading country in the field of measurement reform was France. Beginning in A.D. 650 and continuing under Charlemagne, there were attempts to make the standards of measure somewhat uniform. In 1670, Gabriel Mouton proposed a system of measurement using a scale of 10 and taking as its base length a 1-foot arc on a great circle of the earth. In 1789, the French Academy of Sciences appointed a committee to work out a plan of decimal measure. Several earlier English proposals had suggested that the linear unit be the length of a pendulum beating half-seconds. This suggestion was considered by the French but dropped in favor of an arc of one ten-millionth of a quarter of a meridian. A slight error in measurement caused this proposal to be dropped, but a standard meter of similar length was adopted, and all civilized countries received copies.

The official name for the present version is Système International d'Unités (International System of Units), or simply SI. The following official names and definitions were given to the various basic units of measure:

Meter: The measure of length equal to 1 ten-millionth of the distance from the North Pole to the equator.
Liter: The measure of capacity for both liquids and dry materials whose extent is that of a cube with edges equal to one-tenth of a meter.
Gram: The absolute weight of a volume of pure water equal to a cube with edges equal to one-hundredth part of a meter and at the temperature of melting ice.

In addition, basic units for the area of land (are) and firewood (stere) were defined.

In 1795, the Latin and Greek prefixes by which the decimal subdivisions and multiples (respectively) of the meter, the liter, and the gram are now known were designated. In certain cases, particularly in scientific usage, it becomes convenient to provide for multiples larger than 1,000 and for subdivisions smaller than one-thousandth. Accordingly, the following prefixes have been introduced and are now generally recognized:

tera, meaning 10^{12}	pico, meaning 10^{-12}
giga, meaning 10^{9}	nano, meaning 10^{-9}
mega, meaning 10^{6}	micro, meaning 10^{-6}
*kilo, meaning 10^{3}	*milli, meaning 10^{-3}
hecto, meaning 10^{2}	*centi, meaning 10^{-2}
deka, meaning 10^{1}	*deci, meaning 10^{-1}

Thus, a kilometer is 1,000 meters and a millimeter is 0.001 meter. These prefixes, appropriately applied to all kinds of units, retain their significance, as in kilowatts, picofarads, megacycles, and microinches. A special case is the term *micron* (abbreviated as μ, the Greek letter mu), a coined word meaning one-millionth of a meter (equivalent to one-thousandth of a millimeter). Although these prefixes now replace the millimicron and the micromicron, these terms are still found in the literature. A millimicron (abbreviated as $m\mu$) is one-thousandth of a micron (equivalent to one-millionth of a millimeter), and a micromicron (abbreviated as $\mu\mu$) is one-millionth of a micron (equivalent to one-thousandth of a millimicron, or to 0.000 000 001 millimeter).

LINEAR MEASURE

10 millimeters (mm)	= 1 centimeter (cm)
10 centimeters	= 1 decimeter (dm) = 100 millimeters
10 decimeters	= 1 meter (m) = 1,000 millimeters
10 meters	= 1 dekameter (dkm)
10 dekameters	= 1 hectometer (hm) = 100 meters
10 hectometers	= 1 kilometer (km) = 1,000 meters

AREA MEASURE

100 square millimeters (mm^2)	= 1 square centimeter (cm^2)
10,000 square centimeters	= 1 square meter (m^2) = 1,000,000 square millimeters
100 square meters	= 1 are (a)
100 ares	= 1 hectare (ha) = 10,000 square meters
100 hectares	= 1 square kilometer (km^2) = 1,000,000 square meters

VOLUME MEASURE

10 milliliters (ml)	= 1 centiliter (cl)
10 centiliters	= 1 deciliter (dl) = 100 milliliters
10 deciliters	= 1 liter = 1,000 milliliters
10 liters	= 1 dekaliter (dkl)
10 dekaliters	= 1 hectoliter (hl) = 100 liters
10 hectoliters	= 1 kiloliter (kl) = 1,000 liters

*The most commonly used prefixes.

CUBIC MEASURE

1,000 cubic millimeters (mm^3)	= 1 cubic centimeter (cm^3)
1,000 cubic centimeters	= 1 cubic decimeter (dm^3)
	= 1,000,000 cubic millimeters
1,000 cubic decimeters	= 1 cubic meter (m^3) = 1 stere
	= 1,000,000 cubic centimeters
	= 1,000,000,000 cubic millimeters

WEIGHT

10 milligrams (mg)	= 1 centigram (cg)
10 centigrams	= 1 decigram (dg) = 100 milligrams
10 decigrams	= 1 gram (g) = 1,000 milligrams
10 grams	= 1 dekagram (dkg)
10 dekagrams	= 1 hectogram (hg) = 100 grams
10 hectograms	= 1 kilogram (kg) = 1,000 grams
1,000 kilograms	= 1 metric ton (t)

The metric units can be related to English measure by using the following approximations:

meter = 39.37 inches
liter = 0.908 dry quart, or 1.0567 liquid quarts
gram = 0.035 ounces
kilogram = 2.2046 pounds
inch = 2.54 centimeters
ounce = 28.35 grams
dry quart = 1.101 liters
liquid quart = 0.9464 liter

The government of the French Revolution also decreed a new series of units of time, but they were never accepted by the population, and about 12 years later, they were discarded:

100 seconds = 1 minute
100 minutes = 1 hour
10 hours = 1 day
10 days = 1 week, or decade
3 weeks (30 days) = 1 month
12 months plus
5 or 6 carnival days = 1 year

MEASUREMENT ACTIVITIES

Measurement requires action on the part of the student and the teacher. The very nature of the subject leads to children's discovery. Because of this characteristic of measurement, I have used a slightly different format to develop teaching suggestions. In the material that follows, activities that can be used to develop measurement of length, weight (mass), and temperature are provided. A number of miscellaneous activities are also suggested.

Length Activities (Primary)

1. To develop awareness, children should compare distances: lengths of pencils, their own heights, the widths of books, and so on. Games comparing such measurements can be created. (At first, compare only two objects at a time.)
2. Discuss a new object in terms of measurement; what is its length, its width, the distance around it, its weight?
3. Arrange various objects according to size, from least to greatest in terms of length (or later, weight and volume). Discuss an object as being "shorter than A but longer than B."
4. Compare the lengths of two unmovable objects by using a third object (string, for example). Children should be led to realize that to compare lengths correctly, the end of one object must be aligned with the end of another.
5. Pose a practical problem to be solved ("Can we fit that desk in this space?"). Outlaw moving the desk and using rulers. A few children will try to solve the problem by holding their hands the proper distance apart, but the amount of error is soon recognized. Students will finally try nonstandard units of measurement (books, pencils, and so on). Guide them to discover that even though the desk measured "four books" or "nine pencils," the answer to the *problem* remained the same.
6. Building on the previous activity, lead the students to use the "guess-and-test" approach, measuring their desks with several nonstandard units (matchbooks, handspans, chalk, paper clips, and so on). Discuss what they are doing to measure (comparing lengths).
7. Finally, have all the students measure their desks with their own pencils. Are the numbers reported all different? Are some students measuring incorrectly? "Would it be fair if I gave all you 'one pencil's length' of string candy, using your *own* pencil? How could I be fair?" (Discuss the need for a standard unit of measurement. This is also an excellent opportunity to show that a study of the *history* of measurement uncovers parallel problems and solutions.)
8. Have children measure various small objects by using the centimeter only. Centimeter-squared graph paper can be cut out for this purpose or a Cuisenaire rod, the "white rod" (1 cm^3), can be used. Later, the same objects can be measured by using a centimeter ruler (which, children should understand, is merely a number line marked off in units of length).
9. Students can be led to measure themselves in centimeters, not only height, but head, shoulders, waist and hip circumference, the distance between the eyes, etc. Guide them to look for a relationship between their height and fingertip-to-fingertip length (approximately the same), and between the wrist and neck circumference (approximately 1:2). An activity that both entertains and demonstrates that a circumference is a linear measurement is illustrating how a child would look "unzipped and rolled out"; that is, laying out all the child's measurements horizontally.
10. Have students measure one of their parents, using the same categories. This will give the child extra experience in guessing and measuring, and it will involve the parents in metric activities. Concepts and terminology should be extended from "a little more than 23 cm" to "to the nearest centimeter, it measures 23 cm."
11. Measuring longer objects gives the teacher the opportunity to introduce the meter. Provide students with a meter stick (marked in centimeters), guide discovery of the relationship 1 m = 100 cm, and have the students guess and test various objects in the room (the doorway, windows, desks, length of the room,

etc.). Reinforce the relationship of 1 m = 100 cm by having students measure several objects with length of 1, 2, and 3 meters with both units.

12. Metric mystery boxes: One box is prepared for each group of four to five students. Each (identical) box contains objects of various lengths, including some very close in length to 1 m and 1 cm (for reference measures). The teacher shows a transparency listing various measures (17 cm, 1 m, etc.) which correspond to the objects. Proceeding in order, each group guesses which object corresponds to the given length, tests the guess by measuring, and records the name of the object if the guess was correct. Concluding, the teacher uncovers the answers, and children discuss their results.
13. Have students find as many things at home as they can that measure just about 1 meter or 1 centimeter. Use the best suggestions as reference measures.
14. One student writes the length of a "metric mystery" object on the board. Others try to locate the object by measuring.
15. This activity and the following three can be used independently or simultaneously. If you choose the latter, you can have students begin at one station, complete the activity, and ring a bell for everyone to switch stations. In this manner, the whole class rotates among the stations, and each child participates in every activity. In the first activity, each student receives an activity card or ditto that advertises a "Metric Treasure Hunt." Ten to 15 challenges greet the student, such as "Find something white in the front of the room that is 73 cm long." The student locates the item and records its name.
16. Students at this activity become "Detective Mickey Meter." They locate murder weapons from exhibits A, B, C, D, and E by solving such problems as, "One weapon is 79 cm long. What is it?" and "Another is 8 times as long as it is wide. What exhibit is it?"
17. At this activity, students first guess and test the length of photocopied lines and pictures and then try to draw their own lines freehand, given a stated length. The challenge is to be an accurate estimator. Finally, students investigate an "Optical Illusions" board, proving to themselves that metric measure is more accurate than the eye in some cases!
18. Metric Olympics: Students participate in a class Olympics, measuring their performances in metric length. Events can include the standing broad jump, the straw javelin throw, the paper airplane throw, etc.

Length Activities (Intermediate)

1. Students can make their own meters from a photocopy passed out by the teacher. (This meter can be photocopied onto heavy-duty paper and should *not* include divisions at this point.) Then, utilizing a "guess-and-test" approach and expressions such as "a little more than 2 meters," students are asked to measure various "abouts" in the room. Special attention is given to objects just about 1 meter in length for *reference measures.*
2. Orally presented problems can be utilized, such as, "Olympics are in metric. One U.S. runner did the 100-meter dash in 10 seconds. How far does he average per second?" "An Olympic swimming pool is 50 meters long. In the 200-meter race, how many lengths must someone swim?" Use a hand-held calculator to aid in problem solving.
3. Present a problem that requires some division of a meter (How long is a piece of chalk?). Then students can make their own meter stick with centimeter divisions. Direct them to discover that this meter is, of course, the same length as the undivided meter and, more important, that there are 100 centimeters in 1 meter.

4. Have the students find other words that start with *cent* (cents, century, etc.). Reinforce the relationship of 1 m = 100 cm by having students measure several objects (whose lengths are 1, 2, and 3 meters) in terms of both centimeters and meters.
5. Present a problem ("What is the width of a pencil point?") that necessitates use of smaller divisions of the meter than the centimeter. Guide discovery of a usable unit, the millimeter. Discuss the "magic number" of the metric system, 10, and how this magic number makes conversions easier (10 mm = 1 cm, 30 mm = 3 cm).
6. Measure several objects, lines, etc., in terms of *both* centimeters and millimeters. Be sure to include objects that measure to a "whole centimeters" (18.0 cm) and objects that have to be measured to the nearest centimeter (17.2 cm to the nearest centimeter is 17 cm). Discuss with the students the fact that in the first case, the rule, "Multiply centimeters by 10 to get millimeters," pertains, but, in the second, it would be a mistake to do so, for the accurate measurement would be 172 mm, not 170. This activity stresses the function of smaller divisions: increased precision.
7. Have students find as many things at home as they can that measure just about 1 meter, 1 centimeter, or 1 millimeter. Use the best suggestions as reference measures.
8. Challenge the students to select the most appropriate unit for different measurement problems. For example: "What would you use to measure (1) the length of your arm? (2) the length of the room? (3) the width of a rubber band?" Answers: (1) centimeter, (2) meter, (3) millimeter.
9. After discussion of the term *kilometer* (familiar from British movies or *Star Trek*), children should walk the distance, in the hallways or outside the school. This is best done by using a trundle wheel, an instrument that is rolled along the ground and clicks at every rotation, signaling that 1 meter (its circumference) has been traveled. If this is unavailable, children could tie a knot in a rope every meter and use a 10- to 100-m length for measuring. Also see "Length Activities (Primary)," numbers 5, 6, 7, 9, 10, 12, 15, 16, 17, and 18.

Volume Activities (Primary)

1. Have pupils develop a readiness for volume measurement by finding the volume of a large box in terms of egg cartons (how many fit in the box?), milk containers, blocks, etc. Models built with cubic-centimeter blocks (such as Cuisenaire rods) can also prepare students for standard metric volume measurement.
2. The contents of how many school milk cartons fill a canning bottle? How many canning bottles fill a pail? How many pails fill a sink? (Remember to use a guess-and-test approach!)
3. How many plastic cups fill a milk carton? a bottle? a pail? a sink? a bowl? How would you arrange these objects in order from least to greatest in height? in volume?
4. Introduce the liter, and have the pupils "guess and test" the volume of all the previously used containers in terms of the liter. Direct special attention to those that hold just about 1 liter. Wherever possible, use both dry (sand, macaroni) and liquid (water) measure.
5. Use a calibrated liter container or a 100-ml measuring cup to guide students to discover that 1,000 ml = 1 liter. "Guess and test" small containers with milliliters. Create situations for children to measure several containers of different shapes but with nearly identical volumes.

6. See if the children can discover ways to approximate how much liquid they drink in a day in liters and milliliters.
7. Have students look in their cupboards and refrigerators for commercial items marked in liters and milliliters.
8. Divide the class into teams; hold up various containers and let each team submit a guess as to its volume. The team coming closest gets one point.
9. Adapt volume activities from the next section (intermediate) for your pupils.

Volume Activities (Intermediate)

1. To help children visualize a liter, and to reinforce the internal consistency of the metric system, it is valuable for children to make a liter. A liter is defined as the capacity of a cube measuring 10 cm on all sides. Using centimeter-squared paper pasted on heavy paper, children can cut, fold, and tape to make a liter. (Several patterns are possible. Two are reproduced below.) This liter can be filled with rice, macaroni, or the like and can then be used to fill other containers to measure their capacity. Further, each centimeter of height of the container represents one-tenth of a liter, or 100 milliliters (1,000 ml = 1 liter). Therefore, subdivisions can be marked off and utilized.

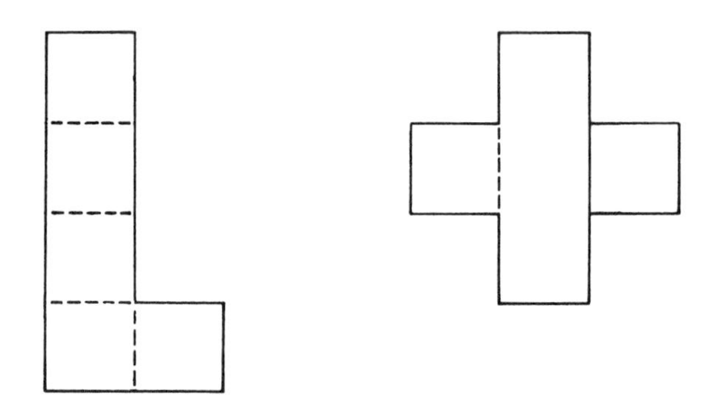

Weight Activities (Primary)

1. Call attention to a familiar object. Have students find objects that are heavier, lighter, and just about as heavy as that object.
2. Order several objects from least to greatest in terms of weight (mass).
3. Have children put an object in each side of a pan balance. Ask them what happened, and why. Use the balance to order several objects in weight.
4. Have students weigh several objects in terms of a nonstandard unit, such as coins or washers. Utilize a guess-and-test approach. Finally, introduce the gram and have students weigh many objects in terms of grams, estimating, measuring, and recording their results.
5. Weigh a ball of clay. Mold the ball into many shapes, and weigh it every time, recording the results. What do we find?
6. Introduce the kilogram, and have students "make" several kilogram weights with nails, books, and the like.

7. Weigh the class in kilograms on a metric scale, and make a chart and graph of the results.
8. Weigh a liter container. Now weigh it filled with water (1 kilogram). What is the relationship between volume and weight (mass)?
9. Adapt activities from the intermediate weight section that follows for use with your children. See especially the "kilogram treasure hunt" (activity 4) and activities 3 and 5.

Weight Activities (Intermediate)

1. After establishing the need for weight measurement, guide the children to see how grams are related to meters and liters by constructing a kilogram. This can be done by lining the class-made liter (a cube 10 cm on all sides) with a plastic bag and filling it with water. Concepts can be further developed by later weighing the liter filled with sand, nails, cotton, and the like. What are the different relationships between volume and mass?
2. Children may point out that a liter of water weighs a *kilo*gram, and since *kilo* means 1,000, what is a gram? If Cuisenaire rods are available, a laboratory lesson might be created that allows the students to discover that a cube 1 cm on all sides is one one-thousandth of a liter; therefore, if it was a cube of water, it would weigh 1 gram. (This concept could be discovered or illustrated in several ways.)
3. Children can make gram weights by cutting sections of tagboard that are 5 cm by 9 cm. Test these gram weights against other objects. For instance, a nickel weighs 5 grams (see the "Reference Measures" section given earlier in this chapter for other examples).
4. Conduct a "treasure hunt" for objects in the room that weigh approximately 1 kilogram. Use these objects for reference measures.
5. Use the "guess-and-test" approach to measure many objects in grams. You can use sugar cubes, paper clips, pieces of chalk, etc., as well as heavier items, especially commercially marked foodstuffs. (*Note:* A soup can's marked weight of 298 grams does not include the mass of the can itself.)

Temperature Activities (Primary and Intermediate)

1. Make a list of things that are hot, warm, cool, cold, etc. Try to put them in order, with coldest at the bottom and hottest at the top. Later, match each with a Celsius temperature reading.
2. Construct a large Celsius thermometer and use it to illustrate temperature readings. Have different students read the inside and the outside temperatures every day. Make a graph of changing temperatures.
3. Students should have wide experience with metric temperature over a long period. However, a good activity to start with is to have them guess and test temperature in several situations and record their results. Include outside and inside temperatures, freezing water (water and ice cubes), water from the cold and hot faucets, water from the drinking fountain, milk, the students themselves (by holding the thermometer in the hand tightly), and so on. For further practice, they can make up mixtures of hot and cold water and guess and test with a friend.

Miscellaneous Measurement Activities

1. Guess-and-Test Game: Three students go to the board. The teacher writes the symbol for a length (20 cm, 1 m, 55 mm). The students try to draw a line, free-hand, that is as close as possible to that length. The class votes on which is closest, number 1, 2, or 3. Three other students go to the board, measure the lines, and record each measurement next to the line. The closest student stays (or selects three new contestants).
2. Organize a "metric center" in your classroom. To be effective, metric measurement must continue throughout the school year. A table or set of shelves can be set up with metric measuring instruments, various objects to measure, activity cards, and answer cards. The students could be directed to participate in the activities, check their answers, and eventually help change the objects every 4–8 weeks, or whenever everyone has participated in the activities.
3. Include metric terms and concepts in practice in word problems and operations. Construct this practice so that familiarity with the units is increased. (Real, practical problem solving is best!)
4. The history of measurement is a fascinating and entertaining research topic, and its study further illustrates the need for standard units.
5. If the situation warrants it, a debate can be staged contrasting the metric and U.S. systems. (The U.S. system *does* have some advantages; for example, it contains units that are of a "handy" size.)
6. Card games: Various activities can easily be created to reinforce metric skills. A form of rummy can be created by writing various lengths on cards (10 cm, 250 cm, 10 mm, 150 mm). Five cards are dealt to each player, the object being to collect any combination of cards that, when added, would produce a length equivalent to 1 meter.
7. "Racing" games can be easily constructed. "Racetrack" can be made or adapted and cards with lines of metric length dittoed. Children choose a card, measure it to the nearest centimeter, and move the equivalent number of spaces (extend to include millimeter).
8. Puzzles that provide interesting measuring practice can be constructed easily, perhaps a pattern of dots, with directions such as, "Start at dot A. Draw a line to a dot 5 cm away. Next, draw a line from this dot to a dot 15 mm away." A picture finally results. These activities are motivating, they increase estimation and measuring skills, and they are self-correcting.
9. Other puzzles and games might include unscrambling metric terms (nearli = linear; trimec = metric; triel = liter; etc.) or completing a crossword puzzle containing metric terms, prefixes, etc.
10. Discuss with the students what the appropriate unit would be to measure gasoline, roadways, meat, machine parts, and any other items of interest to them.
11. A review of our monetary system and practice with metric prefixes can be accomplished simultaneously by showing the relationships in the following table.

Place-Value Chart

1,000	*100*	*10*	*1*	*1/10*	*1/100*	*1/1,000*
Thousand dollars Kilo-	Hundred dollars Hecto-*	Ten dollars Deka-*	Dollar Unit	Dime Deci-*	Cent Centi-*	Mill Milli-

*Remember that hecto-, deka-, and deci- are used basically to illustrate the total structure of the metric system and the relationship between it and our monetary and decimal number systems, and are *not* intended for normal use or memorization.

12. Include metric terms and symbols in spelling activities.

PERIMETER AND AREA

Perimeter

The introduction of *perimeter* is covered in Chapter 2, which concerns the teaching of mathematical vocabulary. The idea of the perimeter of a circle (circumference) can be developed through a laboratory, group-thinking, or pattern-searching lesson.

LABORATORY METHOD

Experiment

Materials: string, ruler.

Use string and a ruler to compare the diameter of and the distance around each circle shown below. Then find the distance around by using only the diameter and ideas you have gained from experimenting. Work in groups of three.

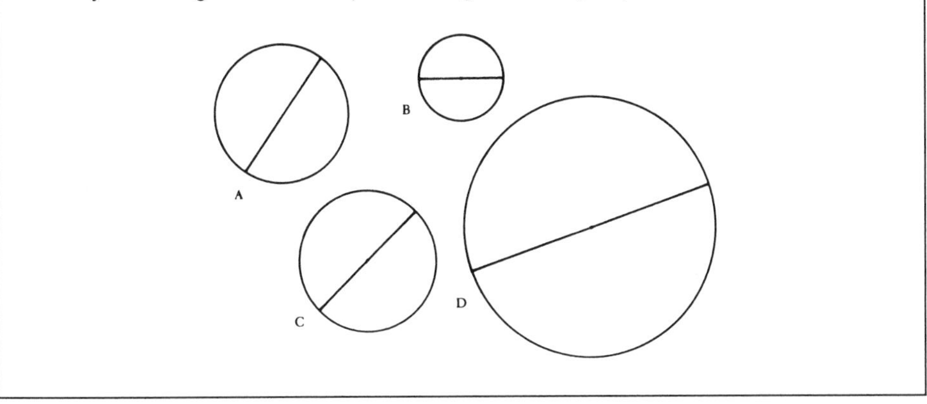

PATTERN SEARCHING

Find the pattern:

	DIAMETER	*CIRCUMFERENCE*
1	8 cm	25.1
2	12 cm	37.7
3	15 cm	47.1
4	18 cm	—
5	23 cm	—
6	40 cm	—

In the intermediate grades, children can be involved in discovery-oriented lessons in which the purpose is to develop formulas for various perimeters. However, place little stress on memorizing these formulas.

The idea of finding a measure of a region (the number of congruent copies of a unit region) can be introduced in terms of rectangular regions at the end of the primary grades. Then the concept of area can be refined in the following grades through problems concerned with finding the areas of triangular, quadrilateral, circular, and irregular regions.

FINDING THE AREA OF A RECTANGULAR REGION

In one class, the topic was introduced with this problem: "I've brought in a new bulletin board for our use. How can we find the number of sheets of construction paper that will be needed to cover it?" The class discussed possible means of determining the number and agreed that the best way would be to lay the bulletin board on the floor and actually cover it with sheets of construction paper. This was done, and the pupils found that 48 sheets of construction paper were required, although the fit was not exact.

Follow-up questions guided the pupils to identify the measure of the rectangular region as 48 units. (A unit represented one sheet of construction paper.) The pupils also generalized that to measure a region, a unit that is itself a region is needed.

The discussion was followed by problems involving the measuring of regions. The pupils were asked to find the measure of the surface of their desks. Several suggestions were made and used in measuring those regions. First, the pupils found the number of "hands" that the desk top contained.

Because of the varying sizes of the pupils' hands, the class decided that although a measure could be taken with a hand, it would be better to have a measure that would not vary. This idea led to a discussion of a possible standard measure of regions. The suggestion most often made was that such a standard region would best be described in terms of centimeters. The pupils were then given a sheet of paper and asked to experiment with measuring that region. Several suggested that the paper be marked off into centimeter squares. When the paper was marked in 1-centimeter-by-1-centimeter regions, these were identified as units of 1 square centimeter.

Next, the teacher asked if any of the pupils knew the name that is usually given to the measure of a region. Several had heard the term *area* and suggested it. The class was referred to textbooks to check the correctness of the term.

The lessons that followed involved using squared paper and finding the area of various rectangular regions marked off on the paper. Familiarity with the array pattern (studied earlier) allowed the pupils to discover quickly that the number of units could be speedily determined by multiplying the number of the length by the number of the width. Mathematical sentences for finding the area were written.

At a later stage of development, the pupils should abstract the general formula for the area of a rectangle: area = length × width; $A = LW$.

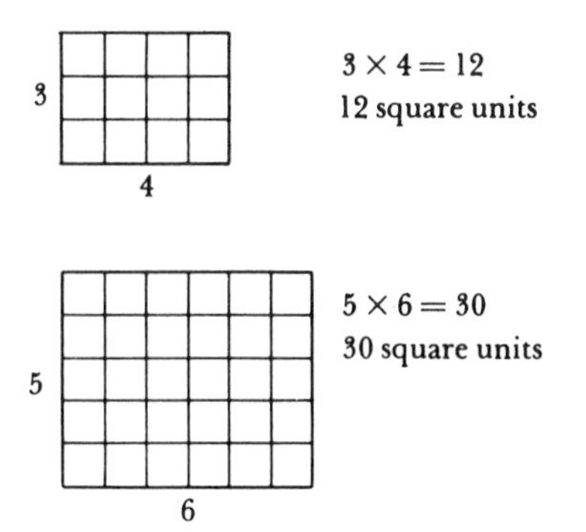

AREAS OF OTHER POLYGONS

Once pupils have developed the basic idea of finding the measure of a region by determining the number of square inches that will cover it, they can discover methods of determining the area of other polygons.

A study of area in the elementary grades should help pupils to understand concepts such as these:

1. The union of a simple closed curve and its interior produce a *plane region.*
2. A standard plane region is used to measure plane regions. The standard plane region for the metric system is the *square meter.* The standard plane region for the English system is the *square foot.*
3. A plane region is measured in terms of the number of unit regions and portions of unit regions that will fit into the plane region. These unit regions must not overlap.
4. The precision of the measurement of the plane region depends on the size of the unit that is used for the measurement. The smaller the unit, the greater the precision of measurement.
5. The area of a plane region is designated by the number of unit regions and is named by this number followed by the designation of the unit region; for example, 5 square centimeters and 9 square yards.
6. The surface (area) of a solid region may be obtained by finding the area of the plane regions needed to cover the surface of the solid.
7. It is possible to develop efficient formulas for obtaining the measure of a region.

Perimeter and Area

Relationships can be studied by studying the specific relationship between perimeter and area. Exercises such as those suggested here for rectangles can be developed for other polygons. The following lab can provide the basis for a bulletin board that emphasizes pupil-doing:

1. What is the area of a rectangular region 12 cm by 1 cm? What is the perimeter of a rectangle 12 cm by 1 cm?
2. What is the area of a rectangular region 9 cm by 2 cm? What is the perimeter of a rectangle 9 cm by 2 cm?
3. What is the largest perimeter you can develop for a region having an area of 24 square cm? (Use only whole centimeters for dimensions.) What is the smallest perimeter you can develop for a region having an area of 24 square cm? (Use only whole centimeters for dimensions.)
4. Charlie has a candy-making machine that makes candy with a perimeter of 16 units. What is the largest candy that can be made by bending the mold? What is the smallest candy that can be made? Use squared paper to experiment.

MEASURING ANGLES

The teacher used a large compass to draw a circle on the board, marked the midpoint of the circular region, and then asked, "How can we mark off this circular region to represent the correct proportions for the various five subjects?" Various suggestions were given by the pupils. They knew that the circle could easily be separated into two equivalent regions by drawing a line segment through the midpoint. The problem then became one of finding $\frac{2}{10}$ and $\frac{2}{10}$ of this region to mark off mathematics and language arts. One pupil suggested that they might try to develop a "unit angle" that could be laid around the center of the circular region ten times. (See below.) This was attempted, and, after a struggle, a reasonable approximation of the circle graph was completed.

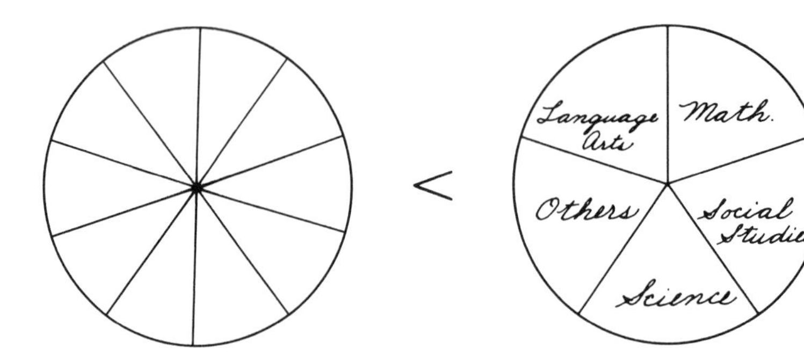

Then the teacher asked, "How could we partition a circular region by using a measure other than tenths?" A class member suggested that it would be necessary to develop some standard unit of measuring angles, just as inches or centimeters are used to measure line segments. Many of the pupils then referred to the fact that older brothers or sisters used protractors to perform this task. The teacher took this lead and provided each pupil with a protractor. Pupils noted that the protractor was marked in degrees and that the semicircle of the protractor contained 180 units. Discussion brought out the idea that the angle measure of an entire circle would be 360 units.

The teacher suggested that pupils experiment with the protractors and attempt to draw angles of various sizes and asked, "How many units would we have in a right angle?"

The children experimented with the protractor for about 5 minutes. Many seemed to be making good progress. Occasionally the teacher stopped to ask a pupil a question such as, "What do you do to make sure that you have the vertex of the angle in the spot that you want?" Then the teacher said, "Let's stop for a few minutes and compare notes. Does anyone know the name given to the units on the protractor?"

Several pupils raised their hands, and one said he thought the unit was called a degree. Discussion continued during the days that followed and was used to bring out the following ideas:

1. The degree is a standard unit for measuring angles.
2. We can now define *acute, right,* and *obtuse* angles in terms of degrees.
3. To measure an angle, the center of the protractor should be placed at the *vertex* of the angle. The number of degrees in an angle is indicated by the spot where the other side of the angle crosses the scale. The degrees should be counted beginning at zero on the base line of the angle.
4. The two scales on the protractor are simply for convenience.

Pupils were then given a variety of experiences measuring the reproducing angles of various sizes at various positions on the paper During the months that followed, pupils measured angles when making mobiles, holiday ornaments, animal feeders, and posters and in doing other art, science, or social-studies projects.

MEASURING TIME

Because of the relativity of time, it is very difficult for children in the primary grades to have developed concepts concerned with "a longer time than," "a shorter time than," "as long a time as." Thus, the 5 minutes that children wait to open a birthday present seems to be much longer than the 5 minutes their mothers spend reading to them before they go to bed.

To develop some concept of time relationships, the teacher should engage in many time comparisons before moving to types of time measure. Questions concerning routine classroom activities can be used to aid in developing these concepts. For example, "Did it take Nancy longer to clean up the tables after our snack than it did Jill to put away the blocks?" "Does it take us longer to put on our boots to go out for recess, or longer to take them off when we come in from recess?" These experiences will lead to discussions of "Which took longer?" which will lead in turn to the need to measure time.

The first experiences with time measure can occur through the use of counting (which proves inaccurate), egg timers, and other nonstandard methods of telling time. From these experiments, the teacher can move to the use of clocks.

Measuring time is one of the more difficult skills for many elementary school pupils. It has been suggested (with tongue in cheek) that the majority of first graders could learn to tell time to the nearest minute if they each received a working watch at the beginning of the school year. There is some justification to this claim, for many pupils who receive watches as holiday gifts return to school with a greatly increased ability to tell time.

Teaching children to tell time is unlike many of the other mathematical tasks, for it is usually done informally. Reference to time and the development of pupils' ability to tell time should be emphasized every day.

The clock can be thought to possess two number lines, the hour number line 1 to 12 and the minute number line 1 to 60. The small hand is used to designate the hour number line, and the large hand is used to designate the minute number line.

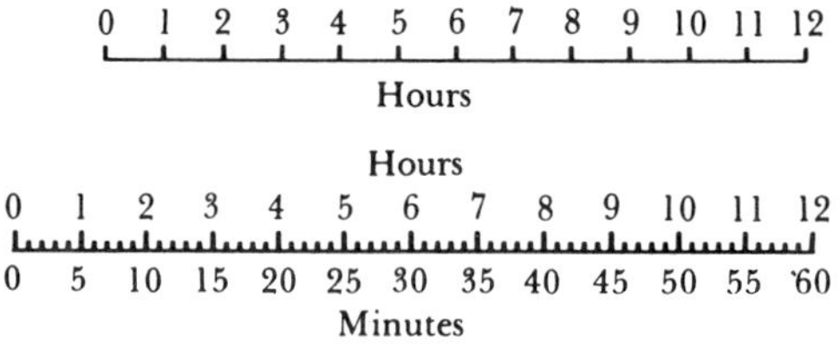

0 1 2 3 4 5 6 7 8 9 10 11 12
Hours

Hours
0 1 2 3 4 5 6 7 8 9 10 11 12
0 5 10 15 20 25 30 35 40 45 50 55 60
Minutes

The teacher can begin with a single-handed clock (the hour hand) by using the story of Grandfather's giving Billy such a clock. Billy thought he could not tell time with it, but Grandfather said, "You can't tell time exactly, but you can get a good idea of the time from this." What did Grandfather mean?

This discussion leads to ideas such as "about 4 o'clock"; "after 4 but before 5"; "closer to 5 than 4"; and so on.

The typical sequence is to move from hours to half hours to quarter hours. However, moving directly from the hour to the minute is usually less confusing to the children. They hear 4:18, 4:55, and the like on the radio and television. Thus, the use of two number lines provides a means of telling "which hour" and "which minute." Figure 14–1 gives an idea of this type of development. Note that several days' activities are combined in this single picture.

Teachers have found the following suggestions helpful in teaching children to tell time:

1. Use a variety of materials and procedures for telling time.
2. As often as possible, use actual clocks for telling time.

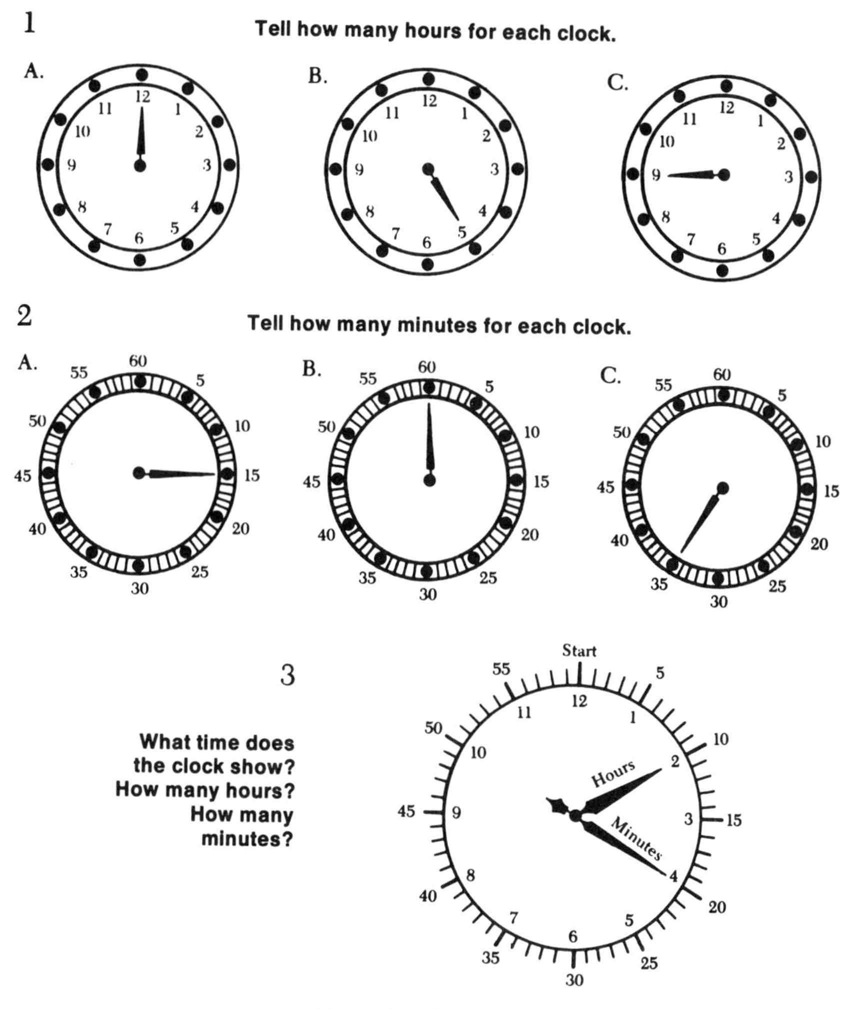

Figure 14–1 What Time?

3. Use "count" activities to develop time concepts. For example, "Let's see how long it takes us to get ready for recess."
4. The usual developmental sequence for telling time begins with identifying the time that particular events occur, such as the time school is out. Then a study of one-half hours is initiated, followed by one-quarter hours. It is debatable whether the next step is to study time by 5-minute or 1-minute periods.
5. Remember that the passage of time is relative. Ten minutes is quite a long time for a person who is waiting anxiously for an overdue airplane. Ten minutes is a short time for people who are engaged in a game of cards. (It can also be a very long time for the nonplaying onlooker.)
6. Telling time is a gradually developing skill. It is better to spend a few minutes every day or two on the development of time concepts than to teach a unit on telling time.

A FINAL NOTE

Measurement is one of the most important social uses of geometry and arithmetic. Instruction in measurement provides an alert teacher with opportunities to make use of historical materials that help pupils see mathematics as a constantly growing and changing subject. Too often in the past, the main emphasis has been on the memorization of tables of weights and measures and of formulas for area, volume, and so on. The teacher should take great care to make the study of measurement stimulating and worthwhile.

THINK ABOUT

1. Plan a series of lessons for fifth grade on the determination of precision in measuring.
2. Working with two or three of your friends, develop a chart of metric "reference distances" between well-known or easily recognized points in your town (e.g., city hall to airport, town limit to first traffic light on a main road, etc.).
3. Compile a set of activity cards on linear measurement for first grade.
4. Discuss ways of introducing and maintaining skill in measuring mass. Include use of nonstandard measures and guess-and-test procedures.
5. Plan at least three motivating activities for learning to measure angles.
6. List at least five activities or experiments for measuring volume for *each* of the following grade levels: K–2, 3–5, and 6–8.
7. Describe two pattern-searching activities for each of four different aspects of measurment (i.e., volume, area, mass, etc.).

KEEPING SHARP

Self-Test: True/False

_______ 1. Since measurement ideas occur in the everyday life of children, they will develop appropriate skill in using measures without carefully planned teaching.

_______ 2. Both the metric and English systems of measurement should be taught, although converting between the systems in not of primary importance.

_______ 3. Mastery of the metric (or English) system includes knowledge of its structure as well as a "feel for" the units being used.

_______ 4. Like the English system, the metric system of measurement is based on arbitrarily determined basic units of length, mass, capacity, etc.

_______ 5. Comparing and measuring common objects helps develop familiarity with unit size.

_______ 6. Encouraging the use of nonstandard units of measurement in early work serves no useful purpose and simply confuses the pupils.

_______ 7. Measurement instruction lends itself well to a laboratory or activity approach, since doing is the best way to learn these skills and concepts.

_______ 8. Precision in measurement is determined by the size of the thing being measured.
_______ 9. Measurement is approximate rather than exact.
_______ 10. Time is a particularly difficult concept for young children.
_______ 11. Measurement is a socially useful tool.

Vocabulary

meter	gram	milli-	reference measures
liter	kilo	centi-	

SUGGESTED REFERENCES

The literature on the metric system has expanded geometrically in the past few years. It is almost impossible to keep up with the number of books and articles written on the topics. It is suggested that the student of elementary mathematics teaching check current issues of *The Arithmetic Teacher, Teacher, Learning, School Science and Mathematics,* and *Instructor* for recent articles. Below are listed several teacher resource books. Books marked with an asterisk (*) contain treatments of the history of measurement.

ACHELIS, ELIZABETH, *Of Time and the Calendar.* New York: Hermitage House, 1955.

BITTER, GARY G., and others, *Activities Handbook for Teaching the Metric System.* Boston: Allyn & Bacon, 1976.

*BUCKINGHAM, B. R., *Elementary Arithmetic: Its Meaning and Practice,* pp. 456–735. Boston: Ginn, 1947.

CLEMENTS, DOUGLAS H., *Computer in Elementary Mathematics Education,* chap. 9, Englewood Cliffs, N.J.: Prentice Hall, 1989.

COBURN, TERRANCE G., and ALBERT P. SHULTE, "Estimation in Measurement." In *Estimation and Mental Computation, 1986 Yearbook,* Reston, Va.: National Council of Teachers of Mathematics, 1986.

DIENES, Z. P. and E. W. GOLDING, *Exploration of Space and Practical Measurement.* New York: Herder and Herder, 1966.

GOLDBECKER, SHERALYN S., *Metric Education.* Washington, D. C.: National Education Association, 1976.

HIGGINS, JOHN L., ed., *A Metric Handbook for Teachers.* Reston, Va.: The National Council of Teachers of Mathematics, 1974.

LAFERLA, VIVIAN R., "Teaching Measurement Estimation through Simulations of the Microcomputer," in *Estimation and Mental Computation, 1986 Yearbook,* Reston, Va.: National Council of Teachers of Mathematics, 1986.

MARKS, JOHN L., ARTHUR A. HIATT, and EVELYN NEUFELD, *Teaching Elementary School Mathematics for Understanding,* 5th ed., chap. 9. New York: McGraw-Hill, 1985.

Measurement in School Mathematics, 1976 Yearbook. Reston, Va.: National Council of Teachers of Mathematics, 1976.

Metric Units of Measure and Style Guide—SI. Boulder, Colo.: U.S. Metric Association, 1976.

O'DAFFER, PHARES G., and STANLEY, R. CLEMENS, *Metric Measurement for Teachers: An Activity Approach.* Menlo Park, Calif.: Addison-Wesley, 1976.

THIESSEN, DIANE, MARGARET WILD, DONALD D. PAIGE, and DIANE L. BAUM, *Elementary Mathematical Methods,* chap. 10. New York: Macmillan, 1989.

WHITMAN, NANCY C., and FREDERICK G. BRAUN, *The Metric System: A Laboratory Approach for Teachers.* New York: John Wiley, 1978.

ZWENG, MARILYN J., "Introducing Angle Measurement through Estimation." In *Estimation and Mental Computation, 1986 Yearbook,* Reston, Va.: National Council of Teachers of Mathematics, 1986.

chapter **15**

EVALUATION OF LEARNING IN ELEMENTARY SCHOOL MATHEMATICS

OVERVIEW

Measuring and evaluating pupil achievement in any area of the curriculum is difficult, and mathematics is no exception. However, there is a tendency to believe that the measurement of educational goals is easier in mathematics than in areas such as science or social studies. This belief is fostered by the ease with which the teacher can obtain an objective measure of certain computational skills that form a portion of the mathematics curriculum. Measurement of the other goals of the mathematics program requires a great deal of thought.

Too often in the past the evaluation of children and the mathematics program has been conducted by using instruments that are not in keeping with the major goals of the program. The learning process should not be limited by the evaluation instruments but be enhanced by them. To measure achievement in a forward-looking program, the National Council of Teachers of Mathematics Standards suggest that the following changes in emphasis be given.[1]

[1]Commission on Standards for School Mathematics, *Curriculum and Evaluation Standards for School Mathematics* (Reston, Va.: National Council of Teachers of Mathematics, 1989).

Emphasis of the Evaluation Standards

Emphasize	De-emphasize
Assessment determines what students know and how they think about mathematics.	Assessment determines only what students do not know.
Assessment is integral to teaching.	Assessment is only unit or semester tests, often used solely for the purpose of assigning grades.
Assessment focuses on a broad range of mathematical tasks and a holistic view of mathematics.	Assessment focuses on a large number of specific skills organized by content-behavior matrix.
Assessment includes problem situations involving the applications of a number of mathematical ideas.	Assessment only includes exercises or word problems requiring one or two skills.
Assessment applies a variety of techniques, including written, oral, and demonstration modes.	Assessment is a written test.
Calculators and computers are used in assessment as appropriate.	Calculators and computers are banned from assessment for other than instructional reasons.
Program evaluation includes systematic collection of information on outcomes, curriculum, and instruction.	Program evaluation is composed of test scores.
Standardized achievement tests are used as one of many indicators of program outcomes.	Standardized achievement tests are used as the only indicator of program outcomes.

The suggestions that follow focus on emerging developments in the measurement of mathematical maturity, as well as on the traditional aspects of elementary school mathematics.

TAKE INVENTORY

Can You:

1. List seven categories (such as problem solving) on which children should be evaluated.

2. Devise ways of measuring in each of the seven categories?
3. Suggest several means of evaluating the attitude and achievement in mathematics of elementary school children?
4. Describe ways to evaluate standardized tests?
5. Suggest methods of reporting to parents?

PURPOSE OF MEASUREMENT

There are a variety of reasons that a teacher would measure or evaluate elementary school mathematics pupils. The teacher might be trying to find answers to questions such as these:

"Where is Bill in multiplication? I need to know to give him appropriate instructional materials."

"How does Bill compare with the others in the class?"

"How does Bill compare with fourth graders in general?"

"Has Bill mastered the basic addition facts?"

"How does Bill feel about elementary school mathematics? Why?"

"Can Bill think creatively in mathematics?"

"Is Bill able to apply information to solve problems?"

Finding answers to questions such as these involve one or more of a number of evaluation techniques, depending upon the purpose of the evaluation or the trait being measured.

The assessment in mathematics of elementary school children should go far beyond paper-and-pencil tests. A variety of techniques must be used so that teachers will have information concerning the child's level within the following categories: (1) integration of mathematical knowledge, (2) problem solving, (3) communication, (4) reasoning, (5) mathematical concepts, (6) mathematical procedures, and (7) mathematical disposition (in addition to attitude, this includes a tendency to think and act in certain ways). In the sections that follow, evaluation techniques will be considered, and then examples of measurement in each of the seven categories will be considered.

EVALUATION TECHNIQUES

Subjective and Informal Procedures

Subjective and informal procedures can be used to measure traits in both the cognitive and the affective domain. The suggestions that follow deal with ideas in both areas. Note that the teacher should have in mind the purpose of each observation or other procedure.

PUPIL OBSERVATION

The phase of a mathematics lesson that is called "study time" is perhaps the single most important means of evaluating pupils' mathematical behavior. The teacher can use this time to move quietly about the room observing the pupils at work, making notes, questioning the children, and making suggestions. The teacher may also very effectively use *class discussion* time to gain insight into the pupils' thinking processes.

I suggest that the following observations be made and that the teacher ask the related questions.

1. *Note the attack used by the pupils.* Is progress consistent? Is Bill always a slow worker? Does he seem to like some phases of mathematics better than others? Is Jean poor at computation, or does she become careless when she gets bored with practice materials?
2. *Limit the observation to some aspect of pupil performance.* How competent are pupils in solving a problem in more than one way? What errors are most common in multiplying numbers when one or both of them contain an internal zero?
3. *Check the depth of pupil thought.* Is the pupil interested in going beyond the lesson? Does Mary consistently try a "just-for-fun" exercise? How often do class members check their work when they are not asked to do so? How many sources does Nancy check when she is preparing her report on the history of decimals?
4. *Observe the emotional climate.* How relaxed is the pupil while working? Does Bob seem to be under pressure when he is doing a geometric construction? Does Alice begin her nervous mannerism of twisting her hair during most mathematics assignments?
5. *Check study habits.* How good or bad are the study habits of the pupils? Does Ken seem to be more interested in the actions of others than in his work? Bill seems to be looking our of the window most of the time. Is he daydreaming, or is he thinking? (He may be doing either.)
6. *Note skill development.* How effective is the numeral writing of the pupils? Will a suggestion now and then help Jerry to write the numeral 2 so that it does not look like a 3?
7. *Observe pupil independence.* Do students really need your help when they raise their hands? Did Chris really know how to work the division exercise she asked about, and was her question a need for reinforcement and praise? How can I give her the needed praise and still develop her independence? How dependent upon the textbook and teacher suggestions are the pupils? Does Joe always look back to the book to check on the method used by the authors to solve an exercise? How many pupils use the teacher's techniques in solving a problem or an exercise?

Because many of the observations that a teacher makes are quickly forgotten, written anecdotal records should be kept. The stock 3-by-5-inch cards or small note pads are easy to use. As the teacher notes an action of significance, it can be recorded briefly on a card or the pad and dropped into the pupils arithmetic folder later.

It is often suggested that daily records of this type be kept on each pupil. Daily records may be unrealistic, but if significant pupil actions are recorded periodically—at least 10 times a year—they will provide perceptions that the teacher may have overlooked. Also, such records are helpful in preparing for parent and pupil conferences. A sample record follows.

CHARLES WASHINGTON—GRADE 3

September 30	Seems able to move from a problem situation to the mathematical idea. Is very interested in puzzle-type problems.
October 10	Makes careless errors in addition and subtraction exercises. Seems to know the addition combinations.
October 15	Charles understands the renaming principle when subtracting multidigit numbers.
October 20	Showed originality in developing several solutions to multiplication problems. Presented his ideas clearly to the class.

INTERVIEWING PUPILS

A study in greater depth may be made by using the interview technique. The teacher may begin by presenting some previously taught material to a child and then ask for an answer and an explanation. As the interview progresses, new material that makes use of the same basic ideas may be interjected. The following sequence could be used to study a pupil's thinking on division combinations:

TEACHER: What is $12 \div 3$?
PUPIL: Four.
TEACHER: How did you get the answer?
PUPIL: I just knew the answer. I guess I could have subtracted 3 from 12 until I got to zero.
TEACHER: What is $18 \div 6$?
PUPIL: I'm not sure. Do you want me to figure it out?
TEACHER: Yes, think out loud as you work the exercise.
PUPIL: I want to find out how many sixes equal 18. I know that 2 sixes equal 12. Twelve plus 6 equals 18. Three sixes equal 18.

Every effort should be made to ease the pressure on the pupil. In fact, it might be better when conducting a first interview to use only material that the teacher feels the pupil will know.

FOLDERS

Manila folders containing samples of the daily work, anecdotal records, and copies of their arithmetic themes are important devices for continuous evaluation of students.

Pupils should periodically be asked to give short answers to questions that require pencil-and-paper computation. Questions of the following variety can be used effectively:

1. Give an example of a problem situation that describes a use of the median; of the mean.

2. Show in at least three ways that the answer to this division is correct:

$$6\overline{)42}\quad 7$$

3. Tell why the quotient is larger than the divisor in this division problem:

$$5 \div \frac{1}{4} = \square$$

The folders serve as a means of systematizing informal evaluative techniques and are convenient repositories for most types of information. Also, the folders provide a fine source of information for parent conferences and can be of great aid to the teachers of these pupils during each succeeding year.

ORAL EVALUATION AND THE TAPE RECORDER

The tape recorder is a unique device for evaluating pupils' mathematical abilities. At least three procedures can be used effectively with tapes.

1. The teacher and the pupil can discuss some aspect of mathematics, with the student using the recorder. The teacher would make comments and ask questions, as follows: "Describe the procedures you would follow to solve this problem." "Tell me why you carry the 1 in addition." "Look at this division exercise, which has been worked incorrectly; can you tell me where the pupil went wrong in his thinking?" "Here is a phase of mathematics that you haven't studied; what would be the first thing you would do to attempt a solution?"

   At the end of the day or in the evening, the teacher can play back the tape and take notes. The pupil's answers are useful in giving insight into his or her thinking pattern. Because many significant bits of the pupil's thinking may have been overlooked in the original discussion, the tape recorded provides a means for a more careful analysis of the situation. This analysis requires a good deal of teacher time, but it is often well worth the time because of the opportunity it affords the teacher to think through a discussion with an individual pupil.
2. Several pupils can work together at the tape recorder and discuss some mathematical concept. Questions such as the following ones can be discussed: "What is the *best* method to add $\frac{2}{3}$ and $\frac{4}{5}$ without using paper and pencil?" "In what ways can you estimate the quotient of $345 \div 23$?" "How can $\frac{1}{3}$ be expressed as a decimal? Why?"
3. A child may use the tape recorder to analyze his or her own thinking. For example, a pupil may explain a problem orally and then play it back on the tape recorder to see how clearly the reasoning has been stated.
4. Short taped interviews can be put to excellent use in parent conferences. Often, parents can gain real insight into their child's thinking and mathematical maturity by listening to a taped interview like those described.

Evaluating With Tests

CHARACTERISTICS OF GOOD TESTS

Any test, whether constructed by an individual teacher or by a team of specialists for a commercial publisher, should meet several criteria, including acceptable validity, reliability, and good format.

Validity. If a test is valid, it will effectively measure the skills, understanding, or knowledge that it was intended to test. Often, a mathematics test that purports to measure understanding measures only computational ability. It may be a good test of computational ability, but if it misses its assigned task of measuring understanding, the test will not be valid.

In checking the validity of a test, the teacher should ask questions such as the following ones. "How well does this test represent the significant behaviors that I want measured?" "Are all the items relevant to these behaviors?" "Is the test a balanced sample of the behaviors I want to assess?"

In general, the best method a teacher has of ascertaining the answers to these questions is to take the test. If it is a standardized test and does not seem to measure the objectives of the mathematical curriculum, its use would be of questionable value. In the case of the teacher-made test, it is probably wise to write the test, let it remain in a desk drawer for a time, and then take the test. This allows the writer to see clearly items that need rewording. It is also helpful for the teacher to have a colleague look at the test and make suggestions.

Reliability. An important factor in the value of a test is the *consistency* of the measurement of a particular achievement. This consistency is usually called reliability. To be considered reliable, a test must consistently and repeatedly measure the same achievement. Fortunately, the nature of mathematics helps make mathematics tests reasonably reliable. The reliability of a carefully constructed teacher-made mathematics test need not be of extreme concern to the teacher.

Good format. The choice of a good format is an important aspect of test selection or construction. A test with good format should meet the following criteria:

1. It can be easily understood by the pupils. The format helps the pupil to understand the basic goals of the test.
2. There are few possibilities of wrong answers because of poor directions.
3. The test can be scored easily.
4. The format does not add unnecessary space to the test.
5. Test questions can be interpreted accurately.

STANDARDIZED TESTS

There is an ever-increasing use of standardized tests in the majority of areas of the elementary school curriculum. With this increased use have come many popular articles of standardized objective tests. When the evidence for and against it is carefully weighed, the findings reveal that the standardized test is helpful if *properly used.* The major difficulty lies in the misuse rather than in the use of such tests. The section that follows will describe some of the features of standardized tests in elementary school mathematics and will make suggestions for their use and interpretation.

Achievement tests. The standardized achievement test differs from the teacher-made test in at least four ways:

1. The objectives and content of standardized tests tend to be based upon those that are common to many school systems, whereas those of teacher-made tests are specific to a given classroom setting.
2. The standardized test surveys a large portion of knowledge; the teacher-made test is usually related to a limited topic.
3. Standardized tests are usually developed by a team of curriculum workers, test editors, and reviewers. The teacher typically works alone.
4. The standardized tests provide norms that are based upon the performance of a large sample of elementary school pupils from representative schools throughout the nation. The teacher-made test usually depends solely upon the performance of from 20 to 40 pupils.

The standardized achievement test in elementary school mathematics may serve many purposes. Some of the more common are these:

1. To provide the teacher with a general picture of achievement in mathematics for individual pupils and for the class as a whole. The teacher can observe from standardized-test results the "spread" in achievement in the class. The test may also give some direction to grouping procedures and the need for special materials.
2. To provide a comparison of these pupils' scores with those made by students in the standardization group. The teacher may see if the class seems on the whole to be above average or below average relative to the pupils on whom the test was standardized.
3. To study pupil growth over a period of months or years. Standardized tests given at regular intervals help the teacher to do this. Some pupils of above-average ability can be expected to make well over a year's growth each year; others will not make a year's growth per year. The teacher can analyze the pattern of growth indicated by the test to ascertain whether the pupil is progressing at a faster or slower pace than previously.
4. To help determine curricular weaknesses. Careful study of the results of a standardized test in mathematics often helps the teacher to see areas of the mathematics curriculum that have been neglected. For example, the results of standardized tests have caused some teachers and school districts to give greater emphasis to basic mathematical concepts and to problem solving.

SELECTION OF STANDARDIZED TESTS

Perhaps the best method of measuring the validity of standardized tests for a given school is to have teachers take the test themselves, comparing the items with the purposes of the local mathematics curriculum.[2] Once tests that measure behaviors important to the curriculum have been identified, several other checks should be made. These are questions that should be asked:

1. Is there good evidence that the test has been well analyzed statistically? Are the norms appropriate? Can raw scores be readily converted to derived scores?
2. Can the information obtained be easily and correctly interpreted?
3. Is the manual adequate? Does it give good suggestions for application of test results?

[2]Leroy Callahan, "Test-Item Tendencies, Curiosity, and Caution," *The Arithmetic Teacher,* 25 (December 1977): 10–13.

4. Can the test be scored by the agency or with a reasonable amount of clerical time?
5. Are these alternate forms so that scores can be compared from year to year? How comparable are these forms?

Teacher-Made Tests

CONSTRUCTION OF TESTS

Teacher-made tests are important for an evaluation of pupil achievement of the curricular goals of the particular school. Well-constructed teacher-made tests also provide a means of motivating and directing student learning. The continual evolution of the elementary school mathematics program forces the teacher to rely heavily on such tests as a means of evaluating the changing program.

When a teacher sits down to develop a test, several questions must be answered. Among the more important are these: (1) What are the purposes or objectives (behavioral and/or experiential) that I am measuring? (2) What mode of presentation—enactive (concrete), iconic (pictorial), or symbolic (abstract)—is in keeping with the instruction on the mathematical topic? (3) How many questions shall I include in the test? (4) What emphasis shall I give to various types of learning? (5) What type of test items shall I use? Suggestions for answering these questions follow.

To answer the first four questions, the teacher must decide the specific purposes of a unit in elementary school mathematics, outline the section, and develop a blueprint for constructing the test.

The teacher may begin by listing the specific learnings of a mathematics unit. Then he or she can decide upon a sample of these learnings on which to develop test items and the number of items for each topic. An illustrative listing appears below.

The suggestions on informal evaluation focused on essay-type evaluation. Therefore, this discussion will focus upon objective test items. There are several types of objective items that can be used: True–false, completion, multiple-choice, and matching items are all possibilities. Probably the two most useful types for mathematics tests are completion and multiple-choice items. The suggestions that follow are primarily concerned with writing good multiple-choice items, but each could be used in writing a completion-type item. In addition, suggestions are given for measuring specific skills.

1. Express your ideas as clearly as possible.
2. Use understandable but precise vocabulary in developing items.
3. Avoid complex or awkward wording.
4. Include all the information necessary for a reasonable response to the item.
5. Eliminate irrelevant sources of difficulty.
6. Unless the item is a very specific mastery item, avoid the type of question that allows a rote response and rewards the rote learner.
7. Adopt the level of difficulty that is appropriate for the group for which the text is used.

CRITERION-REFERENCED TESTS

In recent years, there has been an increased use of criterion-referenced tests by teachers and school districts. With criterion-referenced tests, the pupil is evaluated upon an objective rather than compared with others. The goal of the CRT is to provide the teacher and student with the information necessary to determine the student's mastery of the material. When the material is mastered, the student moves on to other material. Thus, the CRT is concerned with whether the learner has acquired specified objectives.

In theory, all the students will master the material; however, it will take some students a greater number of tries. In many cases, the student will take a CRT several times, until he or she has obtained the determined mastery. The typical teacher-made test falls into this category.

MEASURING EDUCATION GOALS

Integration of mathematical knowledge. Goals in this area include knowledge and understanding of concepts and procedures; ability to apply knowledge to solve mathematical problems and problems in other subjects, ability to reason and analyze, ability to use mathematics to communicate ideas, understanding of the nature of mathematics and disposition toward mathematics, and the extent to which these aspects of the student's mathematical knowledge are integrated.

The sections that follow present an introduction to a foward-looking assessment in the various areas of elementary mathematics instruction. Space does not provide for a thorough development of any one area in which assessment should occur.

Problem solving. Included in this mathematical achievement goal is skill in formulating problems, identifying key features of problems, applying a variety of problem-solving strategies (see Chapter 4), solving problems, verifying and interpreting results of problem solutions, and generalizing solutions to other situations.

A variety of methods should be used to assess the children's ability in this area. Several procedures follow. These suggestions only scratch the surface of the variety of procedures that can be used:

1. Formulate problems. Give the children a page from a toy catalog. Have them use the information on the page to make up several problems. Work in twos and individually. Give the children the grocery sales section of the paper. Suggest that they have a budget of $70 for groceries for the week. Make up five problems concerning the situation.
2. Identify key features of a problem. Give the children a page of problems, three to five, write the problems so that some have too little information to be solved, others have extra information, and others just the amount of information necessary for solution. Have the children indicate the status of each problem and indicate what information is necessary or what information is not needed.
3. Apply strategies to solve problems. Develop problems that lend themselves to a particular approach, such as making a drawing, trial and error, or writing a mathematical sentence. On some occasions, find out if the children can identify

useful strategies. On other occasions, use problems such as this one: "What math sentence most closely corresponds to this problem?" Jill wants to buy a pencil that cost 25 cents. She has 15 cents. How much does she need to buy the pencil? (a) $25 - 15 = N$; (b)$15 + N = 25$ (correct), (c) $N + 15 = 25$.

4. Solve problems. Use a variety of problem settings. Make use of some problems in which data need to be gathered to solve the problem. For example, "Children your age vary in height. What is the tallest fifth grader in school? How many inches taller is this student than the tallest third grader? How could you find out?"
5. Formulate problems; verify and interpret results. For example, "The average fifth grader spends 2 hours a day watching television. Write a question to go with that statement to make a problem. Solve the problem."
6. Write a question to make a problem whose answer is 360 hours.
7. Keep a problem-solving journal. You can keep track of interesting problems that you find outside of class, problems solved in class that you thought were interesting, and ways that you go about solving a problem. This activity also makes a very useful language arts correlation.

Communication. Goals within this area are concerned with expressing mathematical ideas by speaking, writing, and demonstrating pictorially, and by using mathematical vocabulary, notation, and structure to represent ideas, describe relationships, and model situations.

This goal is most often developed by using the kinds of discussion evident in superior science and social studies instruction. Asking questions such as, "I notice that several of you used different methods to solve the problem. Jan, Kim, Tom, and Sally, will you explain how you went about solving the problem?" After each child explained his or her method, the teacher lead the children to discuss the advantages and disadvantages of each approach.

Some teachers make use of the *mathematics theme* to develop and access one phase of the student's mathematical communication. For example, at the end of a unit on fractions, a sixth-grade teacher gave the language arts assignment. "Write a short theme on division of fractions. Tell how you do it, why it works, and perhaps some history." One child's theme follows:

What Are Decimals?

(Annie Lou's paper)

Decimals are based on 10 and multiples of 10. The decimal point separates the whole number(s) from the decimal fraction.

We use decimals because it is easier to get an accurate measure with them, especially with very small measurements.

It is easier for me to use decimals, because there are not many steps to remember when I multiply, divide, add, and subtract them. It is easier for me to read them, too.

When you are adding decimals, you have to remember to have the decimal points in the addends and the sum in a straight column. When you are subtracting decimals, you have to remember to keep the decimal point in the minuend, subtrahend, and difference in a straight column. When you are dividing decimals, you have to remember to move the decimal point in the divisor and the dividend to the right as many places as there are in the divisor. You have to remember to put the decimal point in the quotient right above the one in the dividend after it has been moved. When you multiply a decimal you have to remember to count how many figures there are to the right of the decimal point in the multiplicand and in the multiplier, then you count that many places off in the product.

We use decimals to show someone's batting average, to show how fast someone can swim, to show how high someone can jump, to show how long something is, and to show how much something weighs.

Mr. Riedesel or whoever else reads this theme, I think you would be interested in knowing that the decimal system came into use in the seventeenth century.

In analyzing this procedure, it can be noted that the following benefits can be derived from writing an arithmetic theme:

1. The teacher has a means of evaluating the individual student, because the themes show the variations in the level of understanding among the children.
2. A valuable tool is provided for use in diagnosing the difficulties that individual class members have with a particular arithmetical process.
3. The actual writing focuses attention on a particular skill and makes it necessary for the student to clarify his thinking on a particular arithmetic area.
4. Writing about mathematics entails the ability to develop accurate explanations and descriptions which will help the student in writing about other subjects, particularly in the scientific fields.
5. The children have a new approach to mathematics, a subject in which writing technique is not usually employed.
6. The students are helped to think about the *how* and *why* of a particular arithmetic process.
7. Class unity can be maintained at a high peak, because all of the students are able to participate in the same activity with a feeling of achievement.

Reasoning. There are a variety of mathematical reasoning skills that are important to a student's mathematical success. He or she needs to be able to use inductive reasoning to discover mathematical generalizations and patterns, to analyze situations, and to determine common properties and structures. The student needs to use deductive reasoning to verify conclusions and judge correctness of thinking steps. Here are a few examples:

1. *Reasoning inductively.* One possibility is to find the missing numbers in a pattern. For example.

   "Find the missing numbers:

   (a) 2, 4, 6, ____, 10   (b) 3, 6, 9, 12, ____   (c) 15, 13, 11, ____."

   "Four students have test scores of 78, 80, 95, and 98. Find the average score (you can use a calculator). How much is the average score increased if each students score is increased by

   (a) 1 point,   (b) 6 points,   (c) 8 points?

   "Now, write a sentence or two about your findings. Why do you think you could find answers to a similar problem without computing?"

2. *Reasoning deductively from known facts.* For example, at the third grade level, "Fill in the blanks:

2 cookies	4 cookies	6 cookies	8 cookies	10 cookies
5 cents	10 cents	15 cents	___ cents	___ cents."

3. *Analyze a situation to determine common properties and structure.* For example, "Write the nines time table. What can you say about all of the products (sum of digits add to nine, etc.). Draw an addition table with all of the sums provided." Ask the student to list all of the ideas and patterns they can find from the table.

Mathematical concepts. Concepts are the bread and butter of mathematical knowledge and information. Students develop mathematical power only when they began to understand mathematical properties and use them in a variety of settings. To assess the children's grasp of concepts, they should demonstrate skill in labeling, verbalizing, and defining concepts and should be able to use models, diagrams, and symbols to represent them; translate from one mode of representation to another; recognize the various meanings and interpretations; identify properties of a given concept; and compare and contrast concepts with other related topics. What follows are three examples:

1. Identify example and nonexample of concepts. Use drawings of a fraction such as $\frac{1}{3}$. Have children identify example and nonexamples. Show pictures of several squares and rectangles and quadrilaterals. Ask the students to mark the squares, mark the rectangles, etc.
2. Identify properties of given concepts and compare and contrast concepts. For example, "Study the numbers 3, 5, 7, and 2. Think of multiplication. How are they different than 4, 9, 6, 12?" (The first group has only that number and 1 as factors).

   Give students a sheet containing drawings of various quadrilaterals. Have the students cut them out. Hold up shapes with two pairs of parallel sides. "What are they called?" Hold up shapes with four congruent sides. "What are they called?"
3. Integrate knowledge of concepts. "While riding from Pittsburgh to New York, Pat fell asleep after traveling half of the trip. When she awoke, she still had to travel half the distance that she traveled while sleeping. For what part of the entire trip had she been asleep?"

   "Assuming that the shaded part in each diagram below shows when Pat was asleep, which diagram best depicts the problem?"

This task assesses whether students can use their knowledge of fractions to interpret the problem and to identify the correct representation of its solution. It requires that they consider a fraction in relation to different units. They must first think of the whole trip as a unit, then consider a different unit (the portion of the total trip during which Pat was asleep).

Mathematical procedures. The majority of tests to date measure a child's ability to do mathematical procedures. While it is very important the child be able to execute mathematical procedures in an effective manner, the child must also be able to demonstrate an understanding of the procedures, such as recognizing when it is appropriate to use a procedure, verifying the results of a procedure, and generating a new procedure.

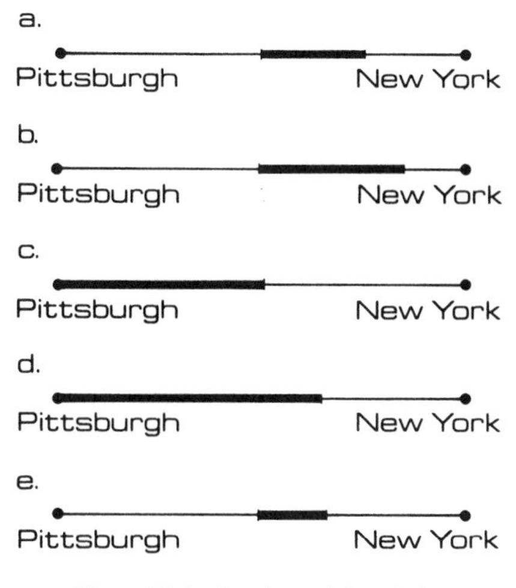

Figure 15–1 Fractions of the whole

Commission on Standards for School Mathematics, *Curriculum and Evaluation Standards for School Mathematics* (Reston, Va.: National Council of Teachers of Mathematics, 1989).

1. Recognize when to use a procedure. Present a variety of single-step word problems. Direct the children to indicate which require addition, subtraction, multiplication, or division. *Note:* some may be correct for both addition and multiplication and others for both division and subtraction.
2. Verify the results of a procedure. A variety of correction procedures using concrete materials and drawings and diagrams can be used for this context. Here are just a few examples: "Solve $64 + 29$ and use tens and one blocks to show your answer is correct.""Solve 9×8 and use squared paper to show your answer is correct.""Solve 23×45 and show that your answer is correct in at least two ways."
3. Generate a new procedure. As you will note this idea is suggested as an introduction to the majority of mathematical ideas presented in the earlier portions of the book.

Mathematical disposition. Assessment of a child's mathematical disposition seeks information about his or her confidence in using mathematics in many contexts; flexibility in exploring mathematical ideas and trying alternative methods in solving problems; willingness to persevere at mathematical tasks; interest, curiosity, and inventivness in doing mathematics; and appreciation of the role of mathematics in our culture and its value as a tool and as a language. A partial model for consideration of mathematics in the affective domain is presented in Figure 15-2.

The best procedures of assessment in this area include discussions, essays, observations, and surveys.

REPORTING TO PARENTS

Methods of reporting vary from school system to school system. Therefore, the form of the written report card depends upon the local school system. However, one of the most valuable means of reporting progress in mathematics to

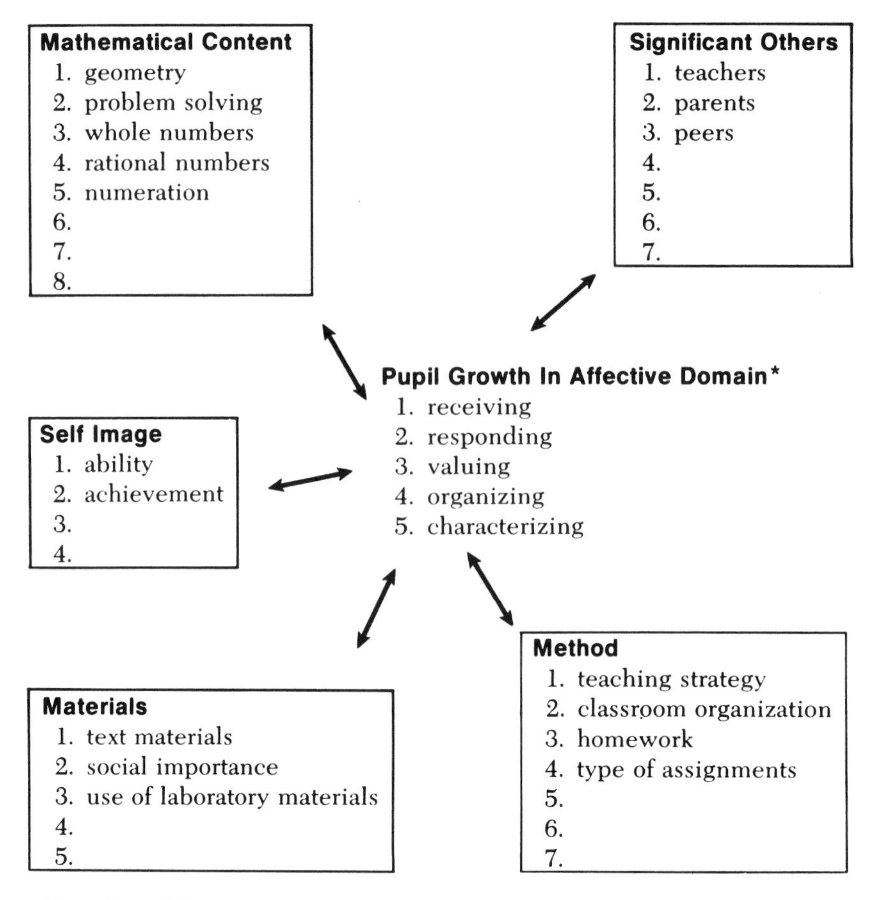

Figure 15–2 A Model for Measuring Mathematical Outcomes in the Affective Domain
Note: The blank spaces are used to indicate that each of these lists is incomplete. The reader should consider the addition of other categories.

parents is through parent–teacher conferences. The teacher should be prepared to give a short overall evaluation of progress in mathematics and to examine with the parent the pupil's arithmetic folder. If a student is having difficulties, the parent will often appreciate suggestions for helping the child at home. In addition, the teacher may effectively use this time to explain the goals of the mathematics program and to clear up any parental confusion concerning the goals and methods of modern mathematics.

FINAL NOTE

Evaluation of pupil progress in elementary school mathematics is a difficult and time-consuming process. But when the dividends to be derived from a good measurement program are assessed, there can be little doubt of its value. Not only does the teacher measure the achievement of a pupil but, in the process, also carefully thinks through his or her own educational objectives, focuses attention on an individual pupil's strengths and weaknesses, and studies the effectiveness of a particular teaching strategy.

SUGGESTED REFERENCES

ANASTASI, ANNE *PsychologicaTesting,* 5th ed. New York: Macmillian, 1982.

BRASWELL, JAMES S., ed. *Mathematics Tests Available in the United States and Canada,* 6th ed., Reston, Va.: The National Council of Teachers of Mathematics, 1988.

BUROS, OSCAR KRISEN, ed. *The Eighth Mental Measurement Yearbook,* Highland Park, N.J.: Gryphon, 1978 (see also 5th, 6th, and 7th Yearbooks).

GRONLUND, NORMAN E., *Measurement and Evaluation in Teaching,* 5th ed. New York: Macmillian, 1985.

KEYSER, DANIEL J., and RICHARD C. SWEETLAND, eds. *Test Critiques,* Kansas City, Mo.: Westport 1985.

MITCHELL, JAMES V., JR., ed. *Tests in Print III: An Index to Tests, Test Reviews, and the Literature on Specific Tests,* Lincoln, Neb.: University of Nebraska Press, 1983.

MITCHELL, JAMES V., JR., ed. *The Ninth Mental Measurements Yearbook,* Lincoln, Neb.: University of Nebraska Press, 1985.

ROMBERG, THOMAS, ed. *Curriculum and Evaluation Standards for School Mathematics,* Reston, Va.: The National Council of Teachers of Mathematics, 1988.

GLOSSARY

Abacus A device used for calculating, usually involving sliding beads or counters along a wire.

Addend Any one of a set of numbers to be added. In the equation $4 + 5 = 9$, the numbers 4 and 5 are addends.

Addition An operation on two numbers called the addends to obtain a third number called the sum.

Additive Identity A number e for which it is true that $a + e = e + a = a$ for every number a. For the set of real numbers, 0 is the additive identity since $a + 0 = 0 + a = a$ for every real number a.

Additive Inverse The additive inverse of a number is a number that, when added to the original number, gives the additive identity. For example, -4 is the additive inverse of 4 since $4 + (-4) = 0$, and the additive inverse of $-\frac{1}{4}$ is $\frac{1}{4}$ since $(-\frac{1}{4}) + \frac{1}{4} = 0$.

Adjacent Sides Two sides of a polygon that share a common vertex are called adjacent sides.

Algorithm A numerical process that may be applied to obtain the solution of a problem; generally used in elementary mathematics to mean one of the various procedures used for computing sums, differences, products, quotients, square roots, etc.; a plan or procedure followed in carrying out a numerical operation. The algorithm for adding $23 + 15$ is in part indicated below:

$$23 = 20 + 3$$

$$15 = 10 + 5$$

$$30 + 8 = 38.$$

Angle The union of two rays that have the same endpoint. The endpoint is called a vertex.

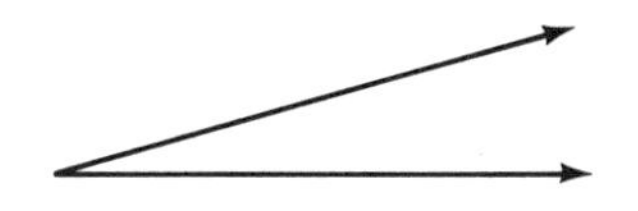

Arc A part of a circle determined by two points on the circle. The arc determined by the points A and B and containing point C is denoted by

Area A numerical measure in terms of a specified unit (usually a square) that is assigned to the plane surface of a region. Note that both number and unit must be given. For example, the area of the table top is 6 square feet.

Arithmetic Mean The arithmetic mean of a set of numbers is the quotient resulting when the sum of the numbers in a set is divided by the number of addends. The arithmetic mean of the set {4, 11, 12} is 9 since the sum 4 + 11 + 12 divided by 3 (the number of addends) is 9. *Synonym:* Average.

Associative Property A binary operation is associative if the result of using it with the ordered pair $[(a,b),c]$ is the same as with $[a,(b,c)]$. The binary operation *, for example, is associative if $(a*b)*c = a*(b*c)$.

Associative Property of Addition The binary operation addition is associative since the sum of more than two numbers that are added in a stated order is independent of the grouping. For example, in finding the sum for 5 + 2 + 8, either the 5 and 2 or the 2 and 8 may be added first, so that $7 + 8 = (5 + 2) + 8 = 5 + (2 + 8) = 5 + 10$. In general, $(a + b) + c = a + (b + c)$.

Associative Property of Multiplication The binary operation multiplication is associative, for, when more than two numbers are multiplied in a stated order, the product is independent of the grouping. For example, in finding the product for $3 \times 4 \times 5$ the first multiplication may be either 3×4 or 4×5 so that $12 \times 5 = (3 \times 4) \times 5 = 3 \times (4 \times 5) = 3 \times 20$. In general, $(a \times b) \times c = a \times (b \times c)$.

Axiom A statement accepted without proof. *Synonym:* Postulate.

Base (of a numeral) In a number written with an exponent, the number being raised to some power is called the base. For example, in 5^3, 5 is the base.

Base (of a numeration system) The number used in the fundamental grouping. For example, 10 is the base of the decimal system, 2 is the base of the binary system. An example of a numeral written in base seven is 563_{seven}, which means $(5 \times 7^2) + (6 \times 7) + 3$.

Base (of a polygon) Any designated side. For example, any side of a triangle may be considered as the base.

BASIC (Beginner's All-Purpose Symbolic Instruction Code) A popular high-level computer language, widely available on most inexpensive microcomputers. While it is often the first language taught to children, there are other languages (see LOGO) that might be more appropriate.

Binary Numeration System A numeration system whose base is two. For example, $11011_{two} = (1 \times 2^4) + (1 \times 2^3) + (0 \times 2^2) + (1 \times 2^1) + (1 \times 2^0) = 27_{ten}$.

Bisect To divide into two congruent parts. For example, to bisect a line segment locate the midpoint.

Bounded Set A set S of numbers is bounded if there is a positive number b such that $|s| < =b$ for all s in S.

CAI (Computer-Assisted Instruction) A method of teaching that uses a computer to present individualized instructional materials. Students interact with the computer in a learning situation that might involve drill and practice, tutorials with different branches of study simulation, or the like. Sometimes referred to as CAL (Computer-assisted learning).

Calculator A device that performs arithmetical operations based on data and instructions that are entered manually.

Cardinal Number The cardinal number of a set is usually thought of as the number of elements in a set, a number designating "how many" as opposed to the order. In the sentence, "There are 25 people in row 5," 25 is a cardinal number and 5 is an ordinal number.

Cartesian Product The Cartesian product of two sets A and B is the set of all ordered pairs, (a, b) where $a \in A$ and $b \in B$. It is the set that consists of all possible matchings of each element or A with each element of B, and it is usually denoted by $A \times B$. For example, let $A = \{e, f, g\}$ and $B = \{1, 2\}$. Then

$$A \times B = \{(e, 1), (e, 2), (f, 1), (f, 2), (g, 1), (g, 2)\}$$

$$B \times A = \{(1, e), (1, f), (1, g), (2, e), (2, f), (2, g)\}$$

Note that $A \times B <> B \times A$.

Chord A line segment that has its endpoints on a given circle. In the following figure, $\overline{BC}$ is a chord.

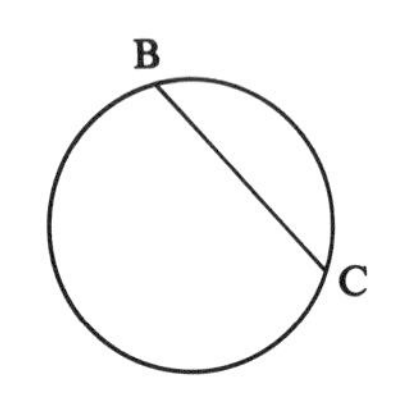

Circle A set of points, all of which are a specified distance from a given point called the center or center point.

Circular Region The union of a circle and its interior.

Circumference The measure of a circle; the distance around a circle.

Closed Curve A curve that has no endpoints (i.e., in drawing a representation, the starting and endpoints are the same).

Closure An operation in a set has the property of closure if the result of the operation on members of that set is a member of the set. Addition has closure in the set of whole numbers since the sum of whole numbers is a whole number. We usually say that the set of whole numbers is closed under addition. The set of whole numbers is also closed under the operation of multiplication, but the set of whole numbers is not closed under subtraction and division, since 3-5 and 3/5 are not whole numbers.

Common Divisor An integer is called a common divisor of a set of integers if it is a divisor of the set of even integers.

Common Factor Common divisor.

Common Multiple An integer is called a common multiple of a set of integers if it is a multiple of each of the members in a given set. For example, 12 is a common multiple of the set $\{2, 4, 6\}$.

Commutative Property A binary operation is commutative if the result of using it with the ordered pair (a, b) is the same as with (b, a). The binary operation *, for example, is commutative if $a*b = b*a$.

Commutative Property of Addition The binary operation addition is commutative since the sum of two numbers is independent of the order in which they are added. For example, $3 + 5$ is the same as the sum $5 + 3$. In general, $a + b = b + a$.

Commutative Property of Multiplication The binary operation multiplication is commutative since the product of two numbers is independent of the order in which they are multiplied. For example, 3×5 is the same as the product 5×3. In general, $a \times b = b \times a$.

Compass A device for drawing models of a circle.

Complement (of a set) The complement of the set A is the set of all those elements in the universal set that do not belong to A. The symbol for the complement of A is usually written A'. For example, if E denotes the set of even integers and the universal set is the set of integers, then E' is the set of odd integers.

Complementary Angles Two angles for which the sum of the measures is 90°.

Composite Number A natural number, different from 1, which is not a prime number (i.e., a natural number, greater than 1, which has a factor other than 1 and itself).

Congruence The relationship between two geometric figures that have the same size and shape.

Coordinate On a number line, the number matched with each point is called the coordinate of the point. In a plane, an ordered pair of numbers is matched with each point, and the ordered pair is called the coordinate of the point.

Coordinate Axes The coordinate axes in a plane are the perpendicular number lines used to match each point in the plane with an ordered pair of numbers.

Corresponding Sides Pairs of sides whose endpoints are paired in a one-to-one pairing of the vertices of two polygons.

Count (1) To name the natural numbers in regular succession. (2) The act of counting the number of elements in a set, which consists of establishing a one-to-one correspondence between the set of objects to be counted and the natural numbers that are used up in the process. The last number used in counting a set of objects is the number of objects in the set. What we do when we count, say, three objects is to show that the set of these objects is equivalent to the set of numbers $\{1, 2, 3\}$.

Counting Numbers The set of natural numbers: $\{1, 2, 3, 4, \ldots\}$.

Cube A rectangular prism such that all faces are squares.

Curve A set of all points that lie on a particular path between two points. It may be thought of as a set of points represented by a pencil drawing made without lifting the pencil off the sheet of paper.

Decagon A polygon with 10 sides.

Decimal Any base 10 numeral.

Decimal Numeration System A numeration system whose base is 10.

Decimal Point The dot that is used in the decimal numeration system to indicate in a decimal (numeral) the separation between the units position and the fractional portion of the numeral. For example, 437.28 represents $(4 \times 10^2) + (3 \times 10^1) + (7 \times 1) + (2 \times 10^{-1}) + (8 \times 10^{-2})$ and is read four hundred thirty-seven and twenty-eight hundredths.

Denominator The number named by the lower numeral in a fractional numeral. For example, the denominator of 2/3 is 3.

Diagonal A segment joining two nonadjacent vertices of a polygon. In the following figure, the diagonal is segment $\overline{AB}$.

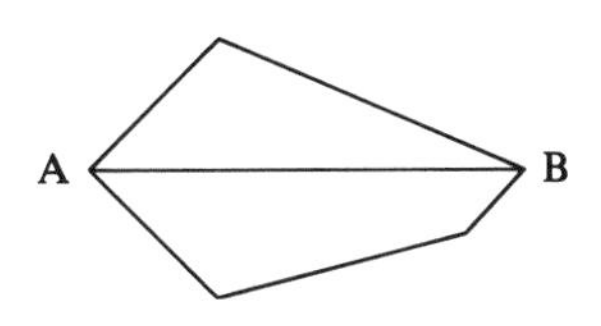

Diameter A chord that passes through the center point of the circle. In the following figure, segment $\overline{AB}$ is a diameter of the circle.

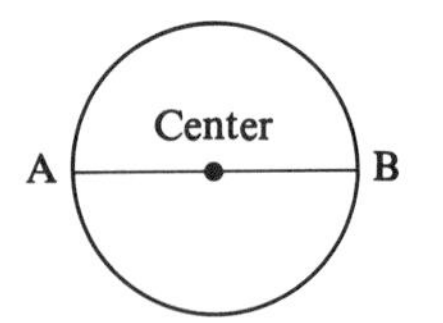

Difference (1) The number resulting from the subtraction operation. (2) The difference $a - b$ of two real numbers a and b is the number c such that $b + c = a$. This definition asserts that the difference $a - b$ is equal to the sum of a and the additive inverse of b; that is, $a - b = a + (-b)$. For example,

$$8 - (-3) = 8 + 3 = 11$$

$$(-7) - 4 = (-7) + (4) = -11$$

$$(-5) - (-2) = (-5) + 2 = -3.$$

Digit The basic Hindu-Arabic symbols used to write numerals. In the base-10 system, these digits are 0, 1, 2, 3, 4, 5, 6, 7, 8, 9.
Disjoint Sets Sets that have no common elements; that is, when S and T are two sets such that $S\ T$ is empty, we say that S and T are disjoint with respect to each other.
Distributive Property a joint property of multiplication and addition that says that multiplication is distributed over or with respect to addition, which means that

$$a \times (b + c) = (a \times b) + (a \times c).$$

For example, $3 \times 7 = 3 \times (5 + 2) = (3 \times 5) + (3 \times 2) = 15 + 6$.
Dividend In a division problem, the number being divided is called the dividend. For example, in the problem $33 \div 7$, 33 is called the dividend.

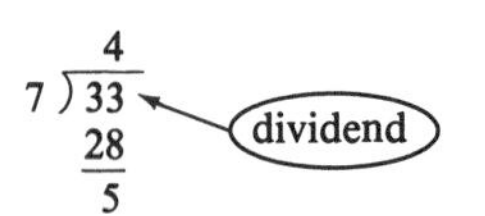

Division The inverse operation of multiplication; an operation on two numbers to obtain a third number called the quotient.
Divisor (1) In a division problem, the number which indicates the size of the divisions is called the divisor. For example, in the problem $33 \div 7$, 7 is called the divisor. (2) If a and b are integers with $a <> 0$ and there exists an integer c such that $b = a \times c$, then we say that a is a divisor of b. We also say that a divides b, b is a multiple of a, and a is a factor of b. For example, 3 is a divisor of 15 since we can find an integer, namely 5, such that $15 = 3 \times 5$.

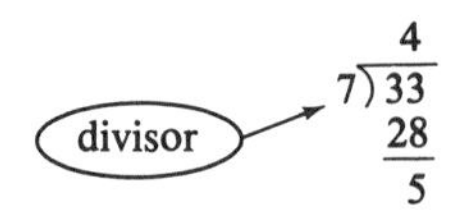

Dodecagon A 12-sided polygon.
Duodecimal Numeration System A numeration system whose base is 12.

Edge An edge of a polyhedron is any one of the segments making up any one of the faces of the polyhedron. In the following figure, segment $\overline{AB}$ is one of the edges of the cube.

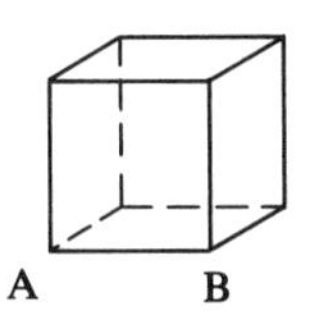

Element A member of a set.

Empty Set The set with no elements; often called the null set. The number zero is associated with the empty set, and the symbol $\emptyset$ is used to denote the empty set.

Equality A relationship between two objects, which means that they are exactly the same. For example, $4 + 5$ and $6 + 3$ are names for the same number, so we may say the $4 + 5$ and $6 + 3$ are equal, and write $4 + 5 = 6 + 3$.

Equal Sets Sets having the same elements.

Equation A mathematical sentence involving the use of the equality symbol. Examples; $5 + 4 = 9$; $7 + 1 = 8$; $n + 3 = 7$.

Equiangular Triangle A triangle with each angle measure equal to 60°.

Equilateral Triangle A triangle in which each side has the same measure.

Equivalence A relation that is reflexive, symmetric, and transitive is called an equivalence relation. That is, if * represents an arbitrary relation on a set $S(* \leqq S \times S)$, then * is an equivalence relation if the following hold:

1. $a * a$ for all a in S
2. if $a * b$, then $b * a$
3. if $a * b$ and $b * c$, then $a * c$

The relation, $=$, "is equal to" is an example of an equivalence relation. Another example is the relation, $\sim$, "is equivalent to" defined for sets, since the relation $\sim$ is reflexive, symmetric, and transitive; that is,

1. for any set, A, $A \sim A$
2. if $A \sim B$, then $B \sim A$
3. if $A \sim B$ and $B \sim C$, then $A \sim C$

Equivalence Class Let A be a set and * an equivalence relation defined on A. If $a \in A$, the subset of A that consists of all elements x of A such that $x * a$ is called an equivalence class. The equivalence class determined by the element a is usually denoted by $[a]$. This definition of the equivalence class $[a]$ can be written formally as follows:

$$[a] = \{x \in A : x * a\}$$

Face A face of a polyhedron is any one of the plane surfaces (polygonal regions) making up the polyhedron. For example, in a cube, each of the square regions is a face of the cube.

Factor (1) Any of the numbers to be multiplied to form a product. For example, in the equation $6 \times 7 = 42$, both 6 and 7 are factors. (2) A divisor of an integer is also called a factor of the integer.

Finite Set A set S is said to be finite if the cardinal number of S is 0 or a natural number; that is, S is finite if S is empty or if a natural number n can be found so that S is equivalent to the set $\{1, 2, 3, \ldots, n\}$. Any set that is not finite is said to be infinite.

Fraction An expression of the form $\frac{a}{b}$ which denotes $b \div a$; $b <> 0$.

Function A function F is a relation where if (a, b) and (a, c) are in F then b must be equal to c. We can think of a function as being a correspondence between two sets X and Y where each element of X corresponds to one and only one element of Y. In other words, F is a function from X to Y if each element of X is associated with a unique element of Y. X is called the domain of F, and Y is called the range of F. The element of Y which is associated with the element x of X is usually denoted by $F(x)$. The notation $F:X$ --> Y is used to denote a function F with domain X and range Y. There are many ways to describe a particular function. For example, the function that associates each natural number with its reciprocal can be written in the following ways:

$$F:N \text{ --> } Q::n \text{ --> } 1/n$$

$$F = \{(n, 1/n) : n \text{ is a natural number}\}$$

$$F(n) = 1/n,\ n \in N.$$

A function F from X to Y is said to be one-to-one if wherever $F(x) = F(y)$, then $x = y$; and F is said to be equal if for every y in Y there is some x in X such that $x = F(y)$.

Greater Than A number a is greater than a number b if $a - b$ is a positive number. We write $a > b$ or $b < a$ to mean that a is greater than b. Examples,

Since $(-3) - (-5) = 2$ is a positive number, $-3 > -5$.
Since $2 - (-3) = 5$ is a positive number, $2 > -3$.
Since $3 - 2 = 1$ is a positive number, $3 > 2$.

Greatest Common Divisor The integer d is called the greatest common divisor (g.c.d.) of the integers a and b, not both zero, if the following conditions are satisfied.

1. d is positive
2. d is a divisor of both a and b
3. Every divisor of both a and b is a divisor of d.

The greatest common divisor of a and b is denoted by (a, b). Note that $(a, 0) = |a|$. Any two integers not both zero always have a g.c.d. Moreover, they can have only one g.c.d.

Hexagon A six-sided polygon.

Hypotenuse The side opposite the right angle in a right triangle.

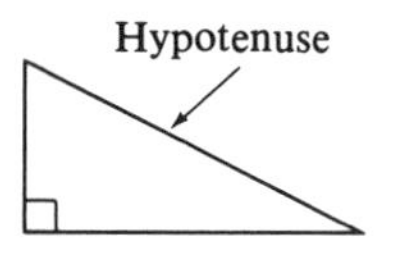

Identity Element Let * denote a binary operation on a set S. Then if there is an element e of S such that $a * e = e * a = a$ for all $a \in S$, e is called the identity element of S under *. For the operation addition on the set R of real numbers, the number 0 is such that $a + 0 = 0 + a = a$ for all $a \in R$, so 0 is called the additive identity of R. For multiplication, $a \times 1 = 1 \times a = a$ for all $a <> 0 \in R$, so 1 is called the multiplication identity of R.

Inequality A relation indicating that two quantities are not the same.

Infinite Decimal A decimal that continues infinitely to the right of the decimal point. To indicate that such a continuation is implied we use three dots to the right of the last written digit in the decimal. For example, the decimal representation for the number pi (π), 3.1415 . . . , is an example of an infinite decimal. The real numbers are the numbers named by the infinite decimals.

Infinite Nonrepeating Decimal An infinite decimal that has no blocks of consecutive digits repeated over and over. The decimal representation for pi (π), 3.1415 . . . , is an example of an infinite nonrepeating decimals.

Infinite Repeating Decimal An infinite decimal in which a certain block of consecutive digits will be repeated over and over in an unending sequence. Examples,

0.272727 . . .
0.463146314631 . . .
29.37854854854 . . .

The bars indicate the repeating block. The rational numbers are the infinite repeating decimals.

Inscribed Angle An angle inscribed in a given circle is an angle whose vertex lies on that circle and whose rays determine chords of that circle. In the following figure, angle ABC illustrates an angle inscribed in a given circle.

Inscribed Circle A circle inscribed in a given polygon is one that lies inside the polygon and has each side of the polygon tangent to it. The circles in the following figures illustrate a circle inscribed in a triangle and a square.

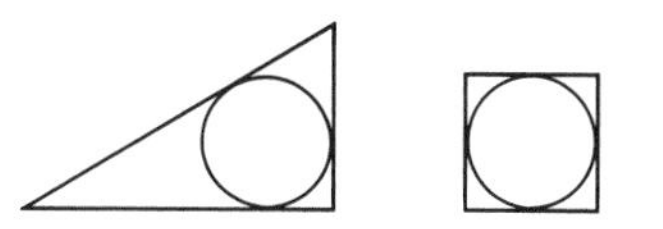

Integers The set J of integers consists of the set N of natural numbers, zero, and for each $n \in N$, the number $-n$ such that $n + (-n) - 0$.

Intersection (of two sets) The set containing those elements and only those elements that are in both of the two sets. The intersection of two sets A and B is denoted by $A \cap B$. For example, if $A = \{c, d, e, f\}$ and $B = \{e, f, g\}$, then $A \cap B = \{e, f\}$.

Inverse Operations The two operations $*$ and $\sim$ on a set S such that $(a * b) \sim b = a$ and $(a \sim b) * b = a$ for all $a, b \in S$. For example, addition and subtraction are inverse operations on the set R of real numbers, since $(a + b - b = a$ and $(a - b) + b = a$. Multiplication and division are also examples of inverse operations since $(a \times b)/b = (a/b) \times b = a$.

Irrational Number Any real number that cannot be expressed in the form a/b where a and b are integers, $b <> 0$; that is, any real number that is not a rational number. The irrational numbers may also be described as the set of infinite nonrepeating decimals.

Least Common Multiple The least common multiple (l.c.m.) of two nonzero integers a and b is the positive integer m with the following two properties:

1. m is a multiple of both a and b
2. every multiple of both a and b is a multiple of m

The least common multiple of a and b is denoted by $[a, b]$. Note that any two nonzero integers always have a least-common multiple and they can have only one.

Leg of a Right Triangle Either of the two sides adjacent to the right angle.

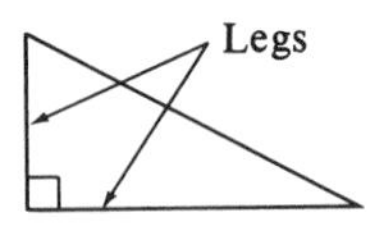

Length A number indicating the measure of one line segment with respect to another line segment called the unit.

Less Than A number a is less than a number b if b is greater than a; that is, if $b - a$ is a positive number, we write $a < b$ or $B > a$ to mean that a is less than b.

Line In geometry, some terms are generally considered undefined. *Line* is such a term. Unless otherwise stated, the word *line* means straight line. A straight line is represented by a straight mark and is named by labeling any two points on it. The line shown here is read "line AB," and we usually write it as

$$\overleftrightarrow{AB}$$

A B

A straight line extends infinitely far in both directions.

Line Segment A definite part of a line; that is, part of a line that begins at one point on a line and ends at another. A line segment has only one dimension-length. A line segment determined by the two points A and B is usually denoted by $\overline{AB}$.

$$\overline{AB}$$

A B

LOGO A computer language designed by Seymour Papert at M.I.T. to be used by grade-school children. Includes "turtle graphics." LOGO can be easily extended. It is said to encourage problem solving by children.

Measure A number assigned to a geometric figure indicating its size with respect to a specific unit.

Multiple If a and b are integers with $a <> 0$ and there is an integer c such that $b = a \times c$, then we say that b is a multiple of a. We also say that a is a divisor of b. The multiples of 3, for example, are the set $\{\ldots, -9, -6, -3, 0, 3, 6, 9\ldots\} = \{3 \times c\colon c$ is an integer$\}$. A multiple of a nonzero integer a is the product of a and an integer.

Multiplication An operation on two numbers called the factors to obtain a third number called the product.

Natural Numbers The set of counting numbers: $\{1, 2, 3, 4\ldots\}$. The subset of the whole numbers that includes all of the whole numbers except zero.

Negative Integers The additive inverses of the natural numbers: $\{-1, -2, -3, -4\ldots\}$.

Negative Rational Numbers A rational number $\frac{m}{n}$ is negative if the integer $m \times n$ is a negative integer. We use the notation $\frac{m}{n} < 0$ to mean $\frac{m}{n}$ is negative.

Negative Real Numbers If we think of the real numbers as infinite decimals, the decimal fraction approximation obtained by neglecting all the decimal places after a particular decimal place is a rational number. Then the negative real numbers are those that have a negative decimal approximation. We indicate that the real number r is negative by the notation $r < 0$.

Noncollinear Points Points not on the same line.

Noncoplanar Points Points not on the same plane.

Null Set The empty set; usually denoted by $\varnothing$.

Number See Cardinal number, Integers, Irrational number, Natural numbers, Ordinal number, Rational number, Real numbers, Whole numbers.

Number Line A model to show numbers and their properties. The model is usually used first for the whole numbers and then the markings and names are extended until finally a one-to-one correspondence is set up between all the points on the line and all the real numbers. The following illustrates the whole-number line:

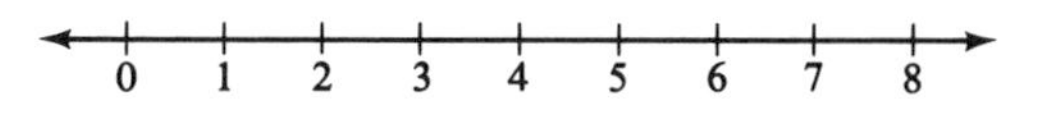

In the following number line some examples of real numbers are labeled:

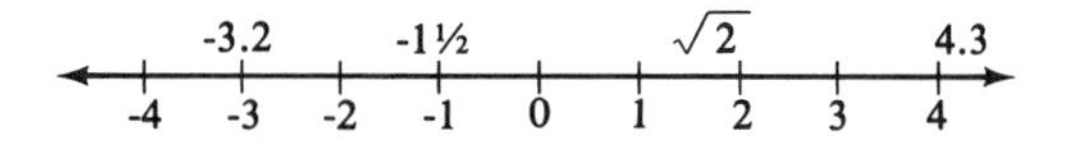

Numeral A name or symbol used for a number.
Numeration System A numeral system for naming numbers.
Numerator The number named by the upper numeral in a fraction. For example, the numerator of $\frac{2}{3}$ is 2.
Odd Integer An integer which is not even. An odd integer can be expressed in the form $(2 \times a) + 1$ where a is an integer. For example, 7 is an odd integer since $7 = (2 \times 3) + 1$, and -7 is odd since $-7 = [2 \times (-4)] + 1$. The set of odd integers $\{\ldots, -1, 1, 1, 3, 5 \ldots\}$ can be described formally as follows:

$$\{(2 \times a) + 1;\ a \text{ is an integer}\}.$$

One-to-one Correspondence Two sets A and B are said to be in one-to-one correspondence if each element of A can be paired to an element of B and each element of B can be paired to an element of A in such a way that distinct elements of A are paired to distinct elements of B and distinct elements of B are paired to distinct elements of A. When two sets A and B can be paired in a one-to-one correspondence we say that A and B are equivalent and we write $A \sim B$. A one-to-one correspondence exists between A and B if there is a one-to-one function from A into B.
Order A property of a set of numbers that permits one to compare any two numbers a and b of the set by saying that $a < b$, $a > b$, or $a = b$ (a is less than b, a is greater than b, or a is equal to b).
Ordered Pair An ordered pair of objects is a set of two objects in which one of them is specified as being first. An ordered pair of objects a and b where a is first is usually written (a, b). The elements of set $A \times B$, the Cartesian product of the sets A and B, are called ordered pairs. Two ordered pairs (a, b) and (c, d) are the same if and only if $a = c$ and $b = d$.
Ordinal Number A number denoting place or position. In the sentence, "There are 25 people in row 5," the number 5 is an ordinal number. The most common elementary forms of the ordinal numbers are first, second, third, etc.
Origin The point of intersection of the perpendicular number lines that constitute the Cartesian coordinate system. The point in the Cartesian coordinate plane assigned to the pair $(0, 0)$.
Parallel Lines Lines in the same plane that do not intersect.
Parallel Planes Planes that do not intersect.
Parallelogram A quadrilateral whose opposite sides are parallel.
Pentagon A five-sided polygon.
Percent Means "per hundred." For example, 3 percent means 3 per hundred and is usually written 3%. $3\% = 0.03 = \frac{3}{100}$.
Perimeter A number denoting the total measure of all sides of a polygon.
Perpendicular Lines Two lines intersecting each other so that right angles are formed.
Pi (π) The ratio of the circumference to the diameter of a circle.

Place Value The value given to a certain position in a numeral. For example, in the decimal numeral 4327, the place value of 4 is four thousand.

Point In geometry, some terms are considered undefined. *Point* is such a term. A point represents an exact location in space. It has no dimensions. We represent a point on paper by a dot.

Polygon A simple closed curve that is the union of three or more line segments. The following figures are representations of some polygons:

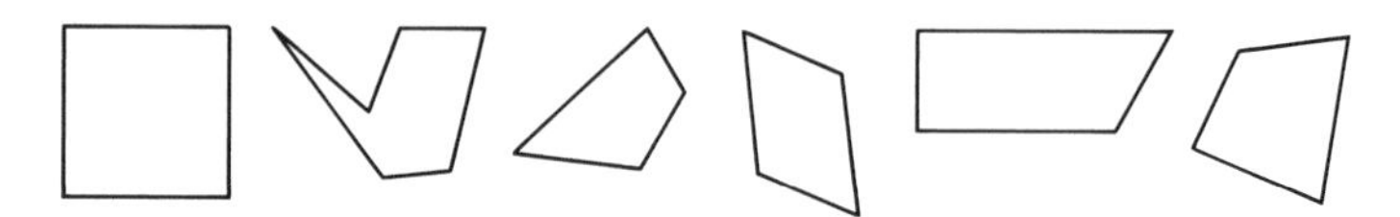

Polyhedron A closed surface made up of four or more intersecting planes. The intersections of the planes are called edges; the portions of the planes included by the edges are called faces; the intersections of the edges are called vertices. Polyhedrons are usually classified according to the number of faces: tetrahedron, four faces; hexahedron, six faces; octahedron, eight faces. Some polyhedrons follow:

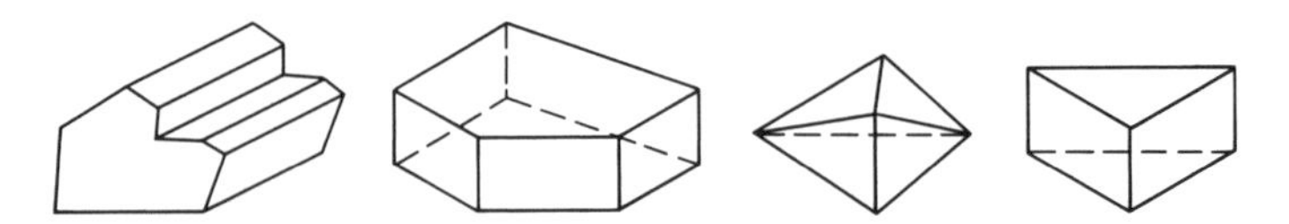

Positional System A numeration system in which the value of a basic numeral depends on the position it occupies in the numeral.

Positive Integers We refer to the natural numbers as positive integers.

Positive Rational Integers A rational number $\frac{m}{n}$ is positive if the integer $m \times n$ is a positive integer; that is, if $m \times n$ is a natural number. We use the notation $\frac{m}{n} > 0$ to mean that $\frac{m}{n}$ is positive.

Positive Real Numbers If we think of the real numbers as infinite decimals, the decimal fraction approximation obtained by neglecting all the decimal places after a particular decimal place is a rational number. Then the positive real numbers are those that have a positive decimal approximation. We indicate that the real number r is positive by the notation $r > 0$.

Postulate A statement accepted without proof. *Synonym:* Axiom.

Power A number shown by means of a base and an exponent. For example, 9 is the second power of 3 since $9 = 3^2$.

Prime Factor A factor that is a prime integer.

Prime Integer A natural number greater than 1 that has, as its only divisors, 1 and itself. 7 is an example of a prime number since the only divisors are 1 and 7.

Prism A polyhedron of which two faces are congruent polygons in parallel planes and the remaining faces are parallelograms.

Product The result of multiplying a pair of numbers. For example, in $3 \times 4 = 12$, 12 is the product.

Program Instructions used to direct a computer to perform a specific set of operations so as to accomplish a given task. A plan for solving a problem, written in a computer language so that it is "understood" by the computer (see also, Software).

Proper Divisor If $1 \leqq a < b$ and a is a divisor of b, then a is called a proper divisor of b.

Proper Subset Any subset of a set except the set itself.

Proportion A statement of equality between two ratios.

Protractor A device used for measuring angles.

Pythagorean Theorem In a right triangle, a and b are measures of the two legs and c is the measure of the hypotenuse, the $a^2 + b^2 = c^2$.

Quadrilateral A four-sided polygon.

Quotient (1) The result of dividing one number called the dividend by a nonzero number called the divisor. (2) The quotient $\frac{a}{b}$ of two real numbers a and b, b<>0, is the number q such that $a = q \times b$. This definition asserts that $\frac{a}{b}$ is equal to the product of a and the multiplicative inverse of b; that is $\frac{a}{b} = a \times b^{-1} = a \times \frac{1}{b}$. For example, $\frac{15}{3} = 15 \times \frac{1}{3} = 5$.

Radius A line segment connecting any point of a circle with its center. In the figure $\overline{OA}$ is a radius.

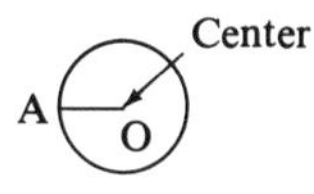

Ratio A relationship between an ordered pair of numbers a and b where b <> 0. The ratio of a and b may be written $a{:}b$ or as a fraction $\frac{a}{b}$.

Rational Number A rational number is a class of ordered pairs of integers. The ordered pairs are written in the form $\frac{m}{n}$, with the restriction that n is never zero. Every rational number can be expressed as an infinite repeating decimal.

Ray A point on a line and all the points on one side of that point that are on that line; that is, the union of a point P on line AB and one of the half-lines determined by P. In the figure that follows, ray PB is the point P and the points on line AB that are on the same side of P as B. The ray PB is denoted by

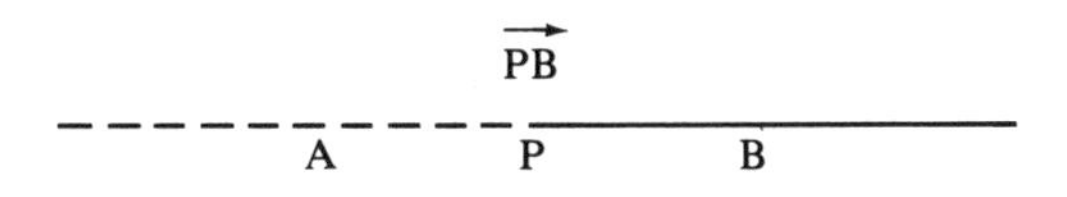

Real Numbers The real numbers are the numbers named by the infinite decimals. We usually think of the set of real numbers as those numbers corresponding to the points on the number line. There is a one-to-one correspondence between the set of infinite decimals and the points on the number line.

Reciprocal The reciprocal of a number is the multiplicative inverse of that number. If m is the reciprocal of n, then n is the reciprocal of m, so we usually refer to a number and its multiplicative inverse as reciprocals. For example, 2 is the reciprocal of $\frac{1}{2}$, and $\frac{1}{2}$ is the reciprocal of 2, so we say that 2 and $\frac{1}{2}$ are reciprocals. Notice that the product of reciprocals is 1 since the product of a number and its multiplicative inverse is the multiplicative identity 1.

Reflection This operation in motion geometry is often, in the classroom, called a "flip." A reflection moves a figure to an upside down position.

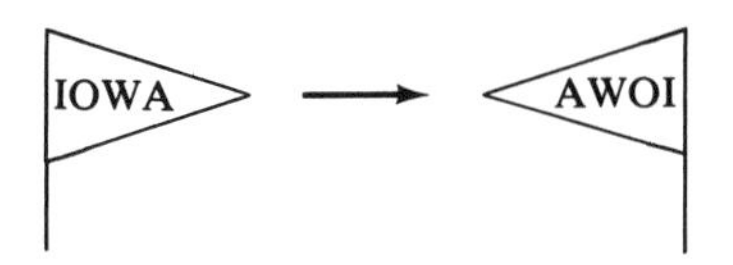

Region A plane region is the union of any simple closed curve and its interior.

Rhombus A parallelogram with four equal sides.

Right Angle An angle that has the measure of 90°.

Right Triangle A triangle that has one right angle. The sides adjacent to the right angle are called legs, and the side opposite the right angle is called the hypotenuse.

Rotation Called in the classroom a "turn"; the rotation moves a figure a given amount around a fixed point.

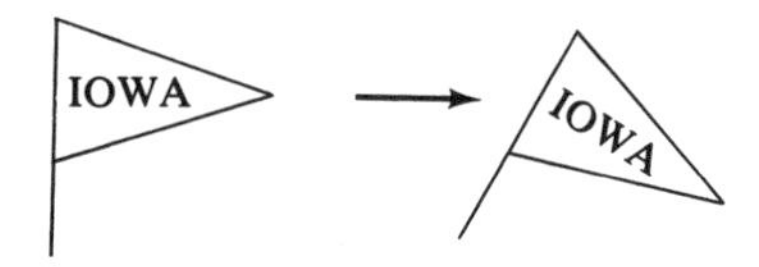

Scalene Triangle A triangle in which no two sides have the same measure.

Scientific Notation A number is said to be written in scientific notation if it is expressed as the product of a number between 1 and 10 and a power of 10. For example, 186,00 in scientific notation is 1.86×10^5, and we write 3.27×10^{-4} as the scientific notation for 0.000327.

Segment (1) A line segment is a definite part of a line determined by two points on the line called the endpoints. It includes the endpoints and all the points on the line between the endpoints. (2) A segment of a circle is the figure formed by the arc of the circle and its chord. The shaded area shown is bounded by a segment. It is composed of a part of the circle, arc $\widehat{ACB}$, and the line segment $\overline{AB}$:

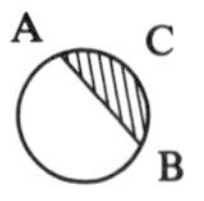

Sequence A sequence is a function having N, the set of natural numbers, as its domain. If x is a sequence, we usually write x_n instead of $x(n)$ for the value of x at n. The value x_n is called the n-th term of the sequence. The function that assigns the rational number $\frac{1}{n}$ to each natural number n is an example of a sequence. The first term of this sequence is 1, the second is $\frac{1}{2}$, the third is $\frac{1}{3}$, the n-th term is $\frac{1}{n}$.

Set A collection.

Similar Polygons Polygons that have their corresponding angles equal and the corresponding sides proportional.

Simple Closed Curve A plane closed curve that does not intersect itself.

Software Computer programs that enable computers to process information and to solve problems. Along with hardware, these constitute the computer system. Although it usually refers to professionally written programs, it sometimes is used to name any computer program. Examples include language translators that allow the use of BASIC, LOGO, PILOT, Pascal, and so on. Another type consists of CAI or CMI programs. A good library of computer software is necessary for worthwhile educational use of microcomputers.

Space The set of all points.

Square A rectangle in which all sides have the same measure.

Square Root A square root of a number is that number that, when taken two times as a factor, results in the given number as the product. For example, the square root of 4 is 2, since $2 \times 2 = 4$. We denote the square root of a number x by $\sqrt{x}$

Subset The set A is a subset of the set B if every element of A is an element of B. We denote this by $A \subseteq B$.

Subtraction The inverse operation of addition; an operation on two numbers to obtain a third number called the difference or remainder.

Successor The successor of a whole number a where $a = n\,(A)$ is the number b such that $b = n(A \cup \{0\})$; that is, the successor of a is $a + 1$. For example, the successor of 4 is 5.

Sum The result of adding a pair of numbers. For example, in $3 + 4 = 7$, 7 is the sum.

Tangent A line that touches a circle at one and only one point.

Ternary Numeration System A numeration system whose base is three.

Theorem A theorem is a statement to be proved. We accept a theorem as true only after it has been proved.

Topology A system of geometry relating to those properties of figures that remain unchanged when the figures are deformed in a specified way. Sometimes called "rubber sheet geometry" in the classroom.

Transitive Property A relation * on a set S is said to be transitive if whenever (a, b) and $(b, c) \in *$, then $(a, c) \in *$; that is, if $a * b$ and $b * c$, then $a * c$. The relation $<$, "is less than," is an example of a relation that is transitive, since $a < b$ and $b < c$ implies that $a < c$.

Translation A motion in a given straight direction over a specified distance. Usually called a "slide" in classroom use.

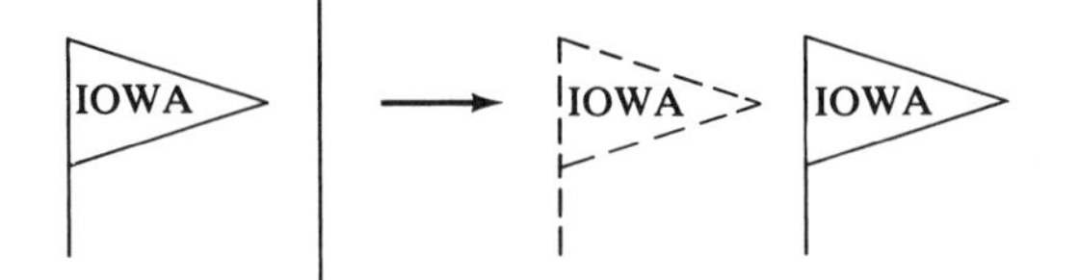

Trapezoid A quadrilateral with a pair of parallel sides.

Triangle A polygon with three sides.

Trichotomy Law If a and b are real numbers, exactly one of the following relationships holds: $a < b$, $a = b$, $a > b$.

Union (of two sets) The union of two sets A and B is the set of all elements that are in A or in B. The union of A and B is denoted by $A \cup B$. For example, if $A = \{e, f, g, h\}$ and $B = \{g, h, i\}$, then $A \cup B = \{e, f, g, h, i\}$. In general, $A \cup B = \{x : x \in A \text{ or } x \in B\}$.

Unit An amount or quantity adopted as a standard of measurement.

Universal Set The set of all things chosen for a particular study.

Variable Any symbol used to represent an unspecified element of a set containing more than one element. For example, when we use the set builder notation to indicate sets, $A + \{x : x \text{ has a certain property}\}$, the symbol x is called a variable.

Venn Diagram A diagram used to illustrate properties of sets. In the Venn diagrams that follow, the shaded areas correspond to the indicated unions and intersections.

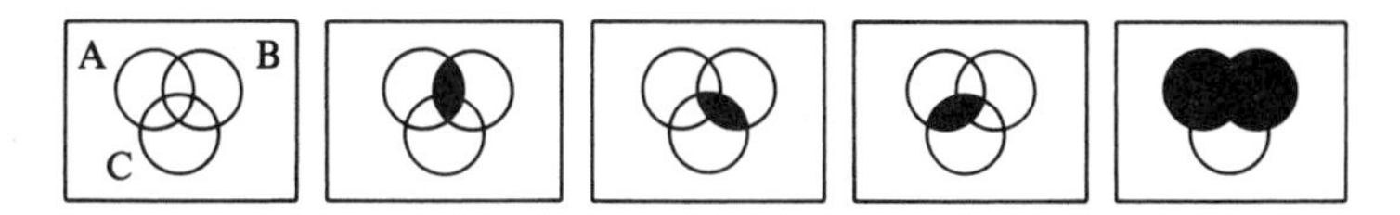

Vertex (1) The vertex of an angle is the point common to the two rays determining the angle. (2) A vertex of a polygon is a point of intersection of two adjacent sides. (3) A vertex of a polyhedron is a point of intersection of the edges.

Volume A numerical measure in terms of a specified unit (usually a cube) assigned to the interior region of a three-dimensional figure. Note that both the number and the unit must be given. For example, the volume of a sphere with a radius of 1 foot is $\frac{4}{3}$ cubic feet.

Whole Number A whole number is the cardinal number of some finite set. The set W of whole numbers is the set $\{0, 1, 2, 3, 4 \ldots\}$.

TEACHER LABORATORY ANSWERS

The answers given will help you think through some of the exercises. There is usually another way of working the exercises.

Chapter 1. Answers will vary.
Chapter 2. Answers will vary.
Chapter 3. Answers will vary.
Chapter 4. 1. Experiment. You will find that you need: 90 clips; 9 erasers; and 1 pencil.
2. Experiment. 4 cows and 31 chickens.
3. Work backwards. Answer: 7 apples.
Chapter 5. Answers will vary.
Chapter 6. 1. Answers will vary.
2. Diagrams will vary.
3. Answers will vary.
Chapter 7. The first problem is shown below. Do the others in the same manner.

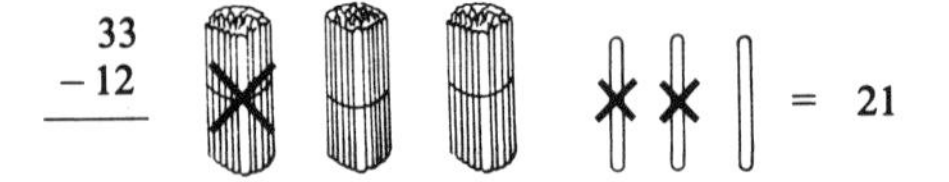

33 − 12 using popsicle sticks.

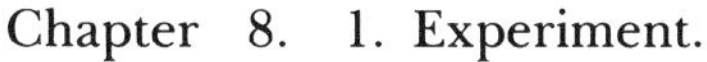

Chapter 8. 1. Experiment.
2. Experiment.
3. Answers will vary.

Chapter 9. 1.

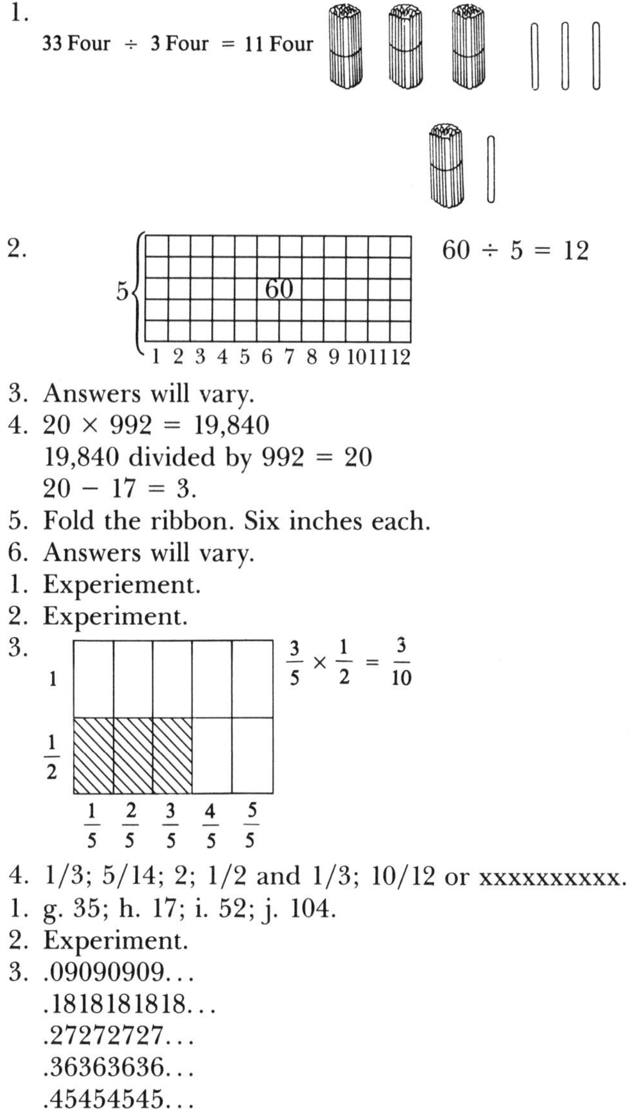

2.

$60 \div 5 = 12$

3. Answers will vary.
4. $20 \times 992 = 19{,}840$
19,840 divided by 992 = 20
$20 - 17 = 3$.
5. Fold the ribbon. Six inches each.
6. Answers will vary.

Chapter 10. 1. Experiement.
2. Experiment.
3.

$\frac{3}{5} \times \frac{1}{2} = \frac{3}{10}$

4. 1/3; 5/14; 2; 1/2 and 1/3; 10/12 or xxxxxxxxxx.

Chapter 11. 1. g. 35; h. 17; i. 52; j. 104.
2. Experiment.
3. .09090909. . .
.1818181818. . .
.27272727. . .
.36363636. . .
.45454545. . .
.54545454. . .
.63636363. . .
.727272. . .
.818181. . .
.90909090. . .
each is the next multiple of 9.

Chapter 12. Answers will vary on each item.
Chapter 13. Experiment.
Chapter 14. 1.–3. Experiment.
4.

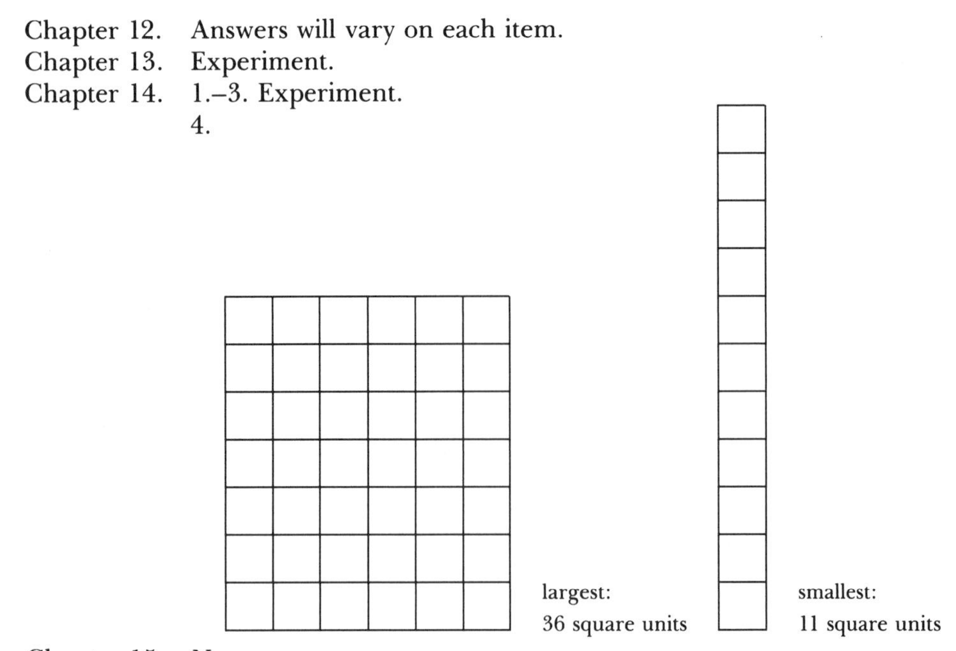

Chapter 15. None.

Answers to Self-Test

Note that true–false questions in the self-test are designed to help you review the chapter. Many of the questions could be answered as either true or false, depending upon the emphasis given.

Chapter 1

1. false (not usually done)
2. true
3. true
4. true
5. false
6. true
7. true
8. false (needs of children also very important)
9. true
10. false
11. false
12. true (in general)

Chapter 2

1. false
2. false
3. false
4. true
5. true (can use everyday materials)
6. false (they benefit from interesting and varied materials)
7. true
8. true
9. false
10. true
11. false (the most important factor is the teacher)
12. true

Chapter 3

1. true
2. false
3. true
4. false
5. true
6. false
7. true
8. true
9. false (use oral presentation)
10. true
11. true
12. true
13. true

Chapter 4

1. false
2. true
3. true
4. true
5. false
6. true
7. true
8. true
9. true
10. true
11. false
12. false
13. true
14. true
15. false
16. true

Chapter 5

1. false
2. false
3. false
4. true
5. true
6. true
7. true
8. true
9. true
10. true (but only teachers with exceptional ideas on teaching)
11. true
12. false
13. true

Chapter 6

1. true
2. false
3. true
4. true
5. true
6. false
7. true
8. false
9. true
10. false
11. false
12. true
13. true

Chapter 7

1. false
2. true
3. true
4. true
5. true (a good thoughtful way—to enrich)
6. false
7. false

Chapter 8

1. true
2. false
3. true
4. true
5. true
6. false
7. false

Chapter 9

1. true
2. false
3. true
4. true
5. false
6. true
7. true
8. true (but they are important)
9. true
10. true

Chapter 10

1. false
2. true
3. true
4. false
5. true
6. true
7. true
8. false
9. true
10. false
11. true
12. true
13. false
14. false

Chapter 11

1. false
2. true
3. true
4. false
5. true
6. false
7. true
8. true

Chapter 12

1. true
2. true
3. true
4. true
5. false
6. true
7. true
8. true
9. false
10. true
11. true
12. true
13. true
14. true
15. false
16. false
17. true
18. true

Chapter 13

1. false
2. false (it is an exact location—a dot is a model)
3. false
4. true
5. false
6. true
7. true
8. false
9. false
10. true
11. true
12. false
13. false
14. false

Chapter 14

1. false
2. true
3. true
4. false (a common basic unit is used)
5. true
6. false
7. true
8. true
9. true
10. true
11. true

Index